Radar Energy Warfare and the Challenges of Stealth Technology

Bahman Zohuri

Radar Energy Warfare and the Challenges of Stealth Technology

Springer

University of New Mexico.
Albuquerque, New Mexico

Bahman Zohuri
Electrical Engineering
University of New Mexico
Albuquerque, NM, USA

ISBN 978-3-030-40618-9 ISBN 978-3-030-40619-6 (eBook)
https://doi.org/10.1007/978-3-030-40619-6

This Springer imprint is published by the registered company Springer Nature Switzerland AG
The registered company address is: Gewerbestrasse 11, 6330 Cham, Switzerland

This book is dedicated to my son Sasha
and grandson Darrius

Preface

When the Wright brothers invented the airplane, they effectively changed the way travel, exploration and warfare forever. Examples of this can be found in World War I and II, a two pilot system was created for successfully identifying and bombing specific targets. In the Battle of Britain this system used by Royal Air Force (RAF) so they could defend their territory against German Luftwaffe and their turn the battle to their advantage. In the Vietnam War American pilots used surface-to-air missile (SAM), where the radar and electronic jamming played a big role, and the use of radar technology is used to this day.

Since the Wright brothers' original design, airplanes have drastically improved upon their design as well as functionality; such as: F-35 on American side and their Russian counterpart Sukhoi Su-57 and even now Chinese playing in this market with their Chengdu J-20. While they all are the sixth-generation (GEN-VI) fighter jets, they have one common technical challenge, that is, they all need to be stealthy in order to avoid any radar detection. However, the question is are they really stealthy as their manufacturers claim by using radar absorbing material (RAM) or reducing their radar cross-section (RCS) against any radar beam detection that is looking for them in the sky.

Targets for radar are still agile within the sixth-generation (GEN-VI) aircraft with an RCS in the 8–12 GHz band between 2 m^2 (head-on) and 100 m^2 (maximum side-on), flying at ground velocities between 150 and 750 m/s. Stealthy aircraft targets may have an RCS as low as 0.005 m^2 in the head-on aspect. Missile targets are likely to have a RCS at least one order of magnitude lower than the non-stealthy aircraft and may travel at 1300 m/s (Mach 4). Ground-based targets include structures such as bridges and buildings or vehicles such as armored vehicles, which may be moving or static. Even today with the announcement of new weaponry systems such as hypervelocity missile or torpedoes by Russian and Chinese, Stealth Scenario has taken a different direction from the technical point of view.

The most pernicious form of electronic countermeasures (ECM) against this new generation of fast-moving jet planes is Digital Radio Frequency Memory (DRFM)-based repeater jammers and transponders. This book suggests that scalar wave (SW) as a countermeasure against a threat such as hypervelocity, and this type of wave is based on scalar longitudinal wave (SLW) derived from more complete equation (MCE) of Ampere's Maxwell's equation using the quantum electrodynamics (QED) approach.

Techniques from SW to radar can replicate the radar's waveform with remarkable integrity and can present a plethora of false targets that may correlate within the victim receiver and the processing chain driven by the fire-control radar (FCR), and all this is described in this book in a very simple and introductory way.

A particular concern might be the ghosting within a medium pulse repetition frequency (PRF) mode in response to a large number of realistic target returns. As we have learned from our basic principle of radar courses in college, FCRs on fast strike aircraft are the quintessential pulse Doppler radars. They must work in a wide variety of air-to-air and air-to-ground modes, they must be lightweight and compact, yet they have to achieve long detection ranges in the presence of extreme clutter scenes and be capable of tracking a large number of agile targets, all of which must be highly automated, especially for single-seat aircraft, to minimize the workload on the aircrew.

Tactical fighter-sized stealth aircrafts must be optimized to defeat higher frequency bands, such as the C, X, and Ku bands—that is just a simple matter of physics. There is a "step change" in an LO aircraft's signature once the frequency wavelength exceeds a certain threshold and causes a resonance effect. Typically, that resonance occurs when a part of an aircraft—such as a tail fin or similar—is less than eight times the size of a particular frequency wavelength. Fighter-sized stealth aircrafts that do not have the size or weight allowances for two feet or more of radar absorbent material coatings on every surface are forced to make trades as to which frequency bands they are optimized for.

It means that radars operating at a lower frequency band, such as parts of the S or L band, are able to detect and track certain stealth aircrafts. Ultimately, in order to counter lower-frequency radars, a larger flying-wing stealth aircraft design like the Northrop Grumman B-2 Spirit or B-21 Raider—which lacks many of the features that cause a resonance effect—is a necessity. However, the UHF and VHF band wavelengths, designers are not trying to make the aircraft invisible—rather engineers hope to create a radar cross-section that will blend in with the background noise that is inherent to low-frequency radars.

Additionally, low-frequency radars can be used to cue fire-control radars, and some US adversaries have started to make an effort to develop targeting radars that operate at lower frequencies. However, those lower-frequency fire-control radars exist only in theory—and are a long way off from being fielded.

The topics above are included throughout this book. Each subject of interest is carefully examined in order to build a solid foundation for its reader that can range from someone with little background knowledge to a reader with more sophisticated reader with solid background in physics of radar and its principles as well the engineers that they do understand since of Stealth Technology.

This book also provides four Appendices to enhance the general knowledge of readers in case they are new to the game of radar energy warfare and stealth technology.

Albuquerque, NM, USA Bahman Zohuri

Acknowledgment

I am indebted to the many people who aided, encouraged, and supported me beyond my expectations. Some are not around to see the results of their encouragement in the production of this book, yet I hope they know of my deepest appreciation. I especially want to thank all my friends to whom I am deeply indebted, for continuously giving their support without hesitation. They have always kept me going in the right direction, especially a true friend, Dr. Patrick J. McDaniel.

Above all, I offer very special thanks to my late mother and father and to my children, in particular, my son Sasha, who always encouraged me while we had him in this world for short time. They have provided constant support and encouragement, without which this book would not have been written. Their patience with my many absences from home and long hours in front of the computer to prepare the manuscript is especially appreciated.

Contents

About the Author

Bahman Zohuri is a Research Associate Professor of Electrical Engineering and Computer Science at the University of New Mexico in Albuquerque, while consulting through his own company, Galaxy Advanced Engineering, Inc., a consulting company that he started himself in 1991 when he left both semiconductor and defense industries after working many years as a chief scientist. After graduating from the University of Illinois in the field of Physics and Applied Mathematics, he joined Westinghouse Electric Corporation where he performed thermal hydraulic analysis and natural circulation for inherent shutdown heat removal system (ISHRS) in the core of a liquid metal fast breeder reactor (LMFBR) as a secondary fully inherent shut system for secondary loop heat exchange. All these designs were used for nuclear safety and reliability engineering for self-actuated shutdown systems. He designed the mercury heat pipe and electromagnetic pumps for large pool concepts of LMFBR for the purpose of heat rejection in this reactor around 1978, when he received a patent for it. He was then transferred to the defense division of Westinghouse, where he was responsible for the dynamic analysis and method of launch and handling of the MX missile out of the canister. He later on worked as a consultant at Sandia National Laboratories after leaving the United States Navy. Dr. Zohuri earned his bachelor's and master's degrees in physics from the University of Illinois and his second master's degree in mechanical engineering as well as his doctorate in nuclear engineering from the University of New Mexico. He has been awarded three patents, has published 26 textbooks, and has numerous other journal publications to his credit.

Recently, he has been involved with cloud computing, data warehousing, and data mining using Fuzzy and Boolean and applying them to artificial intelligence, machine learning, and deep learning logic and has a few books published on these subjects as well as numerous other books that can be found under his name on Amazon or on the Internet.

Chapter 1
Fundaments of Radar

This chapter gives an elementary account of radar's principles and goes over essential of radar as a detecting device. Radar was one of the elements that helped Britain in their air war against German Luftwaffe during the peak of the Battle of Britain, and they managed to survive during such invasion. The subject is simple, the methods are powerful, and the results have broad applications. Radar is a detection system that uses radio waves to determine the range, angle, or velocity of objects, and it can be used to detect aircraft, ships, spacecraft, guided missiles, or motor vehicles or even to be able to forecast the weather formations and terrain. Today with all the stealth technologies that is pushing state of new generation of fighter planes to the stage of sixth generation and new threats hypersonic objects that are flying faster than speed of sound to level 5–15 Mach number; thus, the radar warfare is taking a different innovative level.

1.1 Introduction

The acronym RADAR stands for RAdio Detection And Ranging, and today, the technology is so common that the word has become a standard English noun. As indicated by the name of the system, this device works based on the usage of radio waves, and it is capable of sending out electromagnetic waves in the form of transverse electromagnetic (TEM) form to far distance from the source of radiation. Note that in later chapter we will talk about transverse and longitudinal electromagnetic waves and how they work to detect different objects with different characteristic threats. However, for the time being, a transverse mode of electromagnetic radiation is a particular electromagnetic field pattern of the radiation in the plane perpendicular to the radiation's propagation direction.

A radar system consists of a transmitter producing electromagnetic waves in the radio or microwave domain, a transmitting antenna, a receiving antenna (often the same antenna is used for transmitting and receiving), and a receiver and processor to

B. Zohuri, *Radar Energy Warfare and the Challenges of Stealth Technology*,
https://doi.org/10.1007/978-3-030-40619-6_1

Fig. 1.1 Typical long-range and rotating radar illustrations. (Source: www.wikipedia.com). (**a**) Long-range radar, (**b**) Rotational radar

determine properties of the object(s). Radio waves (pulsed or continuous) from the transmitter reflect off the object and return to the receiver, giving information about the object's location and speed.

Figure 1.1a is an illustration of a long-range radar antenna that is used to track space objects and ballistic missile, while Fig. 1.1b is the type of radar used for the detection of aircraft, and it rotates steadily, sweeping the airspace with a narrow beam.

Radar was developed secretly for military use by several nations in the period before and during World War II. A key development was the cavity magnetron in the United Kingdom, which allowed the creation of relatively small systems with submeter resolution. The term *RADAR* was coined in 1940 by the US Navy as an acronym for *RA*dio *D*etection *A*nd *R*anging [1, 2]. The term *radar* has since entered English and other languages as a common noun, losing all capitalization.

The modern uses of radar are highly diverse, including:

1. Air and terrestrial traffic control and radar astronomy
2. Antimissile systems
3. Marine radars to locate landmarks and other ships and aircraft anti-collision systems
4. Ocean surveillance systems
5. Outer space surveillance and rendezvous systems
6. Meteorological precipitation monitoring
7. Altimetry and flight control systems and guided missile target locating systems
8. Ground-penetrating radar for geological observations

High-tech radar systems are associated with digital signal processing (DSP) and machine learning (ML) integrated into artificial intelligence (AI) in conjunction with deep learning (DL) combined and are capable of extracting useful information from very high noise levels. Radar is a key technology that the self-driving systems are mainly designed to use, along with sonar and other sensors [3].

Other systems similar to radar make use of other parts of the electromagnetic spectrum. One example is LIght Detection And Ranging (LIDAR), which uses predominantly infrared light from lasers rather than radio waves, there are some articles or books that call LIDAR as LADAR, and they both mean the same techniques. With the emergence of driverless vehicles, radar is expected to assist the automated platform to monitor its environment, thus preventing unwanted incidents [4].

LIDAR mechanism of detection is based on surveying method that measures distance to a target by illuminating the target with laser light and measuring the reflected light with a sensor. Differences in laser return times and wavelengths can then be used to make digital 3D representations of the target. The name LIDAR, now used as an acronym of LIght Detection And Ranging [5] (sometimes, *light imaging, detection, and ranging*), was originally a portmanteau or combination or blending of *light* and *radar* [6, 7]. LIDAR sometimes is called 3D laser scanning, a special combination of a 3D scanning and laser scanning. It has terrestrial, airborne, and mobile applications. See Fig. 1.2, in which a Frequency Addition Source of Optical Radiation (FASOR), where it is used at the Starfire Optical Range for LIDAR and laser-guided star experiments, is tuned to the sodium D2a line and used to excite sodium atoms in the upper atmosphere.

Bear in your mind that LIDAR typically uses ultraviolet (UV), visible, or near-infrared (IR) light to image objects. It can target a wide range of materials, including

Fig. 1.2 Typical FASOR demonstration. (Source: www.wikipedia.com)

non-metallic objects, rocks, rain, chemical compounds, aerosols, clouds, and even single molecules [5]. A narrow laser beam can map physical features with very high resolutions; for example, an aircraft can map terrain at 30 c (12 in.) resolution or better [8].

In the case of airborne LIDAR, which is nothing more than *airborne laser scanning*, a laser, while attached to an aircraft during flight, creates a 3D point cloud model of the landscape. This is currently the most detailed and accurate method of creating digital elevation models, replacing photogrammetry. One major advantage in comparison with photogrammetry is the ability to filter out reflections from vegetation from the point cloud model to create a digital terrain model which represents ground surfaces such as rivers, paths, cultural heritage sites, etc., which are concealed by trees. Within the category of airborne LIDAR, there is sometimes a distinction made between high-altitude and low-altitude applications, but the main difference is a reduction in both accuracy and point density of data acquired at higher altitudes. See Fig. 1.3.

In Fig. 1.3 we are observing a schematic diagram of airborne LIDAR performing line scanning resulting in parallel lines of measure points, although there exist other scan pattern methods, but this one is fairly the most common one.

Collection of elevation data using LIDAR has several advantages over most other techniques. Chief among them are higher resolutions, centimeter accuracies, and ground detection in forested terrain [8].

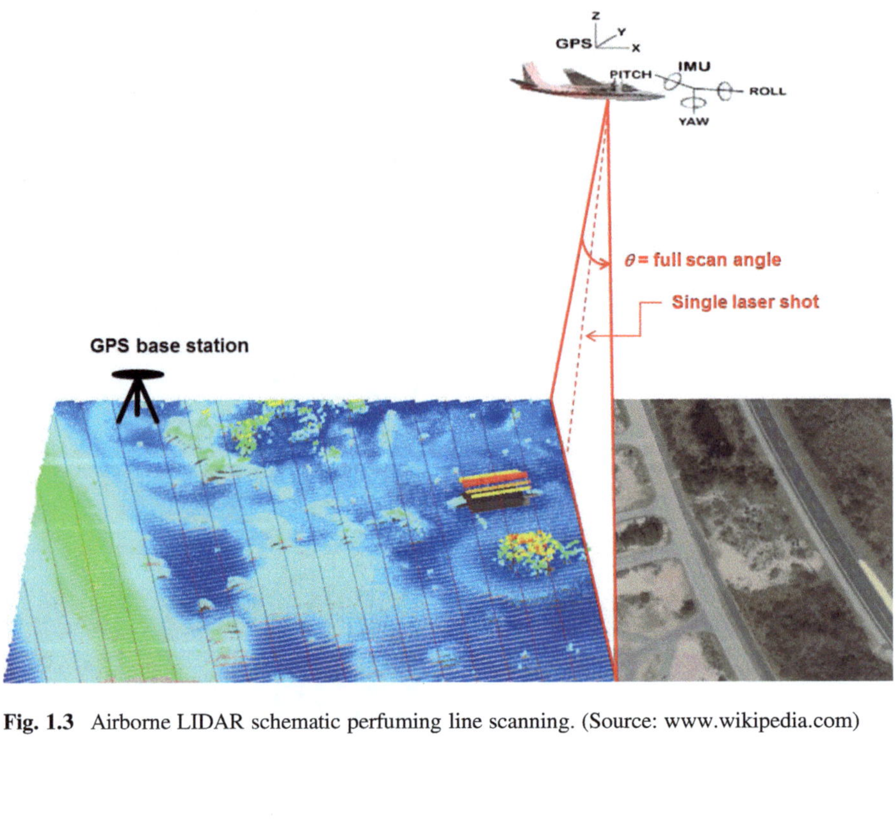

Fig. 1.3 Airborne LIDAR schematic perfuming line scanning. (Source: www.wikipedia.com)

LIDAR has become an established method for collecting very dense and accurate elevation data across landscapes, shallow-water areas, and project sites. This active remote sensing technique is similar to radar but uses laser light pulses instead of radio waves. LIDAR is typically "flown" or collected from planes where it can rapidly collect points over large areas (Fig. 1.3).

Airborne LIDAR can also be used to create bathymetric models in shallow water [9].

1.2 First Principle of Radar Concept and Experiments

Early history of encountering radar principle falls in around 1886, when German physicist Heinrich Hertz showed that radio waves could be reflected from solid objects. Further discovery was done around 1895 by a Russian physics instructor at the Imperial Russian Navy School in Kronstadt, who developed an apparatus using a coherer tube for detecting distant lightning strikes. The next year, he added a spark-gap transmitter. In 1897, while testing this equipment for communicating between two ships in the Baltic Sea, he took note of an interference beat caused by the passage of a third vessel. In his report, Popov wrote that this phenomenon might be used for detecting objects, but he did nothing more with this observation [9].

The German inventor Christian Hülsmeyer was the first to use radio waves to detect "the presence of distant metallic objects." In 1904, he demonstrated the feasibility of detecting a ship in dense fog, but not its distance from the transmitter [10]. He obtained a patent [11] for his detection device in April 1904 and later a patent [12] for a related amendment for estimating the distance to the ship. He also obtained a British patent on September 23, 1904 [13], for a full radar system that he called a *telemobiloscope*. It operated on a 50 cm wavelength and the pulse radar signal was created via a spark-gap. His system already used the classic antenna setup of horn antenna with parabolic reflector and was presented to German military officials in practical tests in Cologne and Rotterdam harbor but was rejected [14].

In 1915, Robert Watson-Watt used radio technology to provide advance warning to airmen [15] and during the 1920s went on to lead the UK research establishment to make many advances using radio techniques, including the probing of the ionosphere and the detection of lightning at long distances. Through his lightning experiments, Watson-Watt became an expert on the use of radio direction finding before turning his inquiry to shortwave transmission. Requiring a suitable receiver for such studies, he told the "new boy" Arnold Frederic Wilkins to conduct an extensive review of available shortwave units. Wilkins would select a General Post Office model after noting its manual's description of a "fading" effect (the common term for interference at the time) when aircraft flew overhead.

Across the Atlantic in 1922, after placing a transmitter and receiver on opposite sides of the Potomac River, US Navy researchers A. Hoyt Taylor and Leo C. Young discovered that ships passing through the beam path caused the received signal to fade in and out. Taylor submitted a report, suggesting that this phenomenon might be

Fig. 1.4 US Naval Research Laboratory experimental antenna configuration. (Source: www.wikipedia.com)

used to detect the presence of ships in low visibility, but the Navy did not immediately continue the work. Eight years later, Lawrence A. Hyland at the Naval Research Laboratory (NRL) observed similar fading effects from passing aircraft; this revelation led to a patent application [16] as well as a proposal for further intensive research on radio-echo signals from moving targets to take place at NRL as illustrated in Fig. 1.4, where Taylor and Young were based at the time [17].

In summary the history of radar as we stated extends to the days of modern electromagnetic theory, where Hertz demonstrated reflection of radio waves in around 1986, and in 1900 T described a concept for electromagnetic detection and velocity measurement during an interview. In 1903 and 1904, the German scientist Hülsmeyer experimented with ship detection via radio wave reflection, an idea advocated again by Marconi in 1922. In that same year, Taylor and Young of the US Naval Research Laboratory (NRL) both demonstrated ship detection by radar, and in 1930 Hyland, also of NRL, first detected aircraft accidentally by radar, setting off a more substantial investigation that led to a US patent for what is known as a continuous-wave (CW) radar in 1934.

The effort of developing radar further took momentum in around the 1930s, and countries like Germany, Russia, Italy, and Japan were the pioneers among the developing countries. In fact, in the United States, R. M. Page of NRL began an effort to develop pulse radar in 1943, with the first successful demonstrations in 1936, and in the same year, the US Army Signal Corps begin active radar work,

Fig. 1.5 A Chain Home tower illustration (Great Baddow, Essex United Kingdom)

leading to an effort in 1938 to its first operational system that is known as SCR-268 anti-aircraft fire-control radar (FCR) system and in 1939 to another version as early warning system (EWS), which we know it as SCR-270, and yet its warning during Pearl Harbor attack by Japanese naval aircraft was tragically ignored.

The British scientist demonstrated pulse radar (PR) the same year and by 1938 established the famous Chain Home (Fig. 1.5) surveillance radar network that helped them during the Battle of Britain and remained active until the end of World War II.

Britain also built the first airborne interceptor radar in 1939. With collaboration between the United States and United Kingdom around 1940 even possibly up to now, most radar work was conducted at high-frequency (HF) and very-high-frequency (VHF) wavelengths (i.e., we have described the radar bandwidth later on in this book), but with the British disclosure of the critical cavity magnetron (CCM) microwave power tube and US formation of the Radiation Laboratory at Massachusetts Institute of Technology (MIT), the groundwork was laid for the successful development of radar at the microwave frequency allowing the system to be built at small size scale near submeter resolution that has predominated ever since [18]. This technique approach was taken based on Watson patent in an article on air defense under Bonnier Corporation in popular science [19].

1.3 Radar Types

As we stated at the introductory of this chapter, the fundamental duty of the RADAR based on its acronym that is given is radio detection and ranging and categorized as:

- **Bistatic**: the transmitting and receiving antennas are at different locations as viewed from the target (e.g., ground transmitter and airborne receiver).
- **Monostatic**: the transmitter and receiver are collocated as viewed from the target (i.e., the same antenna is used to transmit and receive). See Fig. 1.6, where R_r is the receiving range and R_t is the transmitting range, with θ angle between them.
- **Quasi-monostatic**: the transmitting and receiving antennas are slightly separated but still appear to be at the same location as viewed from the target (e.g., separate transmitting and receiving antennas on the same aircraft).

Radar functions are categorized as follows:

- **Normal radar functions**:
  - Range (from pulse delay)
  - Velocity (from Doppler frequency shift)
  - Angular direction (from antenna pointing)
- **Signature analysis and inverse scattering**:
  - Target size (from magnitude of return)
  - Target shape and components (return as a function of direction)
  - Moving parts (modulation of the return)
  - Material composition
- **Radar performance**:
  - The complexity (cost and size) of the radar increases with the extent of the functions that the radar performs.

As we have understood so far based on historical evidences described before, radar technology is one of the most advanced innovative technologies of the nineteenth century, and it is used for measuring objects' distance. Because of this,

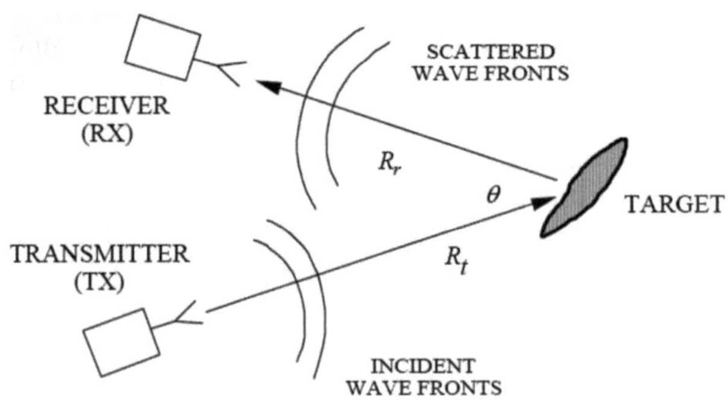

Fig. 1.6 A simple radar configuration and functionality

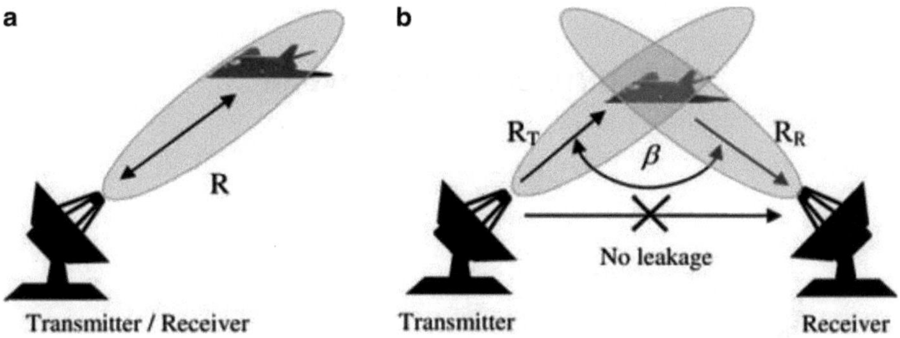

Fig. 1.7 Radar systems: (**a**) monostatic radar and (**b**) bistatic configurations

there has been a variety of radar systems per their application usage and functionality assigned to them as a task for various purposes, and they are classified under various categories. The following list highlights some of the most common radar systems under different functions and used by different commercial and military sectors.

1. **Bistatic Radar**
   Bistatic radar is a radar system that comprises a transmitter and a receiver that are separated by a distance that is equal to the distance of the expected target. A radar in which the transmitter and the receiver are located at the same place is known as a monastic radar. Most long-range surface-to-air and air-to-air missiles employ the use of bistatic radar. In Fig. 1.7 see schematic of monostatic and bistatic radar configuration.
2. **Continuous-Wave Radar**
   A continuous-wave radar is a type of radar where a known stable frequency continuous-wave radio energy is transmitted and then received from any of the objects that reflect the waves. A continuous-wave radar uses Doppler technology which means the radar will be immune to any form of interference by large objects that are stationary or slow moving. See Fig. 1.8.
3. **Doppler Radar**
   A Doppler radar is a special form of radar that employs the use of Doppler effect to produce velocity data about an object at a given distance. This is achieved by sending electromagnetic signals toward a target and then analyzing how the object motion has affected the frequency of the returned signal. This variation has the capacity to give extremely accurate measurements of the radial component of a target's velocity in relation to the radar. Doppler radars have applications in different industries including aviation, meteorology, healthcare, and many others. Figure 1.9 is a presentation of weather Doppler radar.
4. **Monopulse Radar**
   A monopulse radar is a radar system that compares the received signal from a single radar pulse against itself with an aim of comparing the signal as seen in multiple polarizations or directions. The most common form of monopulse radar is the adaptation of conical scanning radar which compares the return from two

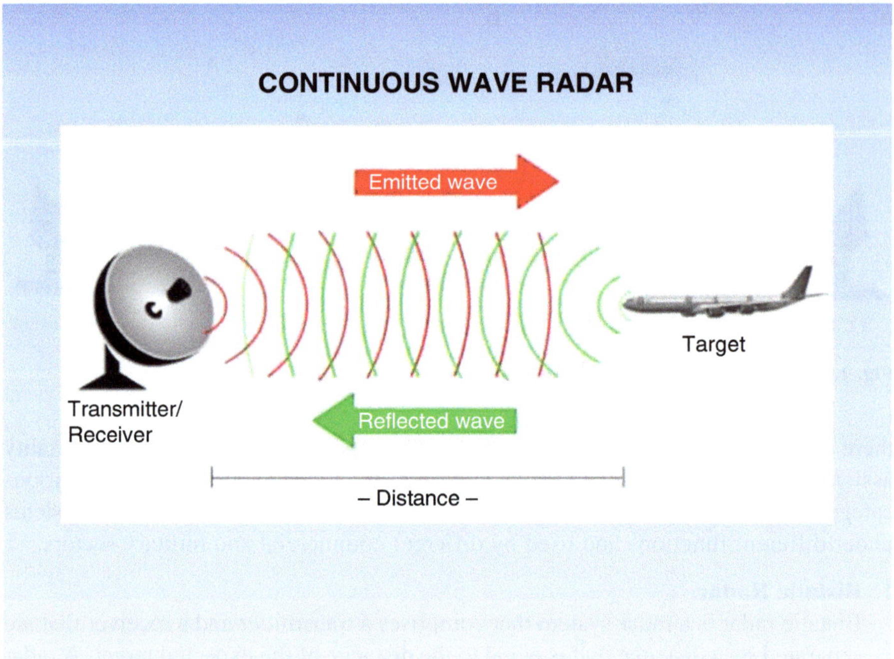

Fig. 1.8 A continuous-wave radar schematic

Fig. 1.9 Weather Doppler radar

directions to directly measure the location of the target. It is important to note that most of the radars that were designed since the 1960s are monopulse radars. Figure 1.10 is an illustration of precision monopulse tracking radar (PMTR).

5. **Passive Radar**

   A passive radar system is a type of radar that is designed to detect and track objects by processing reflections from non-cooperative sources of illumination in the environment. These sources include such things as communications signals and commercial broadcasts. Passive radar can be categorized in the same class of radar as bistatic radar. Figure 1.11 is an image of a civil aviation passive radar.

Fig. 1.10 Precision monopulse tracking radar

Fig. 1.11 Civil aviation passive radar

Fig. 1.12 Test range instrumentation systems

6. **Instrumentation Radar**
Instrumentation radars are radars that are designed to test rockets, missiles, aircrafts, and ammunitions on government and private test ranges. They provide a variety of information including space, position, and time both in the real time and in the post processing analysis.
Figure 1.12 is a presentation of a test range instrumentation system.
7. **Weather Radars**
Weather radars are radar systems that are used for weather sensing and detection. This radar uses radio waves along with horizontal or circular polarization. The frequency selection of weather radar depends on a performance compromise between precipitation refection and attenuation as a result of atmospheric water vapor. Some weather radars are designed to use Doppler shifts to measure the speed of wind and dual polarization to identify precipitation types.
Figure 1.13 is a presentation of an interactive weather radar system.
8. **Mapping Radar**
Mapping radars are used to scan a large geographical region for geography and remote sensing applications. Because of their use of synthetic-aperture radar, they are limited to relatively static objects. There are some specific radar systems that can sense humans behind walls, thanks to the reflective characteristics of humans that are more diverse than the ones found in construction materials. Figure 1.14 is a Titan radar imaging system developed by NASA.
9. **Navigational Radars**
Navigational radars are generally the same as search radars. However, they come with much shorter wavelengths that are capable of reflecting from the Earth and from stones. They are mostly common on commercial ships and other long-distance commercial aircrafts. There are various navigational radars that include

Fig. 1.13 An interactive weather radar system

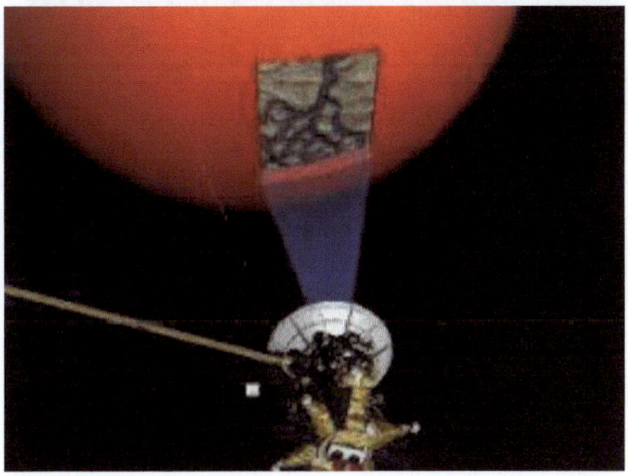

Fig. 1.14 NASA Titan imaging radar

marine radars commonly mounted on ships for collision avoidance and navigational purposes. Figure 1.15 is an image of a navigational radar system in cruise.

There may exist more sub-category types of radar for each of the above categories; however, the above lists are the most common types of radar that are operational today.

Fig. 1.15 Navigational radar cruise system

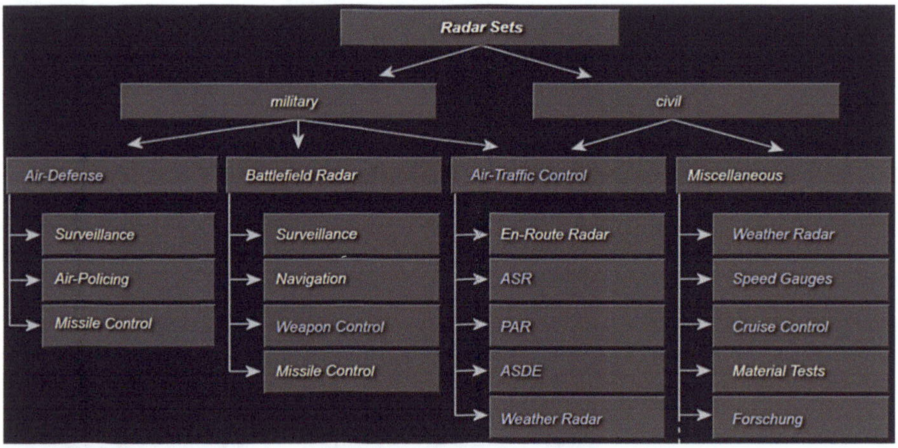

Fig. 1.16 Classification of radar sets based on the designed use. (Source: www.radartutorial.eu)

1.3.1 Radar Classification Sets

Radar systems may be divided into types based on the designed use also. This section presents the general characteristics of several commonly used radar systems as illustrated here in Fig. 1.16.

Radar classification can be described as its functionality in the form of two categories and they are as follows.

1.3.1.1 Multifunction Radars

Active array multifunction radars (MFRs) enable modern weapon systems to cope with saturation attacks of very small radar cross section missiles in a concentrated jamming environment. Such MFRs have to provide a large number of fire-control channels, simultaneous tracking of both hostile and defending missiles, and mid-course guidance commands.

The active phased array antenna as it has been described in Sect. 1.4.2.1 comprises flat sensor panels consisting of arrays of GaAs modules transmitting variable pulse patterns and building up a detailed picture of the surveillance area. A typical fixed array configuration system could consist of about 2000 elements per panel, with 4 fixed panels. Each array panel can cover 90° in both elevation and azimuth to provide complete hemispherical coverage.

1.3.1.2 Multi-target Tracking Radar

Multi-target tracking radar (MTTR) operational functions include:

- Long-range search
- Search for information with a high data rate for low-flying aircraft
- Search for information with high resolution of close-in air targets
- Automatic position and height information
- Simultaneous tracking of a lot of aircraft targets
- Target designation facilities for other systems

Their classifications are briefly described in the following context as:

Air Traffic Control Radar Sets	**En Route Radar**
 Air traffic control radars are used at both civilian and military airports. Airborne radar is designed specially to meet the strict space and weight limitations that are necessary for all airborne equipment. Even so, airborne radar	 "En route" radars operate in L-band mostly and display radar data to controllers in the en route environment at a maximum range up to 450 km.

(continued)

sets develop the same peak power as shipboard and shore-based sets. In fighter aircraft, the primary mission of radar is to aid in the search, interception, and destruction of enemy aircraft.	
Air Surveillance Radar Sets (ASR)  These radar equipments are used for the identification of aircraft, determination of aircrafts approach sequence, and individual aircraft approach controls by Air Traffic Security operators. In the meantime, these radars correlate the data obtained from other radars such as air defense radars or (excluding simple airfields) Mode-4 coordinate data of secondary radar equipment. These radar networks can be used under all weather conditions.	**Precision Approach Radar Sets (PAR)**  The precision approach radar guides aircraft to safe landing under conditions approaching zero visibility. By means of radar, aircraft are detected and observed during the final approach and landing sequence. Guidance information is supplied to the pilot in the form of verbal radio instructions or to the automatic pilot (autopilot) in the form of pulsed control signals.
Surface Movement Radar (SMR)  Surface movement radar (SMR) is the most widely used surveillance system for airport surveillance at present. SMR refers to primary radar that provides surveillance cover for the maneuvering area, which is defined as that used for the takeoff, landing, and taxiing of aircraft, excluding aprons.	**Weather Radar Sets** 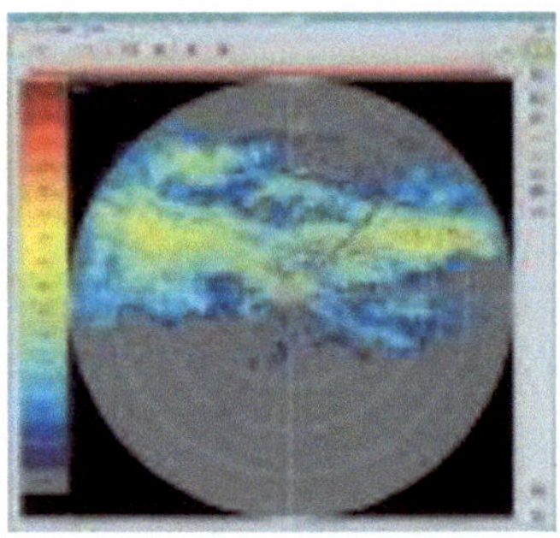 The weather data it finds could be used both for approach support and for feeding into the wider weather data concentration systems. The antenna rotation rate between systems is quite variable (3–6 rpm is common). Assuming multiple elevations are used, the weather picture

(continued)

	gathered might be updated with a frequency of 1 min and upward (this depends on the complexity and number of the elevations required and the antenna rotation rate). Radar in recent years has become an important tool for the measurement of precipitation and the detection of hazardous weather conditions.
Air Defense Radar Sets  Air defense radars can detect air targets and determine their position, course, and speed in a relatively large area. The maximum range of air defense radar can exceed 300 miles, and the bearing coverage is a complete 360° circle. Figure: TAFLIR of the Swiss Air Force	**Air Surveillance Radar Sets**  Air search radar systems initially detect and determine the position, course, and speed of air targets in a relatively large area. The maximum range of air search radar can exceed 300 miles, and the bearing coverage is a complete 360° circle. Air search radar systems are usually divided into two categories, based on the amount of position information supplied. Radar sets that provide only range and bearing information are referred to as two-dimensional, or 2D, radars. Radar sets that supply range, bearing, and height are called three-dimensional, or 3D, radars. Figure: Lockheed Martin's air surveillance radar AN/FPS 117

(continued)

Battlefield Surveillance

The battlefield surveillance radar mission is to alert and/or cue combat troops of hostile and unknown aircraft, cruise missiles, and unmanned aerial vehicles, protect friendly forces from fratricide, and provide air situational data to command and control centers.

Figure: BOR-A 550

Mortar Locating Radar Sets

A mortar locating radar provides quick identification to pinpoint enemy mortar positions in map coordinates, enabling artillery units to launch counter attacks.

Figure: COBRA—Mortar Radar

Air Policing

Another function of an air search radar system is guiding combat air patrol (CAP) aircraft to a position suitable to intercept enemy aircraft. In the case of aircraft control, the guidance information is obtained by the radar operator and passed to the aircraft by either voice radio or a computer link to the aircraft.

In fighter aircraft, the primary mission of radar is to aid in the search, interception, and destruction of enemy aircraft. This requires that the airborne radar system have a tracking feature.

Figure: The nose radar ECR 90 of the Eurofighter EF 2000

Missile Control Radar

A radar system that provides information used to guide a missile to a hostile target is called **guidance radar**. Missiles use radar to intercept targets in three basic ways:

1. Beam-rider missiles follow a beam of radar energy that is kept continuously pointed at the desired target.
2. Homing missiles detect and home in on radar energy reflected from the target; the reflected energy is provided by radar transmitter either in the missile or at the launch point and is detected by a receiver in the missile.
3. Passive homing missiles home in on energy that is radiated by the target.

(continued)

<table>
<tr>
<td>Missiles Guidance and Control

The Patriot is an army surface-to-air, mobile, air defense missile system. Since the mid-1960s, the system has evolved to defend against aircraft and cruise missiles and more recently against short-range ballistic missiles.</td>
<td>Battlefield Radar Sets

Battlefield radars usually have a shorter range and are highly specialized for a particular task. On ships of the Navy, the number of specialized radar antennas is more and more replaced by a multifunction radar.

Figure: Multifunction radar (MFR) “Variant” of Thales Naval Nederland</td>
</tr>
<tr>
<td>Miscellaneous Civil Radar Sets

Radar sets are deployed everywhere where measurements (or positioning) must inevitably be made at certain ranges. As a result of this also for civil purposes, a very wide operation area is developed.</td>
<td>Speed Gauges

Speed gauges are very specialized CW radars. A speed gauge uses the Doppler frequency for measurement of the speed. Since the value of the Doppler frequency depends on the wavelength, these radar sets use a very high frequency in K-band.

Figure: Speed gauge “Traffipax Speedophot”</td>
</tr>
</table>

(continued)

Navigation

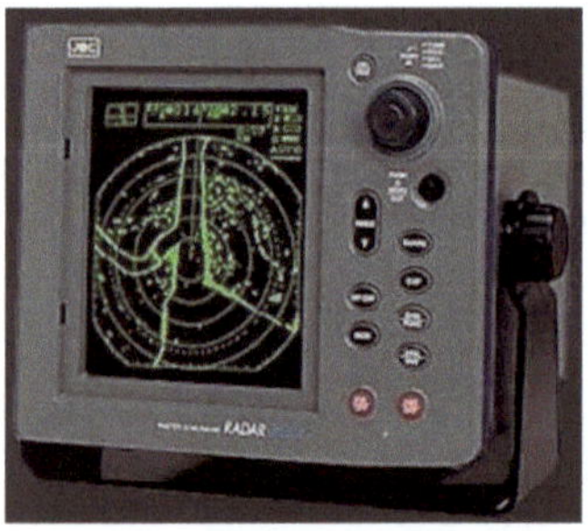

Navigation radars are designed for ship navigation and surface surveillance. When navigating in restricted waters, a mariner most often relies on visual piloting to provide the accuracy required to ensure ship safety. Visual piloting, however, requires clear weather; often, mariners must navigate through the fog. When weather conditions render visual piloting impossible on a vessel, radar navigation provides a method of fixing a vessel's position with sufficient accuracy to allow safe passage.

Radar-Controlled Cruise Control

Here the radiator grille of a Mercedes-Benz SL-Class roadster, the **Distronic** sensor, is being hidden behind the Mercedes star. This future-oriented radar set registers the traffic scenario to a distance of up to 150 m (500 ft) ahead and when necessary applies the brakes automatically.

Ground-Penetrating Radar

Ground-penetrating radar is a geophysical method that has been developed over the past 35 years for shallow, high-resolution, subsurface investigations of the Earth.

Figure: Ground-penetrating radar in action

Non-destructive Material Test

A special radar can be used to penetrate material to detect material defects.

Based on functionality and types of radar, we can also list the following radar type as well.

Radar originally was developed to meet the needs of the military services, and it continues to have critical applications for national defense purposes. For instance, radars are used to detect aircraft, missiles, artillery and mortar projectiles, ships, land vehicles, and satellites. In addition, radar controls and guides weapons; allows one class of target to be distinguished from another; aids in the navigation of aircraft and ships; and assists in reconnaissance and damage assessment.

Military radar systems can be divided into three main classes based on platform: land-based, shipborne, and airborne. Within these broad classes, there are several other categories based mainly on the operational use of the radar system. For the purposes of this report, the categories of military radars will be as described below, although there are some "gray" areas where some systems tend to cover more than one category. There is also a trend to develop multimode radar systems. In these cases, the radar category is based on the primary use of the radar.

Some of the more prominent types of radars are described below. These descriptions are not precise, for each of these radar types usually employs a characteristic waveform and signal processing that differentiate it from other radars.

Land-Based Air Defense Radars These radars cover all fixed, mobile, and transportable 2D and 3D systems used in the air defense mission.

Battlefield, Missile Control, and Ground Surveillance Radars These radars also include battlefield surveillance, tracking, fire-control, and weapons-locating radar systems, whether fixed, mobile, transportable, or man-portable.

Naval and Coastal Surveillance and Navigation Radars These radars consist of shipborne surface search and air search radars (2D and 3D) as well as land-based coastal surveillance radars.

Naval Fire-Control Radars These are shipborne radars that are part of a radar-based fire-control and weapons guidance systems.

Airborne Surveillance Radars These radar systems are designed for early warning, land, and maritime surveillance, whether for fixed-wing aircraft, helicopters, or remotely piloted vehicles (RPVs).

Airborne Fire-Control Radars These include those airborne radar systems for weapons fire control (missiles or guns) and weapons aiming.

Spaceborne Radar Systems Considerable effort has been applied to spaceborne radar (SBR) research for intelligence, surveillance, and reconnaissance missions over the last 30 years. The Department of Defense (DOD) seems to be expressing new interest in SBR.

Military Air Traffic Control (ATC), Instrumentation, and Ranging Radars This type is the most typical radar with a waveform consisting of repetitive short-duration pulses. Typical examples are long-range air and maritime surveillance radars, test range radars, and weather radars. There are two types of pulse radars that use the Doppler frequency shift of the received signal to detect moving targets, such as aircraft, and to reject the large unwanted echoes from stationary clutter that do not have a Doppler shift. One is called moving target indication (MTI) radar and the other is called pulse Doppler radar. Users of pulse radars include the Army, Navy, Air Force, Federal Aviation Administration (FAA), United States Cost Guard (USCG), National Aeronautics and Space Administration (NASA), Department of Commerce (DOC), Department of Energy (DOE), United States Department of Agriculture (USDA), Department of the Interior (DOI), National Science Foundation (NSF), and Department of Treasury (DOT). These include both land-based and shipborne ATC radar systems used for assisting aircraft landing and supporting test and evaluation activities on test ranges.

Simple Pulse Radar This type is the most typical radar with a waveform consisting of repetitive short-duration pulses. Typical examples are long-range air and maritime surveillance radars, test range radars, and weather radars. There are two types of pulse radars that use the Doppler frequency shift of the received signal to detect moving targets, such as aircraft, and to reject the large unwanted echoes from stationary clutter that do not have a Doppler shift. One is called moving target indication (MTI) radar and the other is called pulse Doppler radar. Users of pulse radars include the Army, Navy, Air Force, FAA, USCG, NASA, Department of Commerce (DOC), Department of Energy (DOE), US Department of Agriculture (USDA), Department of the Interior (DOI), National Science Foundation (NSF), and Department of Treasury.

Moving Target Indication (MTI) Radar By sensing Doppler frequencies, an MTI radar can differentiate echoes of a moving target from stationary objects and clutter and reject the clutter. Its waveform is a train of pulses with a low PRR to avoid range ambiguities. What this means is that range measurement at the low PRR is good, while speed measurement is less accurate than at a high PRR. Almost all ground-based aircraft search and surveillance radar systems use some form of MTI. The Army, Navy, Air Force, FAA, USCG, NASA, and DOC are large users of MTI radars.

Airborne Moving Target Indication (AMTI) Radar An MTI radar in an aircraft encounters problem not found in a ground-based system of the same kind because the large undesired clutter echoes from the ground and the sea have a Doppler frequency shift introduced by the motion of the aircraft carrying the radar. The AMTI radar, however, compensates for the Doppler frequency shift of the clutter, making it possible to detect moving targets even though the radar unit itself is in motion. AMTI radars are primarily used by the Army, Navy, Air Force, and USCG.

Pulse Doppler Radar As with the MTI system, the pulse Doppler radar is a type of pulse radar that utilizes the Doppler frequency shift of the echo signal to reject clutter and detect moving aircraft. However, it operates with a much higher PRR than the MTI radar. (A high-PRR pulse Doppler radar, e.g., might have a PRR of 100 kHz, as compared to an MTI radar with PRR of perhaps 300 Hz.) The difference of PRRs gives rise to distinctly different behaviors. The MTI radar uses a low PRR in order to obtain an unambiguous range measurement. This causes the measurement of the target's radial velocity (as derived from the Doppler frequency shift) to be highly ambiguous and can result in missing some target detections. On the other hand, the pulse Doppler radar operates with a high PRR so as to have no ambiguities in the measurement of radial velocity. A high PRR, however, causes a highly ambiguous range measurement. The true range is resolved by transmitting multiple waveforms with different PRRs.

Pulse Doppler radars are used by the Army, Navy, Air Force, FAA, USCG, NASA, and DOC.

High-Range Resolution Radar This is a pulse-type radar that uses very short pulses to obtain range resolution of a target the size ranging from less than a meter to several meters across. It is used to detect a fixed or stationary target in the clutter and for recognizing one type of target from another and works best at short ranges. The Army, Navy, Air Force, NASA, and DOE are users of high-range resolution radars.

Pulse Compression Radar This radar is similar to a high-range resolution radar but overcomes peak power and long-range limitations by obtaining the resolution of a short pulse but with the energy of a long pulse. It does this by modulating either the frequency or the phase of a long, high-energy pulse. The frequency or phase modulation allows the long pulse to be compressed in the receiver by an amount equal to the reciprocal of the signal bandwidth. The Army, Navy, Air Force, NASA, and DOE are users of pulse compression radars.

Synthetic-Aperture Radar (SAR) This radar is employed on an aircraft or satellite and generally its antenna beam is oriented perpendicular to its direction of travel. The SAR achieves high resolution in angle (cross range) by storing the sequentially received signals in memory over a period of time and then adding them as if they were from a large array antenna. The output is a high-resolution image of a scene. The Army, Navy, Air Force, NASA, and NOAA are primary users of SAR radars.

Inverse Synthetic-Aperture Radar (ISAR) In many respects, an ISAR is similar to SAR, except that it obtains cross-range resolution by using Doppler frequency shift that results from target movements relative to the radar. It is usually used to obtain an image of a target. ISAR radars are used primarily by the Army, Navy, Air Force, and NASA.

Side-Looking Airborne Radar (SLAR) This variety of airborne radar employs a large side-looking antenna (i.e., one whose beam is perpendicular to the aircraft's line of flight) and is capable of high-range resolution. (The resolution in cross range is not as good as can be obtained with SAR, but it is simpler than the latter and is acceptable for some applications.) SLAR generates map-like images of the ground and permits detection of ground targets. This radar is used primarily by the Army, Navy, Air Force, NASA, and USCG.

Imaging Radar Synthetic-aperture, inverse synthetic-aperture, and side-looking airborne radar techniques are sometimes referred to as imaging radars. The Army, Navy, Air Force, and NASA are the primary users of imaging radars.

Tracking Radar This kind of radar continuously follows a single target in angle (azimuth and elevation) and range to determine its path or trajectory and to predict its future position. The single-target tracking radar provides target location almost continuously. A typical tracking radar might measure the target location at a rate of ten times per second. Range instrumentation radars are typical tracking radars. Military tracking radars employ sophisticated signal processing to estimate target size or identify specific characteristics before a weapon system is activated against them. These radars are sometimes referred to as fire-control radars. Tracking radars are primarily used by the Army, Navy, Air Force, NASA, and DOE.

Track-While-Scan (TWS) Radar There are two different TWS radars. One is more or less the conventional air surveillance radar with a mechanically rotating antenna. Target tracking is done from observations made from one rotation to another. The other TWS radar is a radar whose antenna rapidly scans a small angular sector to extract the angular location of a target. The Army, Navy, Air Force, NASA, and FAA are primary user of TWS radars.

3D Radar Conventional air surveillance radar measures the location of a target in two-dimensional range and azimuth. The elevation angle, from which target height can be derived, also can be determined. The so-called 3D radar is an air surveillance radar that measures range in a conventional manner but that has an antenna which is mechanically or electronically rotated about a vertical axis to obtain the azimuth angle of a target and which has either fixed multiple beams in elevation or a scanned pencil beam to measure its elevation angle. There are other types of radar (such as electronically scanned phased arrays and tracking radars) that measure the target location in three dimensions, but a radar that is properly called 3D is an air surveillance system that measures the azimuth and elevation angles as just described. The use of 3D radars is primarily by the Army, Navy, Air Force, NASA, FAA, USCG, and DOE.

Electronically Scanned Phased Array Radar An electronically scanned phased array antenna can position its beam rapidly from one direction to another without mechanical movement of large antenna structures. Agile, rapid beam switching permits the radar to track many targets simultaneously and to perform other functions as required. The Army, Navy, and Air Force are the primary users of electronically scanned phased array radars.

Continuous-Wave (CW) Radar Since a CW radar transmits and receives at the same time, it must depend on the Doppler frequency shift produced by a moving target to separate the weak echo signal from the strong transmitted signal. A simple CW radar can detect targets, measure their radial velocity (from the Doppler frequency shift), and determine the direction of arrival of the received signal. However, a more complicated waveform is required for finding the range of the target. Almost all federal agencies used some type of CW radar for applications ranging from target tracking to weapons fire-control to vehicle-speed detection.

Frequency-Modulated Continuous-Wave (FM-CW) Radar If the frequency of a CW radar is continually changed with time, the frequency of the echo signal will differ from that transmitted and the difference will be proportional to the range of the target. Accordingly, measuring the difference between the transmitted and received frequencies gives the range to the target. In such a frequency-modulated continuous-wave radar, the frequency is generally changed in a linear fashion, so that there is an up-and-down alternation in frequency. The most common form of FM-CW radar is the radar altimeter used on aircraft or a satellite to determine their height above the surface of the Earth. Phase modulation, rather than frequency modulation, of the CW signal has also been used to obtain range measurement. The primary users of these radars are the Army, Navy, Air Force, NASA, and USCG.

High-Frequency Over-the-Horizon (HF OTH) Radar This radar operates in the high-frequency (HF) portion of the electromagnetic spectrum (3–30 MHz) to take advantage of the refraction of radio waves by the ionosphere that allows OTH ranges of up to approximately 2000 nautical miles. HF OTH can detect aircraft, ballistic missiles, ships, and ocean-wave effects. The Navy and Air Force use HF OTH radars.

Scatterometer This radar is employed on an aircraft or satellite, and generally its antenna beam is oriented at various aspects to the sides of its track vertically beneath it. The scatterometer uses the measurement of the return echo power variation with aspect angle to determine the wind direction and speed of the Earth's ocean surfaces.

Precipitation Radar This radar is employed on an aircraft or satellite, and generally its antenna beam is scanning at an angle optimum to its flight path to measure radar returns from rainfall to determine rainfall rate.

Cloud Profile Radar Usually employed aboard an aircraft or satellite. The radar beam is oriented at nadir measuring the radar returns from clouds to determine the cloud reflectivity profile over the Earth's surface.

1.3.2 Radar Wave and Frequency Bands

In order to have a better understanding of "radar types," we have to have a better understanding of the frequency bands that operate within that frequency range, consequently we need to look at it from electromagnetic spectrum perspective, and such point of view id well depicted in Fig. 1.17.

The spectrum of electromagnetic waves has frequencies up to 10^{24} Hz. This very large range is subdivided into different subranges due to different physical properties. The subdivision of the frequencies into the different ranges was previously measured according to criteria that were historically developed and are now

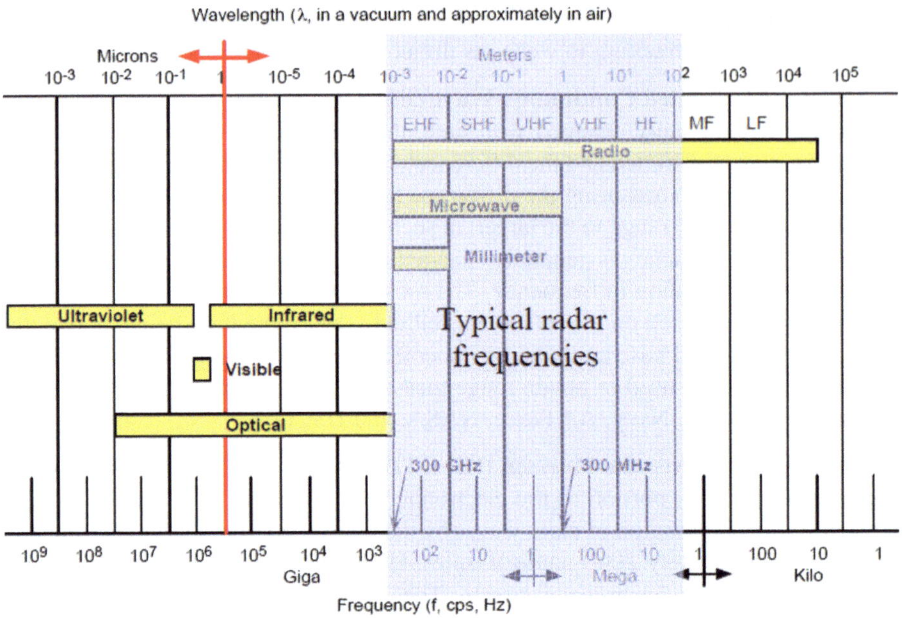

Fig. 1.17 Electromagnetic spectrum ranges

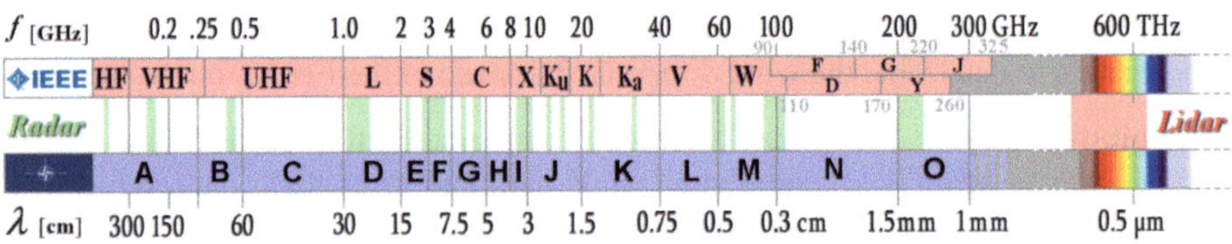

Fig. 1.18 Waves and frequency ranges used by radar

obsolete, and so a new classification of the frequency bands was created. This new classification could not yet fully establish internationally. The traditional frequency band designation is often still used in the literature. In NATO the new subdivision is used.

The following graphics as Fig. 1.18 established by the Institute of Electrical and Electronics Engineers (IEEE) shows an overview of these bandwidths that are used by radar.

Therefore currently there are two valid designation systems for frequency bands which are compared in Fig. 1.1. The IEEE favors the designation system, which originated historically and whose intentionally unsystematic distribution of the letters to the band designation partly originates from the time of World War II. Its selection was initially intended to keep the frequencies used secret.

A newer frequency band classification is used within NATO. Its band boundaries are adapted to the technologies and measurement possibilities in the different frequency ranges. They are almost logarithmically distributed, and the system is open to the high frequencies. In this system, further frequency bands up to the terahertz range can easily be defined in the future. This designation system is also of military origin and is a band division for the electronic war, in which radar equipment finally occupies an essential place.

Since an assignment into the new frequency bands is not always possible without the exact frequency being known, I made use of the traditional band names without comment where they were mentioned in the manufacturer's publications. But be careful! In Germany, for example, companies still use old band names. Radar sets of a so-called C-band family operate with certainty in the new G-band, but radar sets with the letter "L" in the designator (e.g., Signal Multibeam Acquisition Radar for Targeting L (SMART-L)) no longer operate in the L-band but in the D-band.

Note: SMART-L (**S**ignal **M**ultibeam **A**cquisition **R**adar for **T**argeting) is the D-band (former L-band) long-range surveillance radar version of the successful family of SMART multibeam 3D radars. It is designed according to NATO specifications for a volume search radar and designed to fulfill:

- Medium-range detection of the newest generation of small "stealth" air targets.
- Long-range detection of conventional aircraft.
- High electronic counter-countermeasure (ECCM) performance. ECCM techniques are utilized against active deception jamming electronic countermeasure (ECM). Bear in your mind the idea behind the ECCM techniques is that to take electronic warfare to the next level and it is explained in Chap. 2 of this book.
- Guidance support for patrol aircraft.
- Surface surveillance.

Due to its larger power budget, SMART-L is dedicated for the early detection and tracking of very small aircraft and missiles. The accurate 3D target information, gathered by the SMART-L radar, provides an essential contribution to the threat evaluation process, especially in multiple attack scenarios, and it allows the weapon control system to perform the fastest lock-on.

The frequencies of radar sets today range from about 5 MHz to about 130 GHz (130,000,000,000 oscillations per second!). However, certain frequencies are also preferred for certain radar applications. Very-long-range radar systems usually operate at lower frequencies below and including the D-band. Air traffic control radars at an airport operate below 3 GHz air surveillance radar (ASR) or below 10 GHz precision approach radar (PAR).

Note: As it was described before, an airport surveillance radar (ASR) or terminal area radar (TAR) is an air traffic control (ATC) radar system used at airports. It is a midrange primary radar used to detect and display the presence and position of aircraft in the terminal area, the airspace around airports. It usually operates in the frequency range from 2700 to 2900 MHz (E-band), since this frequency range

Fig. 1.19 Air surveillance radar ASR-NG

provides low attenuation due to absorption in heavy rain regions. In addition, this frequency range is still high enough to be able to use highly directional antennas with relatively small dimensions and lower weight. See Fig. 1.19, which is an image air surveillance radar ASR-NG on the test area of the company Hensoldt near Ulm (Germany).

Note: Precision approach radar (PAR) is a primary radar used at aerodromes for approach operations based on specific procedures for the pilot and the controller; however, the use of PARs for civil applications is rapidly decreasing [20]. Precision approach radar offers the possibility of a safe landing even in poor visibility conditions. The radar is placed near the mid-point of the runway (at a distance up to 6000 ft) and works remotely. The radar is particularly important in situations when the pilot has limited sight (because of fog, rain, etc.). In this situation, the radar has to provide the approach controller with maximum quality radar display complemented by computer evaluation of speed, deviations from glide path (or glide slope) and course line, the distance from the previously approaching aircraft, etc. The controller issues azimuth and elevation advisories to the pilot until the aircraft reaches the elevation decision height point, approximately one-half mile from the touchdown point. See Fig. 1.20.

Fig. 1.20 Precision approach radar. (Source: Selex System Integration)

The advantage of the classical method of transmitting the instructions by radio, the so-called talk down, is its general applicability, because no additional equipment in the aircraft is needed.

The technical parameters that a precision approach radar should meet are referred to in a Recommendation of the International Civil Aviation Organization (ICAO). This recommendation includes minimum requirements for technical parameters as well as site conditions [21].

A precision approach always needs height-finder capabilities. An instrument approach and landing, which utilizes lateral guidance, e.g., by a 2D radar but does not utilize vertical guidance, is called non-precise [22, 23]. In this case, depending on the distance measured by the radar, the respective nominal altitude is determined at regular intervals by the air traffic controller from a table and communicated to the pilot by radio.

1.3.2.1 A- and B-Band (HF and VHF Radar)

These radar bands below 300 MHz have a long tradition, as the first radar sets were developed here before and during World War II. The frequency range corresponded to the high-frequency technologies mastered at that time. Later, they were used for early warning radars of extremely long-range, so-called over-the-horizon (OTH), radars. Since the accuracy of angle determination and the angular resolution depend on the ratio of wavelength to antenna size, these radars cannot meet high accuracy requirements. The antennas of these radar sets are nevertheless extremely large and can even be several kilometers long.

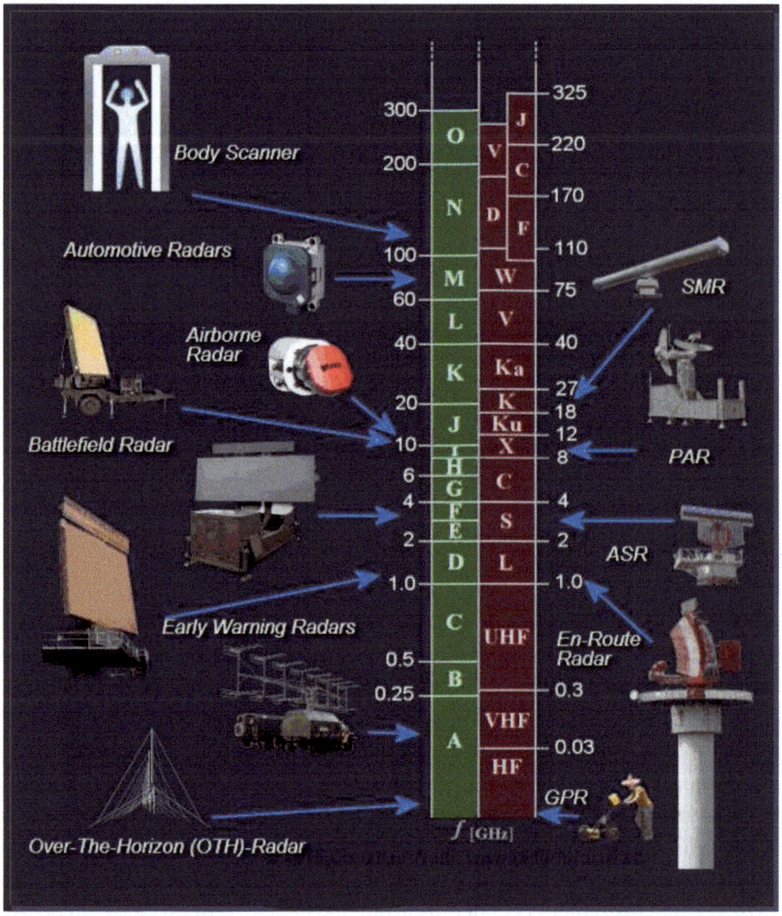

Fig. 1.21 Some radars and its frequency band

Here special abnormal propagation conditions act, which increase the range of the radar again at the expense of the accuracy. Since these frequency bands are densely occupied by communication radio services, the bandwidth of these radar sets is relatively small. See Fig. 1.21.

These frequency bands are currently experiencing a comeback, while the actually used stealth technologies don't have the desired effect at extremely low frequencies.

1.3.2.2 C-Band (UHF Radar)

For this frequency band (300 MHz to 1 GHz), specialized radar sets have been developed which are used as military early warning radar, for example, for the Medium Extended Air Defense System (MEADS) (i.e., Fig. 1.22), or as wind

Fig. 1.22 The MEADS uses a UHF surveillance radar

Fig. 1.23 Ground-penetrating radar in action

profilers in weather observation. These frequencies are damped only very slightly by weather phenomena and thus allow a long range. Newer methods, so-called ultrawideband radars, transmit with very low pulse power from the A- to the C-bands and are mostly used for technical material investigation or partly in archaeology as ground-penetrating radar (GPR) as it is described before. See Fig. 1.23.

1.3.2.3 D-Band (L-Band Radar)

This range is ideally suited for modern long-range air surveillance radars up to a range of 250 nautical miles (≈400 km). Relatively low interference from civil radio communication services enables broadband radiation with very high power. They transmit pulses with high power, wide bandwidth, and an intrapulse modulation to achieve even longer ranges. Due to the curvature of the Earth, however, the range that can be practically achieved with these radar sets is much smaller at low altitudes, since these targets are then obscured by the radar horizon.

In this frequency band, the en route radars or Air Route Surveillance Radars (ARSR) work for air traffic control as it has been described before. In conjunction with a Monopulse Secondary Surveillance Radar (MSSR), these radars operate with a relatively large, slowly rotating antenna (L-band: like large antenna and long range). The designator L-band is good as mnemonic rhyme as **l**arge antenna or **l**ong range.

1.3.2.4 E/F-Band (S-Band Radar)

In the frequency band from 2 to 4 GHz, the atmospheric attenuation is higher than in the D-band. Radar sets require a much higher pulse power to achieve long ranges. An example is the older one military medium power radar (MPR) with up to 20 MW pulse power. In this frequency band, considerable impairments due to weather phenomena are already beginning to occur. Therefore, a couple of weather radars work in E/F-band but more in subtropics and tropic climatic conditions, because here the radar can see beyond a severe storm. Figure 1.24 is a presentation of antenna of MPR without radome.

Special airport surveillance radars (ASR) are used at airports to detect and display the position of aircraft in the terminal area with a medium range up to 50...60 NM

Fig. 1.24 Antenna of MPR

(≈100 km). An ASR detects aircraft position and weather conditions in the vicinity of civilian and military airfields. The designator S-band is good as mnemonic rhyme as **s**maller antenna or **s**horter range (contrary to L-band).

1.3.2.5 G-Band (C-Band Radar)

For this frequency band, mobile military battlefield radars with short and medium range are used. The antennas are small enough to be quickly installed with high precision for weapon control. The influence of weather phenomena is very large, which is why military radar sets are usually equipped with antennas with circular polarization. In this frequency range, most weather radars are also used for moderate climates.

1.3.2.6 I/J-Band (X- and Ku-Band Radars)

Between 8 and 12 GHz, the ratio of wavelength to antenna size has a more favorable value. With relatively small antennas, sufficient angular accuracy can be achieved, which favors military use as airborne radar. On the other hand, the antennas of missile control radar systems, which are very large relative to the wavelength, are still handy enough to be considered as deployable.

This frequency band is mainly used in civil and military applications for maritime navigation radar systems. Small cheap and fast rotating antennas offer sufficient ranges with very good precision. The antennas can be constructed as simple slot radiators or patch antennas.

This frequency band is also popular for spaceborne or airborne imaging radars based on synthetic-aperture radar (SAR) for both military electronic intelligence and civil geographic mapping; see Fig. 1.25. A special application of the inverse synthetic-aperture radar (ISAR) is the monitoring of the oceans to prevent environmental pollution.

Note: A **s**ynthetic-**a**perture **r**adar (**SAR**), or SAR, is a coherent mostly airborne or spaceborne side-looking radar system which utilizes the flight path of the platform to simulate an extremely large antenna or aperture electronically and that generates high-resolution remote sensing imagery. Over time, individual transmit/receive cycles (PRTs) are completed with the data from each cycle being stored electronically.

The signal processing uses magnitude and phase of the received signals over successive pulses from elements of a synthetic aperture. After a given number of cycles, the stored data is recombined (taking into account the Doppler effects inherent in the different transmitter to target geometry in each succeeding cycle) to create a high-resolution image of the terrain being overflown.

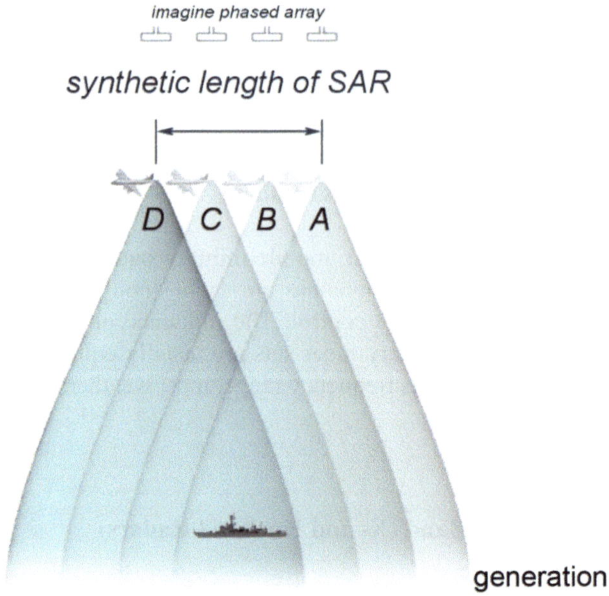

Fig. 1.25 The synthesized expanding beam width. (Source: Christine Wolff 20028)

1.3.2.7 K-Band (K- and Ka-Band Radars)

As the emitted frequency increases, the attenuation in the atmosphere increases, but the possible accuracy and range resolution increase too. Large ranges can no longer be achieved. Radar applications in this frequency range are, for example, airfield surveillance radar, also known as surface movement radar (SMR) or as part of Airport Surface Detection Equipment (ASDE). With extremely short pulses of a few nanoseconds, an excellent range resolution is achieved so that the contours of aircraft and vehicles can be seen on the display.

Note: Surface movement radar (SMR) is the most widely used surveillance system for airport surveillance at the present. SMR refers to primary radar that provides surveillance cover for the maneuvering area, which is defined as that used for the takeoff, landing, and taxiing of aircraft, excluding aprons. Figure 1.26 is a former X-band navy radar used as a surface movement radar on the airfield Logan (USA) in 1995.

SMR provides surveillance of all aircraft and vehicles in this area with a high update rate. SMR antennas are often mounted on the tower, which has good visibility of the maneuvering area. (Very big airfields like the Munich Airport have even a second control tower for its second terminal and the purpose of the airfield taxiway management.) The ground surface environment is quite different from high altitude because of the increased clutter and other physical problems. The quality of surveillance information on the ground is often quite poor and limited by these physical problems.

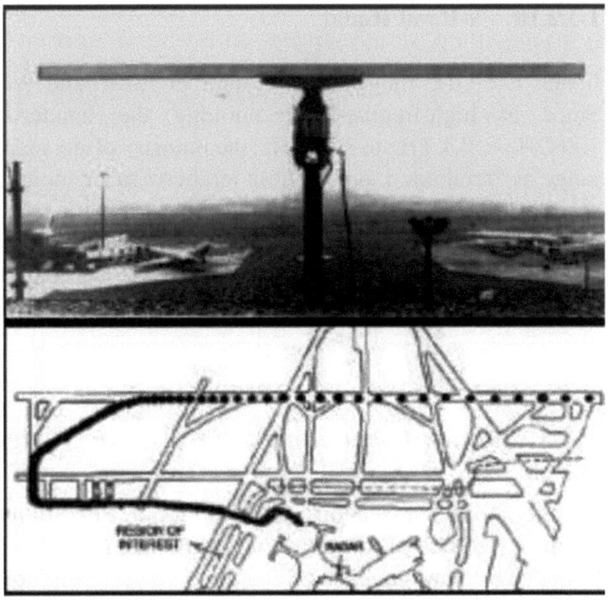

Fig. 1.26 A former X-band navy radar

Use of PSR means that target labeling may not be possible, and hence controllers use visual identification of aircraft (by looking out of the tower window). This is one of the contributing factors to the reduced capacity of airports in low visibility.

1.3.2.8 V-Band Radar

Due to molecular scattering of the atmosphere, the electromagnetic waves suffer a very strong attenuation. Radar applications are limited to a range of a few ten meters.

1.3.2.9 W-Band Radar

Two phenomena of atmospheric attenuation can be observed here: a maximum of attenuation at about 75 GHz and a relative minimum at about 96 GHz. Both frequencies are used practically. At about 75–76 GHz, short-range radar sets are used in automotive engineering as parking aids, brake assist systems, and automatic accident avoidance. This high attenuation through molecular scattering (here through the oxygen molecule O_2) prevents mutual interference through mass use of these radar sets.

There are radar sets operating at 96–98 GHz as laboratory equipments yet. These applications give a preview for a use of radar in extremely higher frequencies as 100 GHz.

1.3.2.10 N-Band Radar

In the 122 GHz range, there is another ISM band for measurement applications. Since in high-frequency technology the terahertz range is defined from 100 GHz = 0.1 THz to 300 GHz, the industry offers radar modules for this frequency range as "terahertz radar." These terahertz radar modules are used, for example, in so-called full-body scanners. Full-body scanners take advantage of the fact that although these terahertz frequencies can easily penetrate dry and non-conductive substances, they cannot penetrate the skin deeper than just a few millimeters due to the moisture of the human skin.

1.3.3 Radar Frequencies, Bands, and Usage

All the above sub-sections and description all different wave and bands can be summarized as it is illustrated in Tables 1.1 and 1.2 here are presenting the radar frequencies, bands and usage and they are very similar to what you are observing in Fig. 1.18.

As far as radar frequencies are concerned, there are no fundamental bounds on radar frequency. Any device that detects and locates a target by radiating

Table 1.1 Radar bands and usage

Band designation	Nominal frequency range	Usage
HF	3–30 MHz	OTH surveillance
VHF	30–300 MHz	Very-long-range surveillance
UHF	300–1000 MHz	Very-long-range surveillance
L	1–2 GHz	Long-range surveillance
S	2–4 GHz	Moderate-range surveillance
		Terminal traffic control
		Long-range weather
C	4–8 GHz	Long-range tracking
		Airborne weather detection
X	8–12 GHz	Short-range tracking
		Missile guidance
		Mapping, marine radar
		Airborne intercept
K_u	12–18 GHz	High-resolution mapping
		Satellite altimetry
K	18–27 GHz	Little use (water vapor)
K_a	27–40 GHz	Very-high-resolution mapping
		Airport surveillance
Millimeter	40–100+ GHz	Experimental

Table 1.2 Standard radar-frequency letter-band nomenclature (Source: IEEE Standard 521 – 1984)

Band designation	Nominal frequency range	Specific frequency range
HF	3–30 MHz	
VHF	30–300 MHz	138–144 MHz
		216–225 MHz
UHF	300–1000 MHz	420–450 MHz
		890–942 MHz
L	1000–2000 MHz	1215–1400 MHz
S	2000–4000 MHz	2300–2500 MHz
		2700–3700 MHz
C	4000–8000 MHz	5250–5925 MHz
X	8000–12,000 MHz	8500–10,680 MHz
K_u	12.0–18 GHz	13.4–14.0 GHz
		15.7–17.7 GHz
K	18–27 GHz	24.05–24.25 GHz
K_a	27–40 GHz	33.4–36.0 GHz
V	40–75 GHz	59–64 GHz
W	75–110 GHz	76–81 GHz
		92–100 GHz
mm	110–300 GHz	126–142 GHz
		144–149 GHz
		231–235 GHz
		238–248 GHz

electromagnetic energy and utilizes the echo scattered from a target can be classed as a radar, no matter what its frequency.

Radars have been operated at frequencies from a few megahertz to the ultraviolet region of the spectrum. The basic principles are the same at any frequency, but the practical implementation is widely different. In practice, most radars operate at microwave frequencies, but there are notable exceptions.

Radar engineers use letter designations, as shown in Table 1.2, to denote the general frequency band at which a radar operates. These letter bands are universally used in radar. They have been officially accepted as a standard by the Institute of Electrical and Electronics Engineers (IEEE) and have been recognized by the US Department of Defense. Attempts have been made in the past to subdivide the spectrum into other letter bands (as for waveguides and for ECM operations), but the letter bands in Table 1.1 are the only ones that should be used for radar.

The International Telecommunication Union (ITU) assigns specific frequency bands for radiolocation (radar) use. These are listed in the third column of Table 2.1. They apply to ITU Region 2, which encompasses North and South America. Slight differences exist in the other two ITU regions. Although L-band, for example, is shown in the second column of the table as extending from 1000 to 2000 MHz, in practice an L-band radar would be expected to be found somewhere between 1215 and 1400 MHz, the frequency band actually assigned by the ITU.

Each frequency band has its own particular characteristics that make it better for certain applications than for others. In the following, the characteristics of the various portions of the electromagnetic spectrum at which radars have been or could be operated are described. The divisions between the frequency regions are not as sharp in practice as the precise nature of the nomenclature.

Extending Tables 1.1 and 1.2 on behalf of radar frequency bands, we can state that the traditional band names originated as code names during World War II and are still in military and aviation use throughout the world. They have been adopted in the United States by the Institute of Electrical and Electronics Engineers (IEEE) and internationally by the International Telecommunication Union. Most countries have additional regulations to control which parts of each band are available for civilian or military use. Table 1.3 here is a more extensive presentation of the above two tables.

Other users of the radio spectrum, such as the broadcasting and electronic countermeasures industries, have replaced the traditional military designations with their own systems.

1.4 Radar Basic, Pulse Repetition Frequency (PRF), and Pulse Repetition Time (PRT)

The pulse repetition frequency (PRF) is the number of pulses of a repeating signal in a specific time unit, normally measured in pulses per second. The term is used within a number of technical disciplines, notably radar.

PRF is the number of times a pulsed activity that occurs every second and this is similar to cycle per second that is used to describe other types of waveform and it is inversely proportional to time period T, which is the property of a pulsed wave and presented as Equation 1.1:

$$T = \frac{1}{\mathrm{PRF}} \tag{1.1}$$

PRF is usually associated with pulse spacing, which is the distance that the pulse travels before the next pulse occurs as demonstrated by Equation 1.2:

$$\text{Pulse Spacing} = (\text{Propagation Speed})/(\text{PRF}) \tag{1.2}$$

In radar, a radio signal of a particular carrier frequency (Fig. 1.27) is turned on and off; the term “frequency” refers to the carrier, while the PRF refers to the number of switches. Both are measured in terms of cycle per second (Fig. 1.28), or hertz. The PRF is normally much lower than the frequency. For instance, a typical World War II radar like the Type 7 GCI radar had a basic carrier frequency of 209 MHz (209 million cycles per second) and a PRF of 300 or 500 pulses per second. A

Table 1.3 Radar frequency bands

Radar frequency bands			
Band name	Frequency range	Wavelength range	Notes
HF	3–30 MHz	10–100 m	Coastal radar systems, over-the-horizon (OTH) radars; "high frequency"
VHF	30–300 MHz	1–10 m	Very long range, ground penetrating; "very high frequency"
P	<300 MHz	>1 m	"P" for "previous," applied retrospectively to early radar systems; essentially HF + VHF
UHF	300–1000 MHz	0.3–1 m	Very long range (e.g., ballistic missile early warning), ground penetrating, foliage penetrating; "ultra-high frequency"
L	1–2 GHz	15–30 cm	Long-range air traffic control and surveillance; "L" for "long"
S	2–4 GHz	7.5–15 cm	Moderate-range surveillance, terminal air traffic control, long-range weather, marine radar; "S" for "short"
C	4–8 GHz	3.75–7.5 cm	Satellite transponders; a compromise (hence "C") between X- and S-bands; weather; long-range tracking
X	8–12 GHz	2.5–3.75 cm	Missile guidance, marine radar, weather, medium-resolution mapping, and ground surveillance; in the United States the narrow range 10.525 GHz $\pm$ 25 MHz is used for airport radar; short-range tracking. Named X-band because the frequency was a secret during World War II
K_u	12–18 GHz	1.67–2.5 cm	High resolution, also used for satellite transponders, frequency under K-band (hence "u")
K	18–24 GHz	1.11–1.67 cm	From German *kurz*, meaning "short"; limited use due to absorption by water vapor, so K_u and K_a were used instead for surveillance. K-band is used for detecting clouds by meteorologists and by police for detecting speeding motorists. K-band radar guns operate at 24.150 $\pm$ 0.100 GHz
K_a	24–40 GHz	0.75–1.11 cm	Mapping, short range, airport surveillance; frequency just above K-band (hence "a"). Photo radar, used to trigger cameras which take pictures of license plates of cars running red lights, operates at 34.300 $\pm$ 0.100 GHz
mm	40–300 GHz	1.0–7.5 mm	Millimeter band subdivided as below. The frequency ranges depend on waveguide size. Multiple letters are assigned to these bands by different groups. These are from Baytron, a now defunct company that made test equipment
V	40–75 GHz	4.0–7.5 mm	Very strongly absorbed by atmospheric oxygen, which resonates at 60 GHz
W	75–110 GHz	2.7–4.0 mm	Used as a visual sensor for experimental autonomous vehicles, high-resolution meteorological observation, and imaging

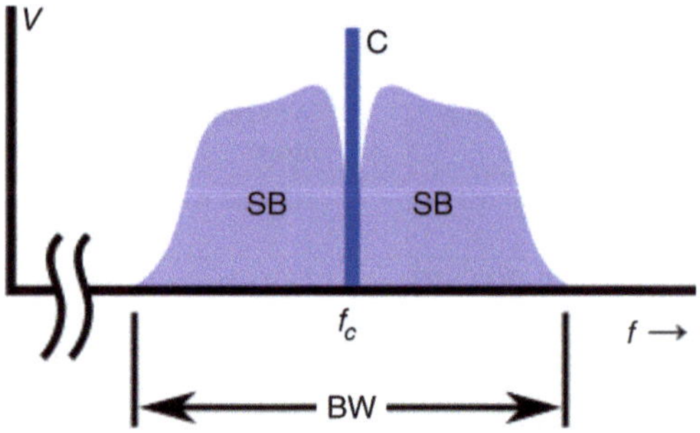

Fig. 1.27 Carrier frequency depiction. (Source: www.wikipedia.com)

Fig. 1.28 A 1000 kilocycle military grade crystal resonator with an octal base. (Source: www.wikipedia.com)

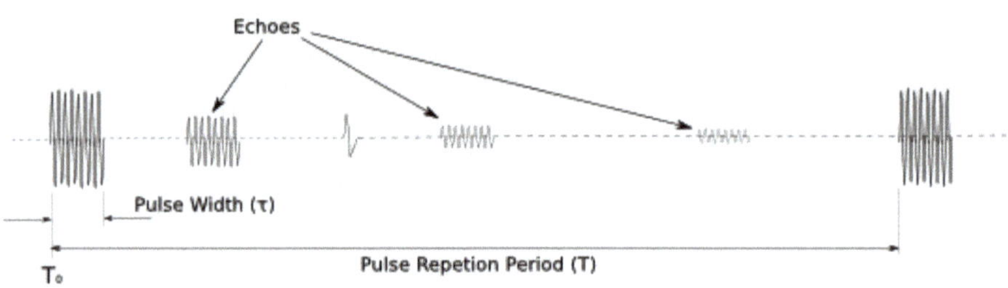

Fig. 1.29 Wave pulse width depiction. (Source: www.wikipedia.com)

related measure is the pulse width (Fig. 1.29), the amount of time the transmitter is turned on during each pulse.

Note: In telecommunications, a *carrier wave*, *carrier signal*, or just *carrier*, is a waveform that is modulated (modified) with an input signal for the purpose of conveying information. This carrier wave usually has a much higher frequency than the input signal does. The purpose of the carrier is usually either to transmit the information through space as an electromagnetic wave or to allow several carriers at different frequencies to share a common physical transmission medium by frequency-division multiplexing. The term originated in radio communication, where the carrier wave is the radio wave which carries the information (modulation) through the air from the transmitter to the receiver. The term is also used for an unmodulated emission in the absence of any modulating signal.

Note: The cycle per second was a once-common English name for the unit of frequency now known as the hertz (Hz). The plural form was typically used, often written cycles per second, cycles/second, c.p.s., c/s, ~, or, ambiguously, just cycles. The term comes from the fact that sound waves have a frequency measurable in their number of oscillations, or *cycles*, per second.

Note: A radar system uses a radio-frequency electromagnetic signal reflected from a target to determine information about that target. In any radar system, the signal transmitted and received will exhibit many of the characteristics described below.

The PRF is one of the defining characteristics of a radar system, which normally consists of a powerful transmitter and sensitive receiver connected to the same antenna. After producing a brief pulse of radio signal, the transmitter is turned off in order for the receiver units to hear the reflections of that signal off distant targets. Since the radio signal has to travel out to the target and back again, the required inter-pulse quiet period is a function of the radar's desired range. Longer periods are required for longer-range signals, requiring lower PRFs. Conversely, higher PRFs produce shorter maximum ranges, but broadcast more pulses, and thus radio energy, in a given time. This creates stronger reflections that make detection easier. Radar systems must balance these two competing requirements.

Using older electronics, PRFs were generally fixed to a specific value or might be switched among a limited set of possible values. This gives each radar system a characteristic PRF, which can be used in electronic warfare to identify the type or class of a particular platform such as a ship or aircraft or, in some cases, a particular unit. Radar warning receivers in aircraft include a library of common PRFs which can identify not only the type of radar, but in some cases the mode of operation. This allowed pilots to be warned when a SA-2 SAM battery had "locked on," for instance. Modern radar systems are generally able to smoothly change their PRF, pulse width, and carrier frequency, making identification much more difficult.

Sonar and LIDAR systems also have PRFs, as does any pulsed system. In the case of sonar, the term pulse repetition rate (PRR) is more common, although it refers to the same concept.

As we stated, the pulse repetition frequency (PRF) of the radar system is the number of pulses that are transmitted per second. See Fig. 1.30.

Radar systems radiate each pulse at the carrier frequency during transmit time (or pulse width (PW)), wait for returning echoes during listening or rest time, and then radiate the next pulse, as shown in the figure. The time between the beginning of one pulse and the start of the next pulse is called pulse repetition time (PRT) and is equal to the reciprocal of PRF as Equation 1.3:

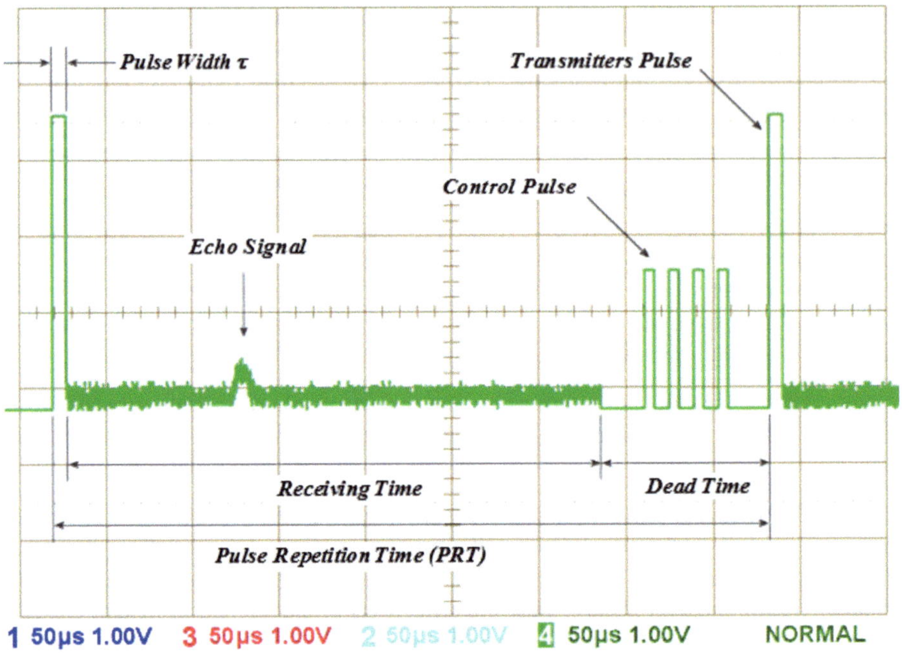

Fig. 1.30 Radar pulse relationships

$$\mathrm{PRT} = \frac{1}{\mathrm{PRF}} \tag{1.3}$$

Definition of context in Fig. 1.14 is described in the following sub-sections.

1.4.1 Receiving Time

Generally, the receiving time is the time between the transmitters' pulses. The receiving time is always smaller than the difference between the pulse repetition period and the length of the transmitter's pulse. It is sometimes also limited by a so-called dead time, in which the receiver is already switched off just before the next transmitting pulse.

In some radars between the transmitting pulse and the receiving time, there is a short recovery time of the duplexer. This recovery time occurs when the duplexer must switch off the receiver response to the high transmitting power. A very low transmitting power, however, can already be received during the transmit pulse also. The receiving time includes transmission time then.

1.4.2 *Dead Time*

If the receiving time ends before the next transmitting pulse, the result is a dead time. During the dead time are carried out system test loops in modern radars generally. Radars that use a phased array antenna urgently need such a dead time. For within this time, the phase shifters of the antenna must be reprogrammed to prepare the antenna for the next direction of the antenna's beam. This can take up to 200 μs, why then the dead time takes quite large values compared with the receive time.

In this dead time, the receiver is already switched off because during the reprogramming the antenna cannot provide received signals. Because during this time, no real data can be processed in any case, this time is used to perform internal testing procedures in the modules of the receive path. This is done in order to verify the operational readiness of certain electronic circuits and to adjust them, if necessary. For this purpose, signals are generated with known size. These signals are fed into the receive paths and their processing in the individual modules is monitored. However, the video processor switches off these pulses, so that they do not appear on the screen. If necessary as a result of the tests, the modules can be automatically reconfigured, and it can be written a detailed error message.

1.4.2.1 Phased Array Antenna

A phased array antenna is an array antenna whose single radiators can be fed with different phase shift. As a result, the common antenna pattern can be steered electrically. The electronic steering is much more flexible and requires less maintenance than the mechanical steering of the antenna. In principle this antenna is functioning based on the effect of interference, which means a phase-dependent superposition of two or usually several radiation sources. See Fig. 1.31.

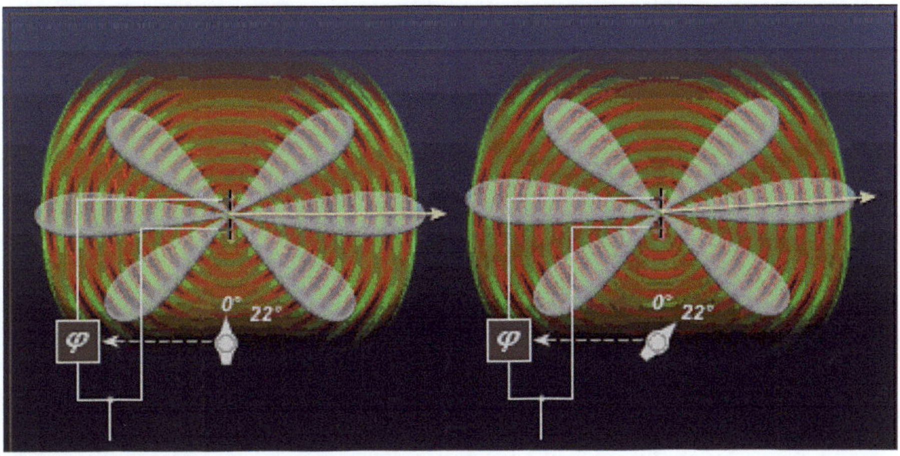

Fig. 1.31 Phased array antenna. (Source: www.radartutorial.eu)

It can be observed that in-phase signals (same color in Fig. 1.31) amplify each other and counter-phase signals cancel each other out. So, if two radiators emit a signal in the same phase shift, a superposition is achieved—the signal is amplified in the main direction and attenuated in the secondary directions. Here in the left radiator group in Fig. 1.31, both radiators are fed with the same phase. The signal is amplified in the main direction, therefore.

In this figure, left two antenna elements fed in phase; right two antenna elements fed out of phase.

In the second graphic in Fig. 1.31, the signal from the upper radiator is transmitted phase-shifted by 22 (i.e., slightly delayed) than from the lower radiator. Therefore, the main direction of the signal emitted in common is slightly steered upward.

Figure 1.31 shows radiators without reflectors. Therefore, the back lobe of the antenna pattern is as big as the main lobe. However, the back lobe has also steered upward.

If the signal to be transmitted is now routed through a phase-regulating module, the direction of radiation can be controlled electronically. However, this is not possible indefinitely, because the effectiveness of this antenna arrangement is greatest in a main direction perpendicular to the antenna field, while extreme tilting of the main direction increases the number and size of the unwanted side-lobes while at the same time reducing the effective antenna area. The sine theorem can be used to calculate the necessary phase shift. See Fig. 1.32, where it presents animation of the electronically steered beam.

Any type of antenna can be used as a radiator in the phased array antenna. Significantly, the single radiators must be controlled with a variable phase shift, and thus the main direction of the radiation can be changed continuously. To achieve high directivity, many radiators are used in the antenna field. The antenna of the

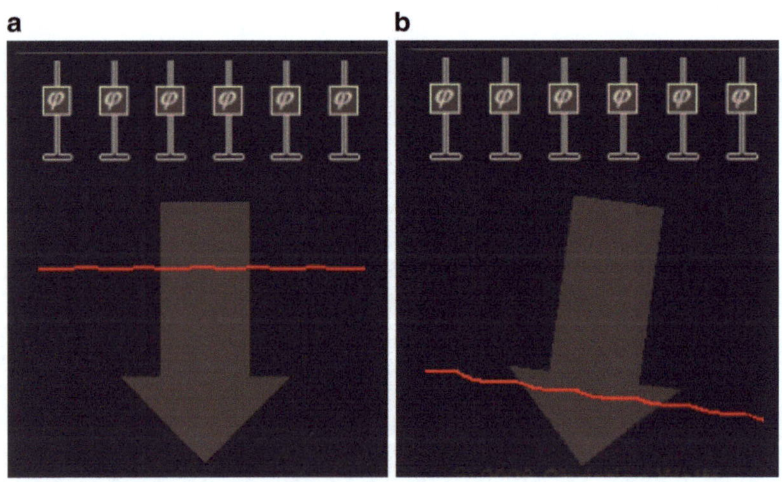

Fig. 1.32 Animation of the electronically steered beam. (**a**) Perpendicular, (**b**) oblique

RRP-117, for example, consists of 1584 radiators whose received signal is still combined in an analogue way to the antenna pattern. More modern multifunction radar sets, on the other hand, use the digital beamforming during the reception.

1.4.2.2 Phased Shifter

Phase shifters switching different detour lines are faster than regulators. In Fig. 1.33, a 4-bit switching phase shifter which is used in radar unit is shown. Different detour lines are switched to the signal way. It created therefore 16 different phase angles between 0° and 337.5° in steps with a distance of 22.5.

The inductivities (the thin meander wires as low-pass filters) also can be recognized in the switching voltage supplies for the altogether 24 pin diodes.

Since this phase shifter module works both for the transmitting way and for the reception way, branching between these two paths is attached with pin diode switches on the ceramic strap at the entrance and exit of the module.

The same data word must be used for the reception time and for the transmitting moment. It is easy to understand: This one radiator, transmitting the latest phase shift, first receives the echo signal. Its phase shifter must have the largest detour line for diagram forming in a decided direction. The same detour line is needed for a summation of the received energy. See Fig. 1.34.

The phase shifter routes the microwave signal that is supplied to each radiating element through cables of varying length. The cables delay the wave, thereby shifting the relative phase of the output. The illustration shows the three basic delays each phase shifter can introduce. The switches are fast pin diode switches. A central computer calculates the proper phase delay for each of the radiating elements and switches in the appropriate combination of phase-shifters pathways.

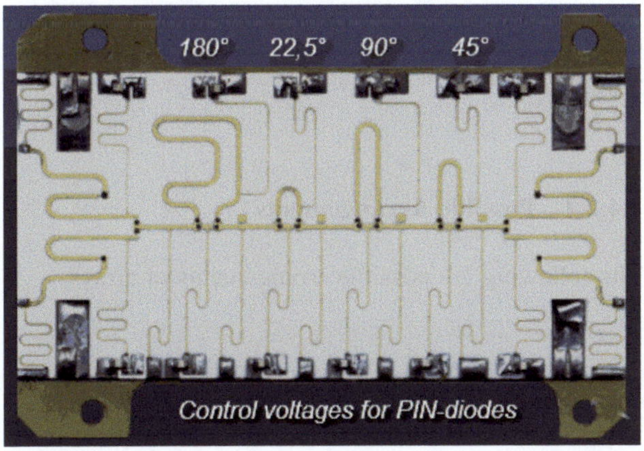

Fig. 1.33 Circuit board with a phase shifter wiring with switched detour lines. (Source: www.radartutorial.eu)

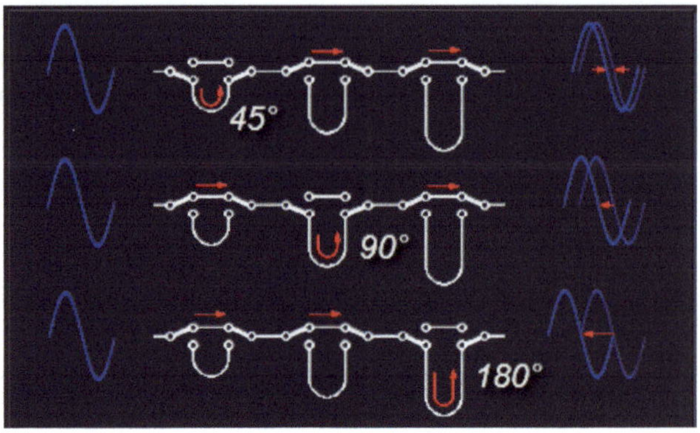

Fig. 1.34 Wiring of the phase shifters' delay lines. (Source: www.radartutorial.eu)

Table 1.4 Phased array antenna advantages and disadvantages list

Advantages	Disadvantages
• High antenna gain with large side-lobe attenuation • Very fast change of beam direction (in range of microseconds) • High beam agility • Arbitrary space scanning • Freely selectable dwell time • Multifunction operation by simultaneous generation of multiple beams • Failure of some components does not result in a complete system failure	• Limited scanning range (up to max. 120 in azimuth and elevation)[a] • Deformation of the antenna pattern during beam steering • Low-frequency agility • Very complex structure (computer, phase shifter, data bus to each radiator) • High costs (still)

[a]Note: *The limitation of the scanning range can be overcome with a three-dimensional radiator distribution. This arrangement of the radiators got the name crow's nest antenna*

1.4.2.3 Advantages and Disadvantages

The advantages and disadvantages of phased array antenna (PAA) are listed below in Table 1.4:

1.4.2.4 Possible Arrangements

The following are possible arrangements of arrays.

Linear Array

These phased array antennas consist of lines, which are commonly controlled by a single phase shifter. (Only one phase shifter is needed per group of radiators in this

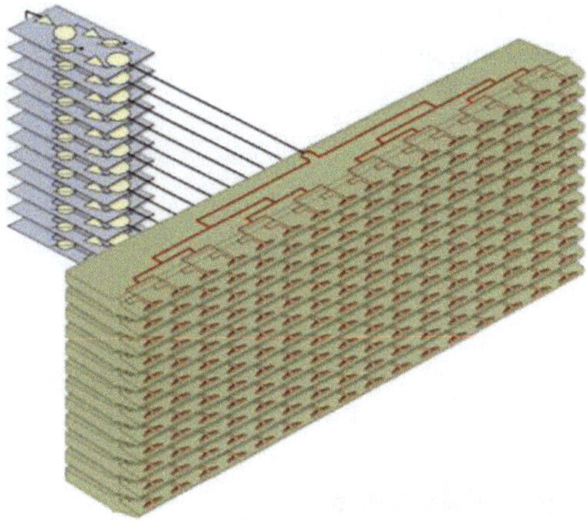

Fig. 1.35 Linear array of a phased array antenna. (Source: www.radartutorial.eu)

line.) A number of linear arrays arranged vertically on top of each other form a flat antenna. See Fig. 1.35.

- **Advantage**: simple arrangement
- **Disadvantage**: beam steering only in a single plane (**Source**: www.radartutorial.eu)
- Examples given:
  - PAR-80 (horizontal beam deflection)
  - FPS-117 (vertical beam deflection)
  - Large vertical aperture (LVA), an SSR antenna with fixed beam pattern

Planar Array

These phased array antennas consist completely of single elements with a phase shifter per element. The elements are arranged like a matrix; the flat arrangement of all elements forms the entire antenna. See Fig. 1.36.

- **Advantage**: Strahlschwenkung in zwei Ebenen möglich
- **Disadvantage**: a large number of phase shifters
- Examples given: AN-FPS-85 and Thomson Master-A

Frequency Scanning Array

The frequency scanning array is a special case of the phased array antenna, in which the beam steering is controlled by the transmitter's frequency without use of any phase shifter. The beam steering is a simple function of the frequency. This type of phased array antenna was often used in older radar sets. See Fig. 1.37.

A vertical antenna array is fed serially. At the main frequency $\mathbf{F_1}$, all radiators get a part of the power of the same phase through structurally identical detours, which

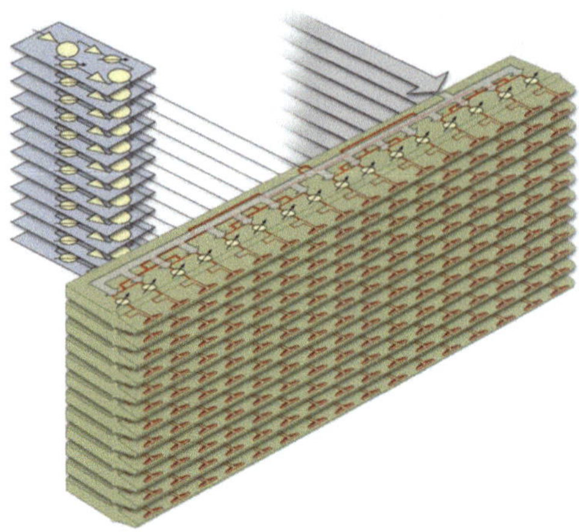

Fig. 1.36 Planar array of a phased array antenna. (Source: www.radartutorial.eu)

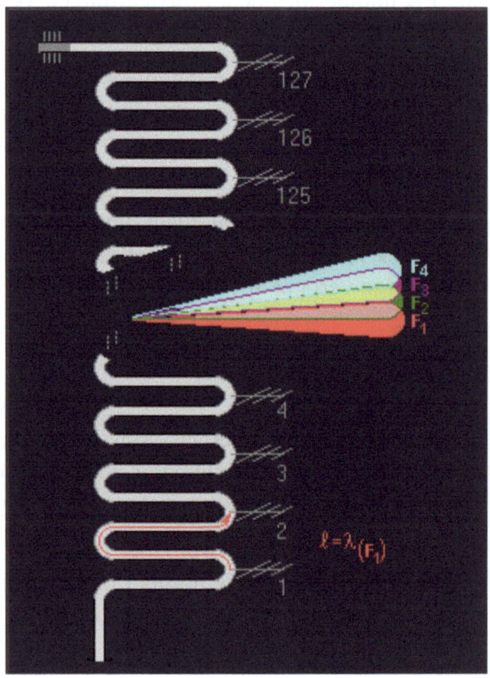

Fig. 1.37 Frequency scanning array. (Source: www.radartutorial.eu)

cause a phase shift of $n \cdot 360°$. All radiators therefore radiate with the same phase. The resulting beam is thus perpendicular to the antenna's plane.

If the transmitter's frequency is increased by a few percent, however, the constructively defined length of the detour lines is no longer correct. At a higher frequency, the wavelength decreases and the detour line is now a bit too long. There appears a phase shift from one radiator to the next radiator. The first radiator radiates this few percent earlier than the next neighboring radiator, etc. The resulting beam for the $\mathbf{F_2}$ frequency is thus steered upward by the angle Θ_s.

Although this type of beam steering is very simple, it is limited to a few permanently installed frequencies. In addition to the susceptibility to interference, there are even more limitations to be accepted, e.g., this radar set cannot use pulse compression because its bandwidth is too low. See description of feeding systems of phased array as defined in this reference by Radartutorial.eu [24].

1.5 Calculation of the Phase Shift

How large must be the phase shift $x = \Delta\varphi$ from one radiator to the next radiator to achieve a desired deflection angle?

A linear arrangement of isotropic single radiators is considered. See Fig. 1.38.

Between the radiators, between the respective beam of the deflection angle and the applied phase shift, a right-angled triangle can be drawn, whose shorter side lies on the beam. The hypotenuse is the distance between two radiators. The third side is an auxiliary line perpendicular to the beam direction of the previous radiator.

$$x = d \cdot \sin \Theta_s \tag{1.4}$$

This distance x can be set in relation to the wavelength as:

$$\frac{360^\circ}{\Delta\varphi} = \frac{\lambda}{x} \tag{1.5}$$

where:

$\Delta\varphi$ = phase shift between two successive elements
d = distance between the radiating elements
Θ_s = beam steering

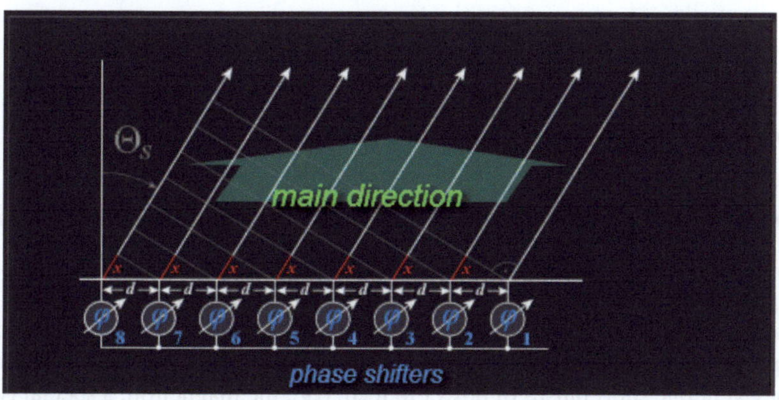

Fig. 1.38 Graphic derivation of the formula. (Source: www.radartutorial.eu)

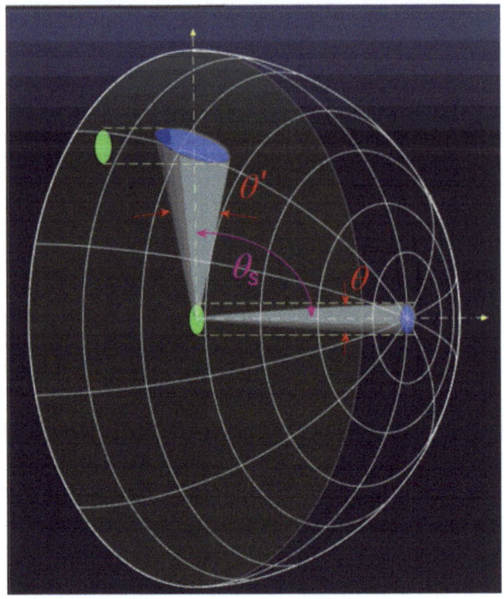

Fig. 1.39 Headlight model of a phased array antenna [25]. (Source: www.radartutorial.eu)

Both Equations 1.3, 1.4 and 1.5 together are the solution of Equation 1.6 as:

$$\Delta\varphi = \frac{360^{\circ} \cdot d \cdot \sin \Theta_s}{\lambda} \tag{1.6}$$

Note that Fig. 1.30 also shows the reason why a phased array antenna focuses worse at larger angles as depicted in Fig. 1.39 [3].

The auxiliary line perpendicular to the adjacent radiator is always smaller than the radiator distance d at an angle that differs from the main beam direction. If the distance "seen" from the deflected beam direction is smaller than the optimum distance d, the antenna quality must deteriorate, which results in a wider antenna pattern.

1.5.1 Modulators

Modulators act to provide the waveform of the RF pulse. There are two different radar modulator designs:

- High-voltage switch for non-coherent keyed power oscillators [26]. These modulators consist of a high-voltage pulse generator formed from a high-voltage supply, a pulse-forming network, and a high-voltage switch such as a thyratron. They generate short pulses of power to feed, e.g., the magnetron, a special type of vacuum tube that converts direct current (DC) (usually pulsed) into microwaves.

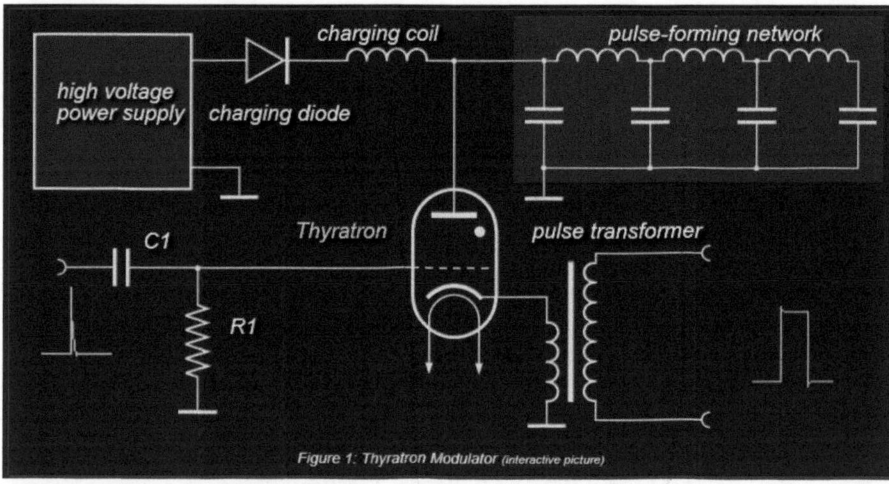

Fig. 1.40 Thyratron modular. (Courtesy of Radartutorial.eu)

This technology is known as pulsed power. In this way, the transmitted pulse of RF radiation is kept to a defined and usually very short duration.

Furthermore, radio-frequency energy in radar is transmitted in short pulses with time durations that may vary from 1 to 50 μs or more. In order to generate this short pulse of high power, a special modulator is required which generates a high voltage for the transmitter tube at the moment of transmission. This radar modulator switches on the anode voltage for the high-power tube for the duration of the pulse. Therefore, it is sometimes called "keyed on/off" radar modulator. See Fig. 1.40.

However, high-power amplifiers using cross-field amplifiers (amplitron) [27] also require such a radar modulator, as they may only get the anode voltage for the duration of the transmission pulse.

The schematic modular presented in Fig. 1.40 uses a pulse-forming network for energy storage. This pulse-forming network is charged to twice the voltage of the high-voltage power supply unit during charging using the magnetic field of the charging coil. This charging coil simultaneously limits the charging current. A charging diode is inserted so that the pulse-forming network is not discharged via the internal resistance of the power supply after charging.

The hydrogen thyratron [28] operates as an electronic switch and is controlled by a short trigger. The R-C combination separates the thyratron input from the pre-amplifier's bias voltage. The pulse transformer is used to adjust the impedances during the discharging.

- Hybrid mixers [29], fed by a waveform generator and an exciter for a complex but coherent waveform. This waveform can be generated by low-power/low-voltage input signals. In this case the radar transmitter must be a power amplifier, e.g., a klystron or a solid-state transmitter. In this way, the transmitted pulse is intrapulse-modulated, and the radar receiver must use pulse compression techniques.

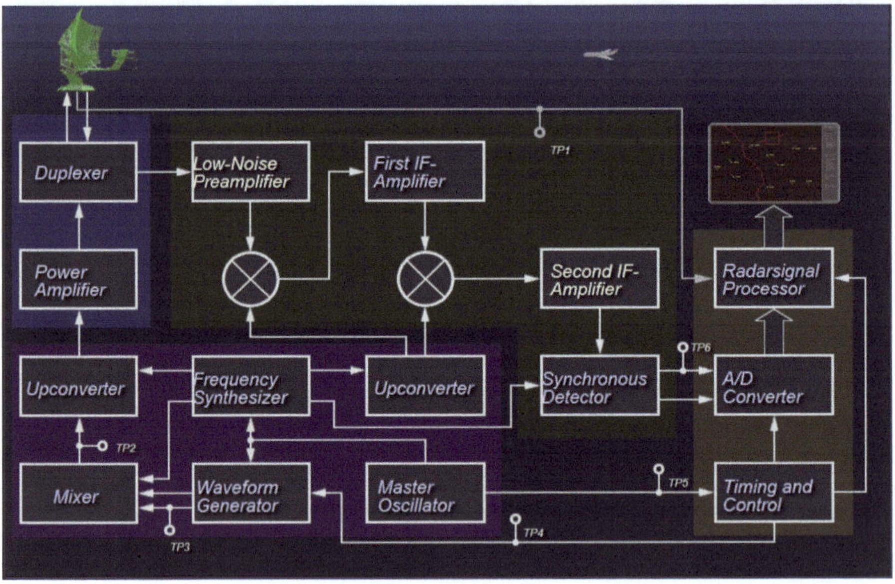

Fig. 1.41 A simplified block diagram of a fully coherent radar. (Courtesy of Radartutorial.eu)

Fig. 1.42 Physical construction of a TWT. (Courtesy of Radartutorial.eu)

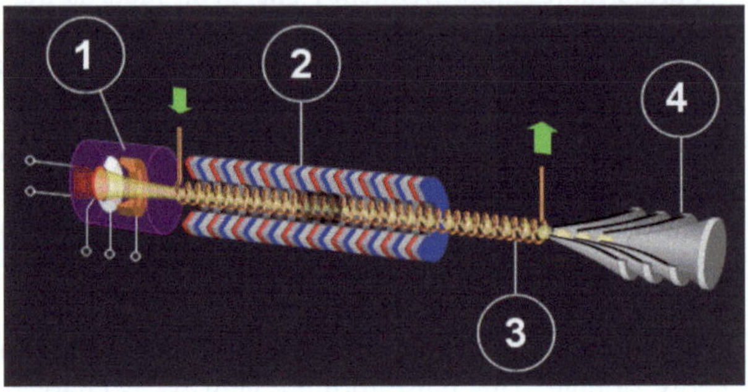

At a fully coherent radar, all the necessary clocks, pulses, gates, and frequencies are derived from the highly stable oscillation of a master oscillator and are synchronous with its oscillation. All derivative frequencies have a fixed phase relationship to this one master oscillator. See Fig. 1.41.

The block diagram on the figure illustrates the principle of a fully coherent radar. The fundamental feature is that all signals are derived at low level and the output device serves only as an amplifier. All the signals are generated by one master timing source, usually a synthesizer, which provides the optimum phase coherence for the whole system. The output device would typically be a klystron, traveling-wave tube (TWT) (i.e., Fig. 1.42), or solid-state. Fully coherent radars exhibit none of the drawbacks of the pseudo-coherent radars, which we studied as before.

Figure 1.42 shows the physical construction of a typical TWT, which consists of four basic elements:

1. Electron gun which produces and then accelerates an electron beam along the axis of the tube.
2. Magnetic electron beam focusing system which provides a magnetic field along the axis of the tube to focus the electrons into a tight beam.
3. Slow-wave structure as RF interaction circuit, e.g., a coiled wire (helix) at the center of the tube, that provides a low-impedance transmission line for the RF energy within the tube.
4. Collector. The electron beam is received at the collector after it has passed through the slow-wave structure.

All components of the TWT are held under a very high vacuum. The RF input and output may couple onto and removed from the helix by waveguide directional couplers that have no physical connection to the helix [30].

1.5.2 Burst Mode

The distribution of the dead time does not have to be uniform. It can be also be transmitted a number of pulses in rapid succession one after the other with each a short receive time before dead time appears. For example, if several pulse periods are oriented in the same direction as like necessary for pulse pair processing [27] and moving target detection, then a dead time is not needed. This has advantages for the time budget [31] of the radar. A random unwanted change in the phase angle of the generator is not likely after a shorter time (Fig. 1.43).

Therefore, the radar will be more accurate in the distance measurement. Simultaneously, the pulse repetition frequency changes in this short period of time: it is very higher than the average. The higher the pulse repetition frequency, the better is the unambiguous measurement of the velocity (see Doppler ambiguity) [32].

The burst mode is mostly used in didactical radars [33]. These radars do not require large receiving time for the extremely short distances within a training room. However, they require a longer dead time to transfer the data of the echo signals over a relatively narrowband serial cable to the computer. For example, they transmit ten pulses per second only, which corresponds to an average pulse repetition frequency of 10 Hz. These 10 pulses are transmitted but within 200 μs.

For the calculation of an unambiguous Doppler frequency that corresponds to a pulse repetition frequency of 50 kHz, the dead time which follows is almost a full second. During this time the data are transferred via USB using a sampling rate of up to 280 Mbit/s.

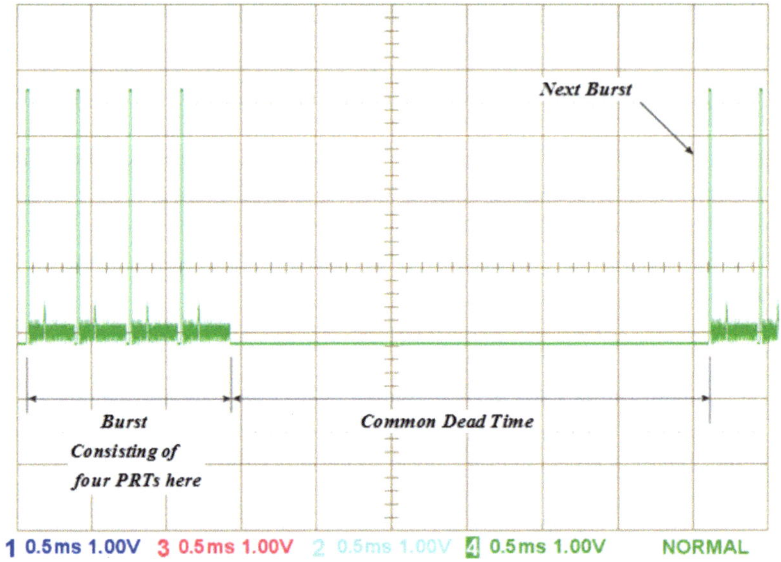

Fig. 1.43 Burst mode of a pulse radar

1.5.3 Ambiguity Range

Pulse repetition frequency (PRF) is crucial to perform measurements for certain physics phenomenon. For example, a tachometer may use a strobe light with an adjustable PRF to measure rotational velocity. The PRF for the strobe light is adjusted upward from a low value until the rotating object appears to stand still. The PRF of the tachometer would then match the speed of the rotating object. Other types of measurements involve distance using the delay time for reflected echo pulses from light, microwaves, and sound transmissions. The devices that measure distance as part of PRF systems are:

- Radar
- Laser range finder
- Sonar

Different PRFs allow systems to perform very different functions. A radar system uses a radio-frequency electromagnetic signal reflected from a target to determine information about that target.

PRF is required for radar operation. This is the rate at which transmitter pulses are sent into air or space.

In range ambiguity principle works based on a radar system determines range through the time delay between pulse transmission and reception by the relation as Equation 1.7:

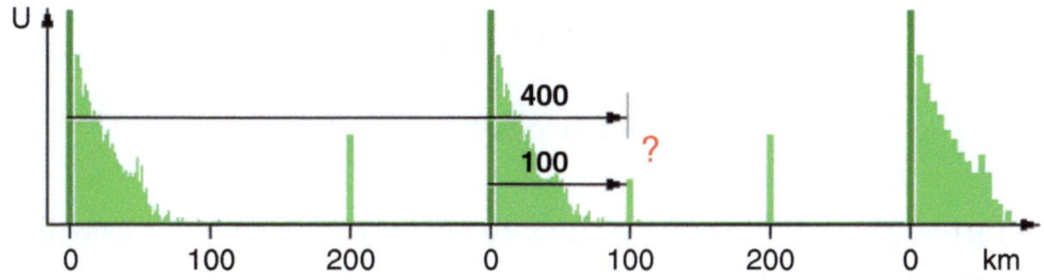

Fig. 1.44 Target being echoed illustration

$$\text{Range} = \frac{c\tau}{2} \tag{1.7}$$

For accurate range determination (Fig. 1.44), a pulse must be transmitted and reflected before the next pulse is transmitted. This gives rise to the maximum unambiguous range limit as Equation 1.8:

$$\text{Max Range} = \frac{c\tau_{\text{PRT}}}{2} = \frac{c}{2\text{PRF}} \Leftrightarrow \boxed{\tau_{\text{PRT}} = \frac{1}{\text{PRF}}} \tag{1.8}$$

In Fig. 1.44, a real target in 100 km or a second-sweep echo in a distance of 400 km?

The maximum range also defines a range ambiguity for all detected targets. Because of the periodic nature of pulse radar systems, it is impossible for some radar system to determine the difference between targets separated by integer multiples of the maximum range using a single PRF. More sophisticated radar systems avoid this problem through the use of multiple PRFs either simultaneously on different frequencies or on a single frequency with a changing PRT.

The range ambiguity resolution process is used to identify true range when PRF is above this limit.

1.5.3.1 Low Pulse Repetition Frequency

Systems using PRF below 3 kHz are considered low PRF because direct range can be measured to a distance of at least 50 km. Radar systems using low PRF typically produce unambiguous range.

Unambiguous Doppler processing becomes an increasing challenge due to coherency limitations as PRF falls below 3 kHz.

For example, an L-band radar with 500 Hz pulse rate produces ambiguous velocity above 75 m/s (170 mile/h) while detecting true range up to 300 km. This combination is appropriate for civilian aircraft radar and weather radar.

$$\begin{cases} 300\ \text{km range} = \dfrac{c}{2 \times 500} \\ 75\ \text{m/s velocity} = \dfrac{500 \times c}{2 \times 10^9} \end{cases} \tag{1.9}$$

Low PRF radars have reduced sensitivity in the presence of low-velocity clutter that interferes with aircraft detection near terrain. Moving target indicator is generally required for acceptable performance near terrain, but this introduces radar scalloping issues that complicate the receiver. Low PRF radars intended for aircraft and spacecraft detection are heavily degraded by weather phenomenon, which cannot be compensated using moving target indicator.

1.5.3.2 Medium Pulse Repetition Frequency

Range and velocity can both be identified using medium PRF, but neither one can be identified directly. Medium PRF is from 3 to 30 kHz, which corresponds with radar range from 5 to 50 km. This is the ambiguous range, which is much smaller than the maximum range. Range ambiguity resolution is used to determine true range in medium PRF radar.

Medium PRF is used with pulse Doppler radar, which is required for look-down/shoot-down capability in military systems. Doppler radar return is generally not ambiguous until velocity exceeds the speed of sound.

A technique called ambiguity resolution is required to identify true range and speed. Doppler signals fall between 1.5 and 15 kHz, which are audible, so audio signals from medium PRF radar systems can be used for passive target classification.

For example, an L-band radar system using a PRF of 10 kHz with a duty cycle of 3.3% can identify true range to a distance of 450 km (30 × *c*/10,000 km/s). This is the instrumented range. Unambiguous velocity is 1500 m/s (3300 mile/h).

$$\begin{cases} 450\ \text{km range} = \dfrac{c}{0.033 \times 2 \times 10,000} \\ 1500\ \text{m/s velocity} = \dfrac{10,000 \times c}{2 \times 10^9} \end{cases} \tag{1.10}$$

The unambiguous velocity of an L-band radar using a PRF of 10 kHz would be 1500 m/s (3300 mile/h) (10,000 × *c*/(2 × 10^9)). True velocity can be found for objects moving under 45,000 m/s if the band-pass filter admits the signal (1500/0.033).

Medium PRF has unique radar scalloping issues that require redundant detection schemes. Bear in your mind that scalloping is a radar phenomenon that reduces sensitivity for certain distance and velocity combinations.

The name is derived from the appearance of areas that are scooped out of graphs that indicate radar sensitivity.

Moving objects cause a phase shift within the transmit pulse that produces signal cancelation. This phenomenon also has detrimental effect on moving target indicator systems, where the detection scheme subtracts signals received from two or more transmit pulses.

1.5.3.3 High Pulse Repetition Frequency

Systems using PRF above 30 kHz function better known as interrupted continuous-wave (ICW) radar because direct velocity can be measured up to 4.5 km/s at L-band, but range resolution becomes more difficult.

High PRF is limited to systems that require close-in performance, like proximity fuses and law enforcement radar.

For example, if 30 samples are taken during the quiescent phase between transmit pulses using a 30 kHz PRF, then true range can be determined to a maximum of 150 km using 1 μs samples (30 × *c*/30,000 km/s). Reflectors beyond this range might be detectable, but the true range cannot be identified.

$$\begin{cases} 450\ \text{km range} = \dfrac{30 \times c}{02 \times 30{,}000} \\ \\ 4500\ \text{m/s velocity} = \dfrac{30{,}000 \times c}{2 \times 10^9} \end{cases} \tag{1.11}$$

It becomes increasingly difficult to take multiple samples between transmit pulses at these pulse frequencies, so range measurements are limited to short distances.

1.5.3.4 Sonar

Sonar systems operate much like radar, except that the medium is liquid or air, and the frequency of the signal is either audio or ultra-sonic. Like radar, lower frequencies propagate relatively higher energies longer distances with less resolving ability. Higher frequencies, which damp out faster, provide increased resolution of nearby objects.

Signals propagate at the speed of sound in the medium (almost always water), and maximum PRF depends upon the size of the object being examined. For example, the speed of sound in water is 1.497 m/s, and the human body is about 0.5 m thick, so the PRF for ultrasound images of the human body should be less than about 2 kHz (1.497/0.5).

As another example, ocean depth is approximately 2 km, so sound takes over a second to return from the sea floor. Sonar is a very slow technology with very low PRF for this reason.

1.5.3.5 Laser

Light waves can be used as radar frequencies, in which case the system is known as LIDAR, which is short for "LIght raDAR" or basically LIDAR.

Laser range or other light signal frequency range finders operate just like radar at much higher frequencies. Non-laser light detection is utilized extensively in automated machine control systems (e.g., electric eyes controlling a garage door, conveyor sorting gates, etc.), and those that use pulse rate detection and ranging are at heart, the same type of system as a radar—without the bells and whistles of the human interface.

Unlike lower radio signal frequencies, light does not bend around the curve of the Earth or reflect off the ionosphere like C-band search radar signals, and so LIDAR is useful only in line of sight applications like higher-frequency radar systems.

1.5.4 Unambiguity Range

In this section we describe the unambiguity range for both *single* and *multiple* pulse repetition frequency (PRF).

1.5.4.1 Single Pulse Repetition Frequency

In simple systems, echoes from targets must be detected and processed before the next transmitter pulse is generated if range ambiguity is to be avoided. Range ambiguity occurs when the time taken for an echo to return from a target is greater than the pulse repetition period (T); if the interval between transmitted pulses is 1000 μs, and the return time of a pulse from a distant target is 1200 μs, the apparent distance of the target is only 200 μs. In sum, these "second echoes" appear on the display to be targets closer than they really are.

Consider the following example: if the radar antenna is located at around 15 m above sea level, then the distance to the horizon is pretty close (perhaps 15 km). Ground targets further than this range cannot be detected; thus, the PRF can be quite high; a radar with a PRF of 7.5 kHz will return ambiguous echoes from targets at about 20 km or over the horizon. If, however, the PRF was doubled to 15 kHz, then the ambiguous range is reduced to 10 km, and targets beyond this range would only appear on the display after the transmitter has emitted another pulse. A target at 12 km would appear to be 2 km away, although the strength of the echo might be much lower than that from a genuine target at 2 km.

The maximum non-ambiguous range varies inversely with PRF and is given by:

$$\text{Range}_{\text{max unambiguous}} = \left(\frac{c}{2\text{PRF}}\right) \tag{1.12}$$

where c is the speed of light. If a longer unambiguous range is required with this simple system, then lower PRFs are required, and it was quite common for early search radars to have PRFs as low as a few hundred Hz, giving an unambiguous range out to well in excess of 150 km. However, lower PRFs introduce other problems, including poorer target painting and velocity ambiguity in pulse Doppler systems.

1.5.4.2 Multiple Pulse Repetition Frequency

Modern radars, especially air-to-air combat radars in military aircraft, may use PRFs in the tens to hundreds of kilohertz and stagger the interval between pulses to allow the correct range to be determined. With this form of staggered PRF, a *packet* of pulses is transmitted with a fixed interval between each pulse, and then another *packet* is transmitted with a slightly different interval. Target reflections appear at different ranges for each *packet*; these differences are accumulated, and then simple arithmetical techniques may be applied to determine true range. Such radars may use repetitive patterns of *packets*, or more adaptable *packets* that respond to apparent target behaviors. Regardless, radars that employ the technique are universally coherent, with a very stable radio frequency, and the pulse *packets* may also be used to make measurements of the Doppler shift (a velocity-dependent modification of the apparent radio frequency), especially when the PRFs are in the hundreds-of-kilohertz range. Radars exploiting Doppler effects in this manner typically determine relative velocity first, from the Doppler effect, and then use other techniques to derive target distance.

1.5.5 Maximum Unambiguous Range

At its most simplistic, maximum unambiguous range (MUR) for a pulse stagger sequence may be calculated using the total sequence period (TSP).

TSP is defined as the total time it takes for the pulsed pattern to repeat. This can be found by the addition of all the elements in the stagger sequence. The formula is derived from the speed of light and the length of the sequence.

$$\mathrm{MUR} = (c \times 0.5 \times \mathrm{TSP}) \tag{1.13}$$

where c is the speed of light, usually in meters per microsecond, and TSP is the addition of all the positions of the stagger sequence, usually in microseconds. However, in a stagger sequence, some intervals may be repeated several times; when this occurs, it is more appropriate to consider TSP as the addition of all the unique intervals in the sequence.

Also, it is worth remembering that there may be vast differences between the MUR and the maximum range (the range beyond which reflections will probably be too weak to be detected) and that the maximum *instrumented* range may be *much* shorter than either of these. A civil marine radar, for instance, may have user-selectable maximum *instrumented* display ranges of 72 or 96 or rarely 120 nautical miles, in accordance with international law, but maximum unambiguous ranges of over 40,000 nautical miles and maximum detection ranges of perhaps 150 nautical miles. When such huge disparities are noted, it reveals that the primary purpose of staggered PRF is to reduce "jamming," rather than to increase unambiguous range capabilities.

1.6 Staggered Pulse Repetition Frequency (PRF)

Staggered PRF is a transmission process where the time between interrogations from radar changes slightly, *in a patterned and readily discernible repeating manner.* The change of repetition frequency allows the radar, on a pulse-to-pulse basis, to differentiate between returns from its own transmissions and returns from other radar systems with the same PRF and a similar radio frequency.

Consider a radar with a constant interval between pulses; target reflections appear at a relatively constant range related to the flight time of the pulse. In today's very crowded radio spectrum, there may be many other pulses detected by the receiver, either directly from the transmitter or as reflections from elsewhere. Because their apparent "distance" is defined by measuring their time relative to the last pulse transmitted by "our" radar, these "jamming" pulses could appear at any apparent distance. When the PRF of the "jamming" radar is very similar to "our" radar, those apparent distances may be very slow-changing, just like real targets. By using stagger, a radar designer can force the "jamming" to jump around erratically in apparent range, inhibiting integration and reducing or even suppressing its impact on true target detection.

Without staggered PRF, any pulses originating from another radar on the same radio frequency might appear stable in time and could be mistaken for reflections from the radar's own transmission. With staggered PRF the radar's own targets appear stable in range in relation to the transmit pulse, while the "jamming" echoes may move around in apparent range (uncorrelated), causing them to be rejected by the receiver.

Staggered PRF is only one of several similar techniques used for this, including jittered PRF (where the pulse timing is varied in a less-predictable manner), pulse-frequency modulation, and several other similar techniques whose principal purpose is to reduce the probability of unintentional synchronicity. These techniques are in widespread use in marine safety and navigation radars, by far the most numerous radars on planet Earth today.

In summary, staggered pulse repetition frequency (PRF) is where the time between interrogations from radar changes slightly. The change of repetition frequency allows the radar, on a pulse-to-pulse basis, to differentiate between returns from itself and returns from other radar systems with the same frequency. Without stagger any returns from another radar on the same frequency would appear stable in time and could be mistaken for the radar's own returns. With stagger the radar's own targets appear stable in time in relation to the transmit pulse, while the "jamming" echoes are moving around in time (uncorrelated), causing them to be rejected by the receiver.

1.7 Multiple Pulse Repetition Frequency (PRF)

Modern radars frequently use PRFs in the hundreds of kilohertz and stagger the interval between pulses to allow the correct range to be determined. With a staggered PRF, a "packet" of pulses is transmitted, each pulse a slightly different interval after the last (or viewed a different way; delayed variable amounts from the reference trigger). At the end of the packet, the timing returns to its original value, in sync with the trigger. See Fig. 1.45 as presentation of staggered multiple pulse repetition frequency radar.

In Fig. 1.45, the top sequence represents the conventional equally spaced transmission strategy (PRP = pulse repetition period). The respective second and third sequences represent dual- and triple-PRF schemes; pairs of pulses are emitted with delays changing sequentially. These pulse pairs are used to generate several Doppler images V_{D_i}, which are combined to disambiguate the Doppler velocity field.

All these mean that the second and subsequent echoes will appear in the receiver's processing circuits at slightly different times, relative to the current transmitter pulse. These echoes can then be correlated with their associated T_0 pulse in the packet to build up a true range value. Echoes from other T_0 triggers, e.g., *ghost echoes*, will therefore recede from the display or be canceled in the signal processor, leaving only the true echoes which can then be used to calculate range.

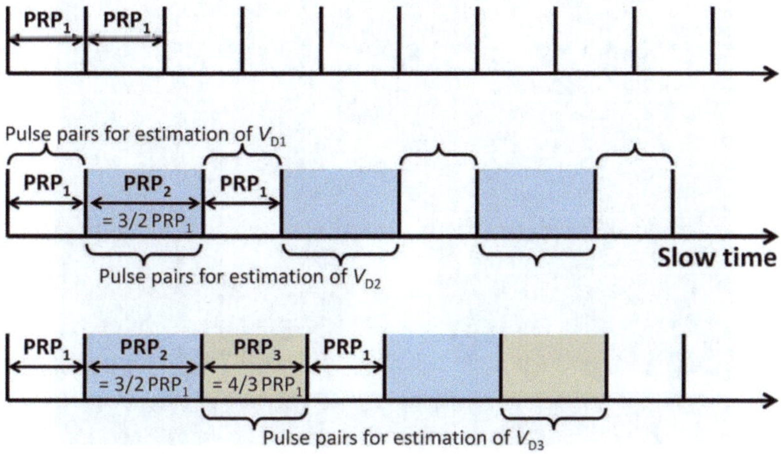

Fig. 1.45 Staggered multiple PRF sequences

Thus, maximum unambiguous range (MUR) for a pulsed stagger radar is calculated using the total sequence period (TSP). TSP is defined as the total time it takes for the pulsed pattern to repeat. This can be found by the addition of all the elements in the stagger sequence. The formula is:

$$\text{MUR} = (c \times \text{TSP}) \tag{1.14}$$

where c is the speed of light and TSP is the addition of all the positions of the stagger sequence usually in microseconds.

Furthermore, in the radar signal in the frequency domain perspective, pure CW radars appear as a single line on a spectrum analyzer display, and when modulated with other sinusoidal signals, the spectrum differs little from that obtained with standard analogue modulation schemes used in communications systems, such as frequency modulation, and consists of the carrier plus a relatively small number of sidebands. See Fig. 1.46.

When the radar signal is modulated with a pulse train as shown above, the spectrum becomes much more complicated and far more difficult to visualize.

Basic Fourier analysis shows that any repetitive complex signal consists of a number of harmonically related sine waves. The radar pulse train is a form of square wave, the pure form of which consists of the fundamental plus all of the odd harmonics. The exact composition of the pulse train will depend on the pulse width and PRF, but mathematical analysis can be used to calculate all of the frequencies in the spectrum. When the pulse train is used to modulate a radar carrier, the typical spectrum shown on the left will be obtained.

Examination of this spectral response shows that it contains two basic structures: the coarse structure (the peaks or *lobes* in the diagram on the left) and the fine

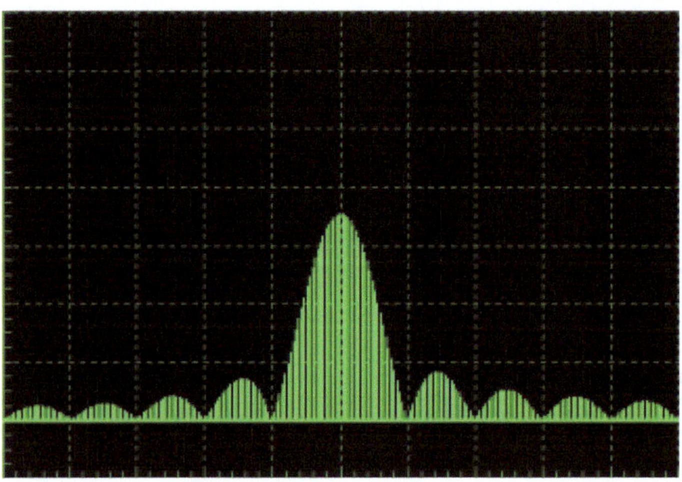

Fig. 1.46 Radar signal modulated with a pulse train

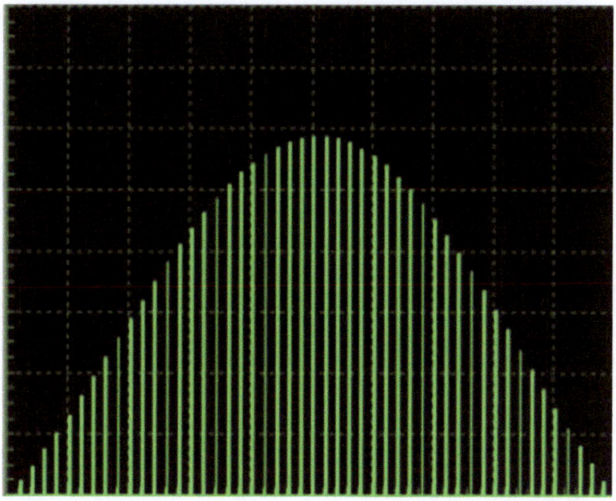

Fig. 1.47 Pulse width or lobe illustration

structure which contains the individual frequency components as shown below. The envelope of the lobes in the coarse structure is given by:

$$\text{Coarse Structure} = \frac{1}{\pi f \tau} \tag{1.15}$$

Note that the pulse width as per Fig. 1.47 appears on the bottom of this equation and determines the lobe spacing. Smaller pulse widths result in wider lobes and therefore greater bandwidth.

Examination of the spectral response in finer detail, as shown on the right, shows that the fine structure contains individual lines or spot frequencies. The formula for the fine structure is given by $T/\pi f \tau$, and since the period of the PRF (T) appears at the top of the fine spectrum equation, there will be fewer lines if higher PRFs are used. These facts affect the decisions made by radar designers when considering the trade-offs that need to be made when trying to overcome the ambiguities that affect radar signals.

1.8 What Is Radar Energy

As we have learned so far, *radar* is a detection system that uses radio waves to determine the range, angle, or velocity of objects. It can be used to detect aircraft, ships, spacecraft, guided missiles, motor vehicles, weather formations, and terrain.

The waves used in the radar are radio waves or microwaves, where the radar is usually used to detect and track space objects and ballistic missiles (this is a long-range radar antenna) and also to detect aircraft at all altitudes, when the antenna rotates at a steady rate, sweeping the local airspace with a narrow vertical fan-shaped beam, to detect aircraft at all altitudes.

Also, it detects motor vehicles, weather formations, and terrain. A radar system consists of a transmitter producing electromagnetic waves in the radio wave range or microwaves domain.

In these days modern radars are used in different fields using radio waves or microwaves according to the programed goal, such as astronomy, air traffic control marine radar to locate landmarks for geological observation, etc., as it is illustrated in Fig. 1.48, which is a presentation of conical scanning radar beam.

And Fig. 1.49 shows depiction of a monopulse radar. Monopulse radar is a radar system that uses additional encoding of the radio signal to provide accurate directional information.

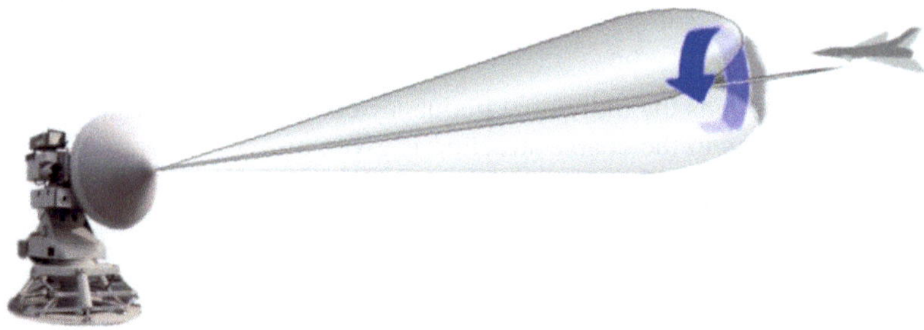

Fig. 1.48 Conical scanning radar signal

Fig. 1.49 A typical monopulse radar beam

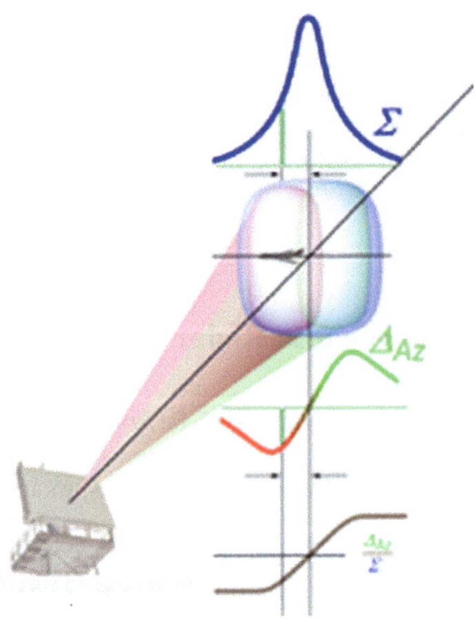

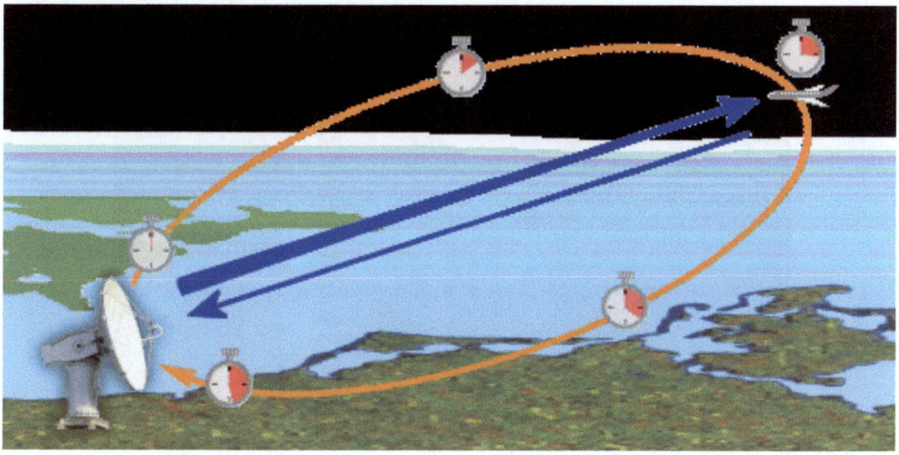

Fig. 1.50 Basic principle of radar

Given that the holistic information about the basic foundation of radar that is using electromagnetic wave in microwave range, as we said, allows a radar system to be able to send signal via its transmitter that emits a radio waves that is called *radar signals* in a predetermined direction either in a continuous wave (CW) or in the form of pulse wave (PW), therefore to understand radar energy, we need to understand electromagnetic (EM) wave and its transmission wave accordingly. Figure 1.50 is an illustration a physical fundamentals of radar principle.

Finally, we need to ask what electromagnetic energy is and it has a simple answer in a simple form as that electromagnetic energy is a form of energy that is reflected or emitted from objects in the form of electrical and magnetic waves that can travel through space. Examples are radio waves, microwaves (i.e., radar beam in general), infrared radiation, visible light (all colors of the spectrum that we see), ultraviolet light, X-rays, and gamma radiation.

In summary, electromagnetic energy is a term used to describe all the different kinds of energies released into space by stars such as the sun. These kinds of energies include some that you will recognize and some that will sound strange. They include:

- Radio waves
- TV waves
- Radar waves
- Heat (infrared radiation)
- Light
- Ultraviolet light (this is what causes sunburns)
- X-rays (just like the kind you get at the doctor's office)
- Short waves
- Microwaves, like in a microwave oven
- Gamma rays

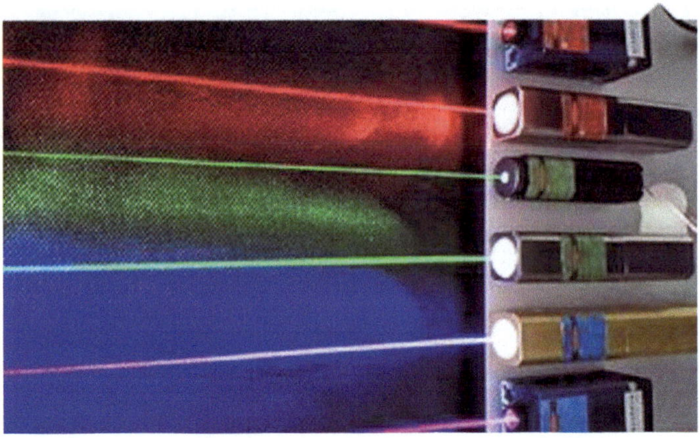

Fig. 1.51 Illustration of phone energy radiation. (Source: www.wikipedia.com)

All these waves do different things (e.g., light waves make things visible to the human eye, while heat waves make molecules move and warm up, and X-rays can pass through a person and land on film, allowing us to take a picture inside someone's body), but they have some things in common.

They all travel in waves, like the waves at a beach or like sound waves, and also are made of tiny particles. Scientists are unsure of exactly how the waves and the particles relate to each other. The fact that electromagnetic radiation travels in waves lets us measure the different kinds by wavelength or how long the waves are. That is one way we can tell the kinds of radiation apart from each other.

Although all kinds of electromagnetic radiation are released from the Sun, our atmosphere stops some kinds from getting to us. For example, the ozone layer stops a lot of harmful ultraviolet radiation from getting to us, and that's why people are so concerned about the hole in it.

We humans have learned uses for a lot of different kinds of electromagnetic radiation and have learned how to make it using other kinds of energy when we need to. However, electromagnetic energy is a physical phenomenon whose utility and benefits are constantly being utilized in ever-new and creative ways. Like all new and constantly improving uses of energy, there is always a time where skepticism of safety becomes an important issue.

In physics, electromagnetic radiation (EM radiation or EMR) refers to the waves or their quanta, photons (i.e., Fig. 1.51) of the electromagnetic field, propagating (radiating) through space, carrying electromagnetic radiant energy. Bear in your mind that the photon is a type of elementary particle, the quantum of the electromagnetic field including electromagnetic radiation such as light and radio waves, and the force carrier for the electromagnetic force.

It includes:

- Radio waves
- Microwaves

- Infrared
- Visible light
- Ultraviolet
- X-rays
- Gamma rays

Classically, electromagnetic radiation consists of electromagnetic waves (EMWs) as illustrated in Fig. 1.52, which are synchronized oscillations of electric and magnetic fields.

In a vacuum electromagnetic wave travels at the speed of light, commonly denoted c. In homogeneous, isotropic media, the oscillations of the two fields are perpendicular to each other and perpendicular to the direction of energy and wave propagation, forming a transverse wave. See Fig. 1.53. Where A linearly polarized sinusoidal electromagnetic wave, propagating in the direction $+\mathbf{z}$ through a homogeneous, isotropic, dissipation-less medium, such as vacuum. The electric field (blue arrows) oscillates in the $\pm\mathbf{x}$-direction, and the orthogonal magnetic field (red arrows) oscillates in phase with the electric field, but in the $\pm\mathbf{y}$-direction.

The wavefront of electromagnetic waves as illustrated in Fig. 1.54 emitted from a point source such as a light bulb is a sphere. The position of an electromagnetic wave within the electromagnetic spectrum (e.g., the electromagnetic spectrum is the range of frequencies of electromagnetic radiation and their respective wavelengths and photon energies) can be characterized by either its frequency of oscillation or its wavelength.

Electromagnetic waves of different frequency are called by different names since they have different sources and effects on matter. In order of increasing frequency and decreasing wavelength, these are radio waves, microwaves, infrared radiation,

Fig. 1.52 Electromagnetic wave depiction. (Source: www.wikipedia.com)

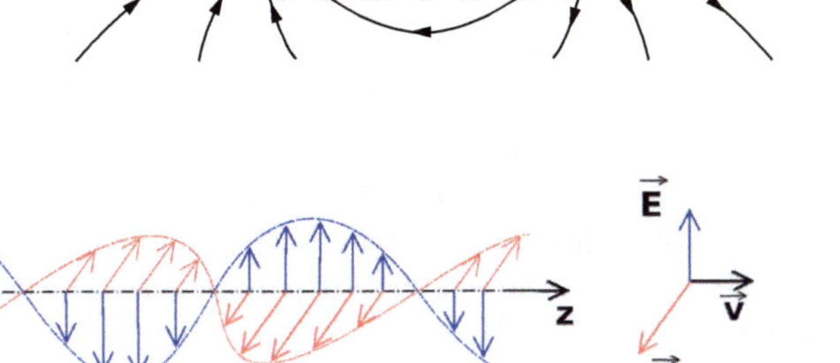

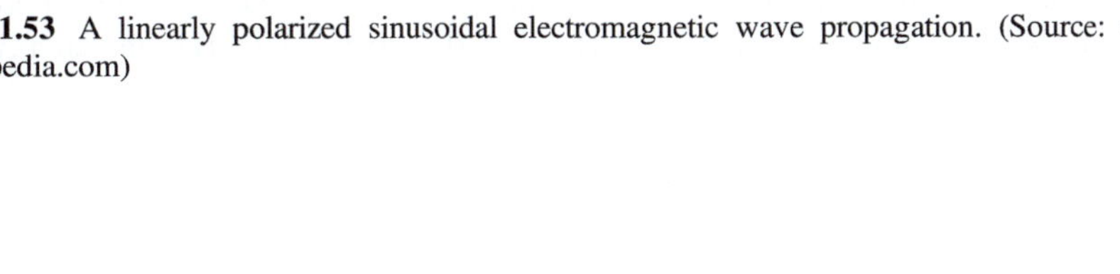

Fig. 1.53 A linearly polarized sinusoidal electromagnetic wave propagation. (Source: www.wikipedia.com)

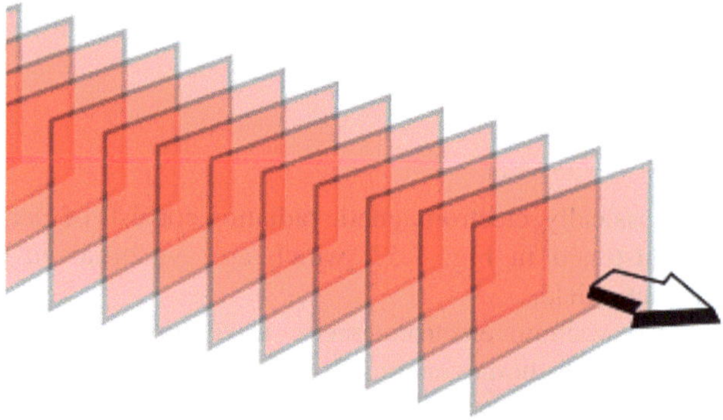

Fig. 1.54 Wavefront Depiction. (Source: www.wikipedia.com)

Fig. 1.55 Visible light depiction. (Source: www.wikipedia.com)

and visible light as depicted in Fig. 1.55, ultraviolet radiation and X-rays as depicted in Fig. 1.56, and gamma rays as depicted in Fig. 1.57.

Note: The visible spectrum is the portion of the electromagnetic spectrum that is visible to the human eye. Electromagnetic radiation in this range of wavelengths is called visible light or simply light. A typical human eye will respond to wavelengths from about 380 to 740 nm. In terms of frequency, this corresponds to a band in the vicinity of 430–770 THz.

Note: X-rays make up X-radiation, a form of high-energy electromagnetic radiation. Most X-rays have a wavelength ranging from 0.01 to 10 nm, corresponding to frequencies in the range 30 PHz to 30 EHz (3×10^{16} to 3×10^{19} Hz) and energies in

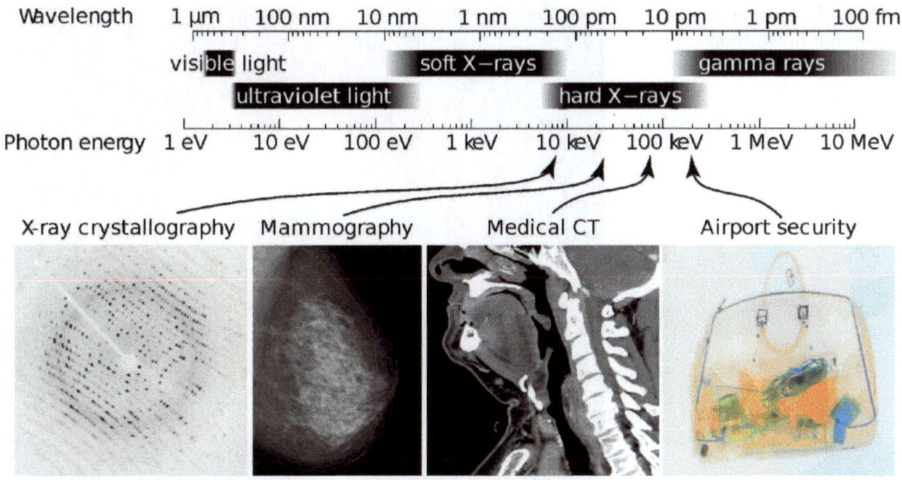

Fig. 1.56 X-ray spectrum illustration. (Source: www.wikipedia.com)

Fig. 1.57 Gamma ray radiation depiction

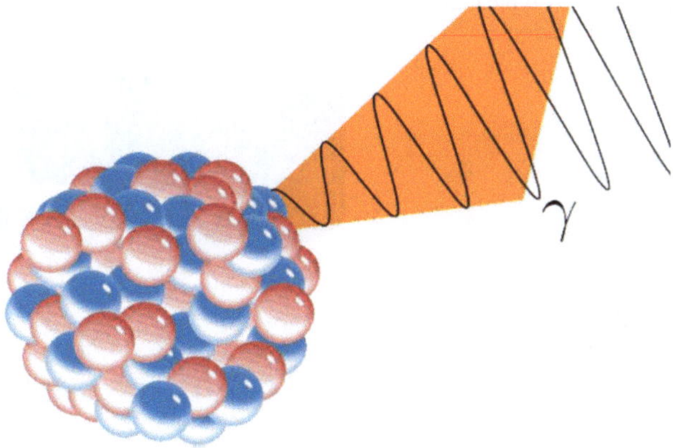

the range 100 eV to 100 keV. X-ray wavelengths are shorter than those of UV rays and typically longer than those of gamma rays.

Note: A **gamma ray**, or **gamma radiation** (symbol γ), is a penetrating electromagnetic radiation arising from the radioactive decay of atomic nuclei as illustrated in Fig. 1.58. It consists of the shortest wavelength electromagnetic waves and so imparts the highest photon energy.

Note: Gamma rays are emitted during nuclear fission in nuclear explosions.

At this stage it is worth to mention that so far we have been discussing two-dimensional (2D) radars as is illustrated in Fig. 1.59, where it shows a typical diagram of 2D radar rotating cosecant squared antenna pattern. However, three-dimensional (3D) radar provides for radar coverage in three dimensions; unlike the

Fig. 1.58 Nuclear atomic explosion illustration

Fig. 1.59 A typical 2D radar illustration

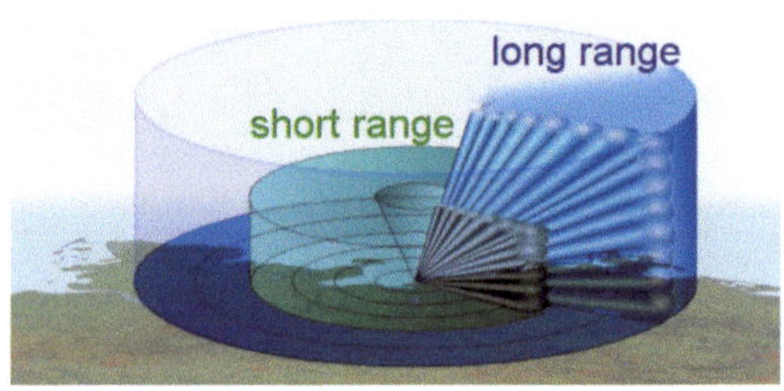

Fig. 1.60 A typical 3D radar illustration

more common 2D radar which provides range and bearing, the 3D radar also provides elevation.

Applications include weather monitoring, air defense, and surveillance as depicted in Fig. 1.60, where it shows a diagram of a typical 3D radar, a judicious mix of vertical electronic beam steering, and mechanically horizontal movement of a pencil beam.

The information provided by 3D radar has long been required, particularly for air defense and interception. Interceptors must be told the altitude to climb to before making an intercept. Before the advent of single unit 3D radars, this was achieved with separate search radars giving range and azimuth and separate height-finding radars that could examine a target to determine altitude. These had little search capability, so were directed to a particular azimuth first found by the primary search radar.

Techniques that are implemented in the above scenarios fall under two categories, and they include:

1. **Steered beam** radars steer a narrow beam through a scan pattern to build a 3D picture. Examples include NEXRAD Doppler weather radar, which uses a parabolic dish and the AN/SPY-1 passive electronically scanned array radar employed by the Ticonderoga class of guided missile cruisers and other ships so equipped with the Aegis Combat System.
2. **Stacked beam** radars emit and/or receive multiple beams of radio waves at two or more elevation angles. By comparing the relative strengths of the returns from each beam, the elevation of the target can be deduced. An example of a stacked beam radar is the Air Route Surveillance Radar.

With the above information, this section about radar energy is pretty much covered from a holistic perspective; however, we encourage those readers who need more detailed and granular information should refer to other technical textbook under subject of radar principles or do search on the Internet.

1.9 Propagation of Electromagnetic (EM) Energy and Pulse Volume

The radar transmits a stream or "beam" of energy in discrete pulses, which propagate away from the radar antenna at approximately the speed of light (~3×10^8 m s^{-1}). The volume of each pulse of energy will determine how many targets are illuminated. This directly determines how much energy (power) is returned to the radar. The shape of the radar antenna; the wavelength, λ, of the energy transmitted; and the length of time the radar transmits determine the shape and volume of each radar pulse.

The Weather Service Radar-1988 Doppler (WSR-88D) transmits a narrow, conical-shaped beam of pulses with each pulse resembling a truncated cone. The radar pulse volume is illustrated in Fig. 1.61.

The angular width of the radar beam is defined as that region of transmitted energy that is bounded by one-half (−3 dB) the maximum power. The maximum power lies along the beam centerline and decreases outward.

These "half-power" points for the WSR-88D result in an angular width of less than 1°. However, the actual physical width increases with increasing range; thus the physical length remains constant such that the pulse volume increases with increasing range (Fig. 1.62).

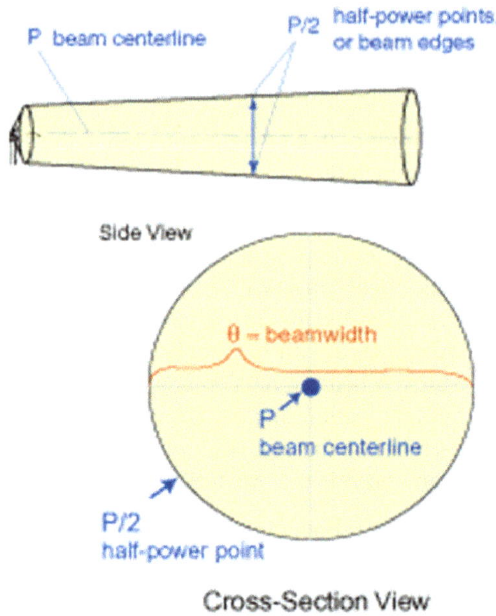

Fig. 1.61 The radar pulse volume depiction

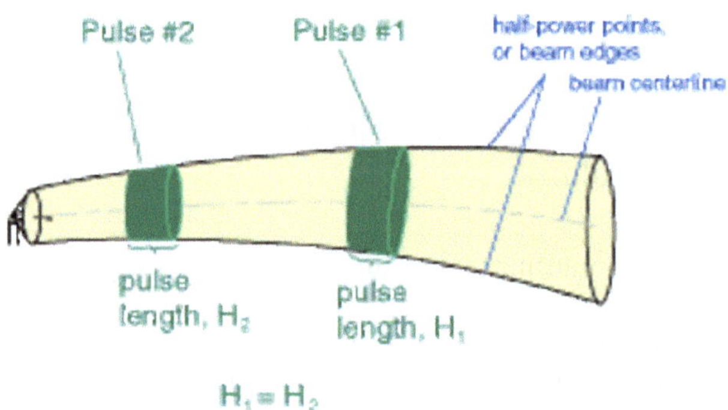

Fig. 1.62 The pulse volume increasing range illustration

Since the amount of transmitted power is fixed, a radar pulse's power density decreases with increasing range. Pulsed transmission also allows for obtaining target range information.

Essentially, bear in your mind that an electromagnetic wave is propagated by the oscillations of the electric and magnetic fields. A changing electric field produces a changing magnetic field, and a changing magnetic field produces a changing electric field. Thus, an electromagnetic wave is self-propagating and does not need a medium to travel through.

Also you should know that a wave describes a mechanism of how energy is transferred from one place to another without any matter being transferred. It is the disturbance that is propagated only. Waves travel with well-defined speeds determined by the properties through which they travel. For example, a disturbance on a

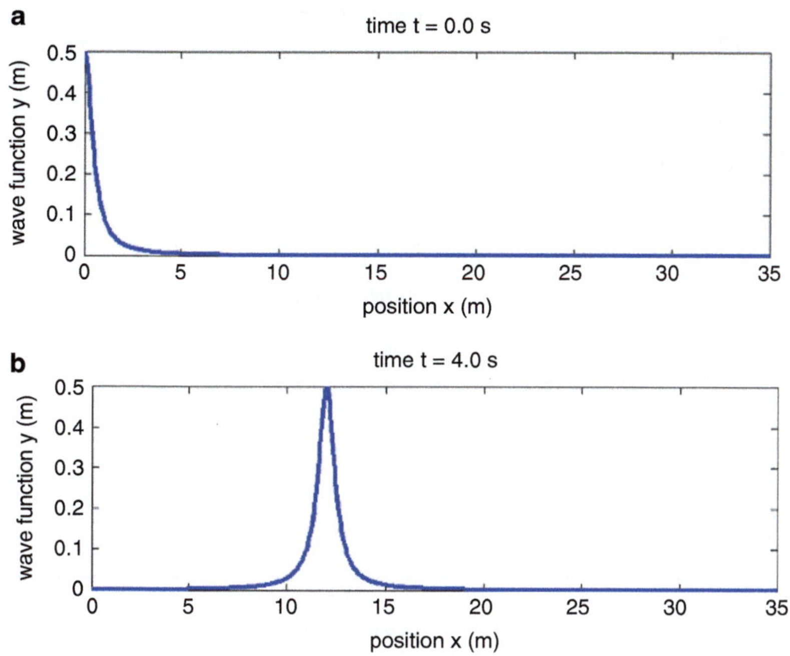

Fig. 1.63 (**a**) Wave position after $t = 0.0$ second traveling (at origin), of a pulse along a string. (**b**) Wave position after $t = 4.0$ second traveling of a pulse along a string

string (e.g., guitar string) propagates along the string. Figure 1.63a, b shows an animation of a pulse traveling along a string.

As the pulse propagates, the string moves up or down but the energy in the form of the kinetic energy and potential energy is transferred as shown in Fig. 1.64a, b.

Important parameters describing a wave are its amplitude A, wavelength λ, period T, frequency f, and speed of propagation ν. The period and frequency are the reciprocals of each other as Equation 1.16:

$$f = \frac{1}{T}, \quad T = \frac{1}{f} \tag{1.16}$$

The speed of propagation through a given medium is constant and depends upon its wavelength and frequency or period as presented by Equation 1.17:

$$\nu = f\lambda = \frac{\lambda}{T} \tag{1.17}$$

Figure 1.65 shows an animation of a sinusoidal wave traveling along a string. Study the animation carefully and before you look at the answers below, determine the amplitude, wavelength, period, frequency, and speed of the wave.

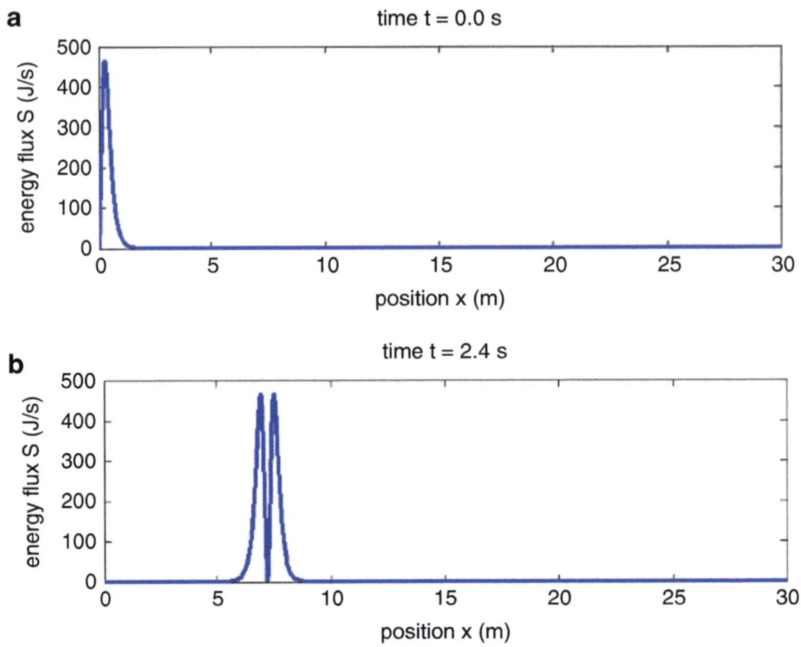

Fig. 1.64 (**a**) Wave energy at $t = 0.0$ second, as the transfer of energy (K.E. + P.E.) along a string. (**b**) Wave energy at $t = 2.4$ s, as the transfer of energy (K.E. + P.E.) along a string

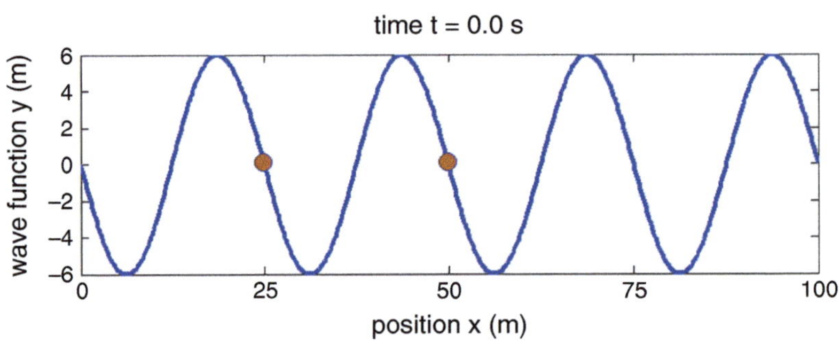

Fig. 1.65 A sinusoidal wave traveling along a string

Notice that it is the shape of the disturbances that advances while parts of the string move up and down, each segment of the strings executing simple harmonic motion.

From the animation:

Amplitude $A = 6.0$ m
Wavelength $\lambda = 25.0$ m
Period $T = 20.0$ s

Fig. 1.66 Water waves entering shallow water from left

Frequency $f = 0.050$ Hz
Speed $\upsilon = 1.25$ m/s

A disturbance on water produces water waves. Figure 1.66 shows a water wave propagating from the left to the right, and then the wave travels into shallower water and is slowed down which produces a change in direction of propagation. This phenomenon is called refraction.

Note that in Fig. 1.66, the frequency does not change but the speed and wavelength are reduced. This type of bending of the waves is known as refraction.

Given the information about wave characteristic, we are able to say that electromagnetic waves in physics of classical electrodynamic world fall into following conditions as:

- Transverse wave.
- An electromagnetic wave is propagated by the oscillations of the electric and magnetic fields. A changing electric field produces a changing magnetic field and a changing magnetic field produces a changing electric field. Thus, an electromagnetic wave is self-propagating and does not need a medium to travel through.
- Can travel through vacuum; speed is $c = 3.0 \times 10^8$ m s^{-1}
- When electromagnetic waves are emitted or absorbed by an atom, done so in quanta of energy as Equation 1.18 presented below as:

$$E = hf \tag{1.18}$$

where:

E = energy of photon (J) electron volt 1 eV = 1.6×10^{-19} J
f = frequency of electromagnetic wave (Hz)
$h = 6.62606876 \times 10^{-34}$ J s Planck's constant (J s)

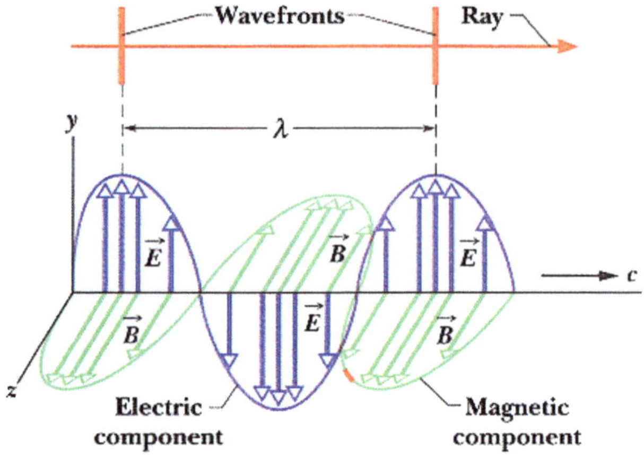

Fig. 1.67 Typical electromagnetic wave depiction

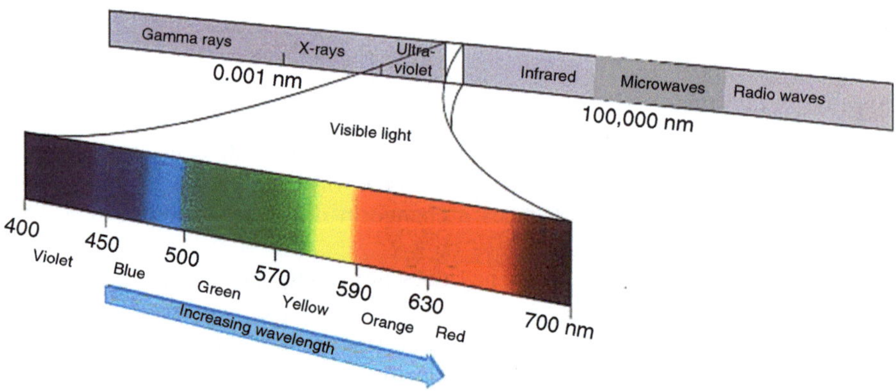

Fig. 1.68 The electromagnetic spectrum

Note that the particles associated with electromagnetic waves are called **photons**. The energy of a single photon is given by Equation 1.18.

Figure 1.67 is a depiction of an electromagnetic wave in the form of transverse electromagnetic (TEM).

The electromagnetic spectrum is depicted in Fig. 1.68.

A progressive electromagnetic wave is a self-supporting, energy-carrying disturbance that travels free of its source. The light from the Sun travels through space (no medium) for only 8.3 min before arriving at Earth. Each form of electromagnetic radiation (radio waves, microwaves, infrared, light, ultraviolet, X-rays, and γ-rays) is a web of oscillating electric and magnetic fields inducing one another. A fluctuating electric field (electric charges experience forces) creates a magnetic field (moving charges experience forces) perpendicular to itself, surrounding and extending beyond it. That magnetic field sweeping off to a point further in space is varying there, and so generates a perpendicular electric field that spreads out. Nothing actually is displaced in space like a water wave where the water oscillates up and down and sideways.

All electromagnetic waves propagate in vacuum at exactly the speed of light:

$$c = 2.997\ 924\ 85\ \text{m s}^{-1}$$

This is a tremendous speed, and light travels 1 m in only 3.3×10^{-10} s.

"There are only two fundamental mechanisms for transporting energy and momentum: a streaming of particles and a flowing of waves. And even these two seemingly opposite conceptions are subtly intertwined—there are no waves without particles and no particles without waves."

Considering the above description of electromagnetic wave and energy, bear in your mind that particles such as photons are behaving also as waves. Waves are a mechanism for transferring energy via some kind of vibration without any matter being transferred. One characteristic of waves, but not of particles, is that diffraction/ interference is observed as shown in Fig. 1.69 when a wave passes through an aperture.

However, in experimental arrangements analogous to the two-slit interference for light, when a beam of electrons is incident upon a biprism (mimics two slits for light as the electrons can travel in two paths around a filament) and is detected upon a screen, an interference pattern is observed. When a few electrons hit the screen, no notice pattern is discerned as shown in Fig. 1.70.

However, for much longer exposures involving 80,000 plus electrons, a very distinctive two-slit diffraction pattern is clearly observed as shown in Fig. 1.71.

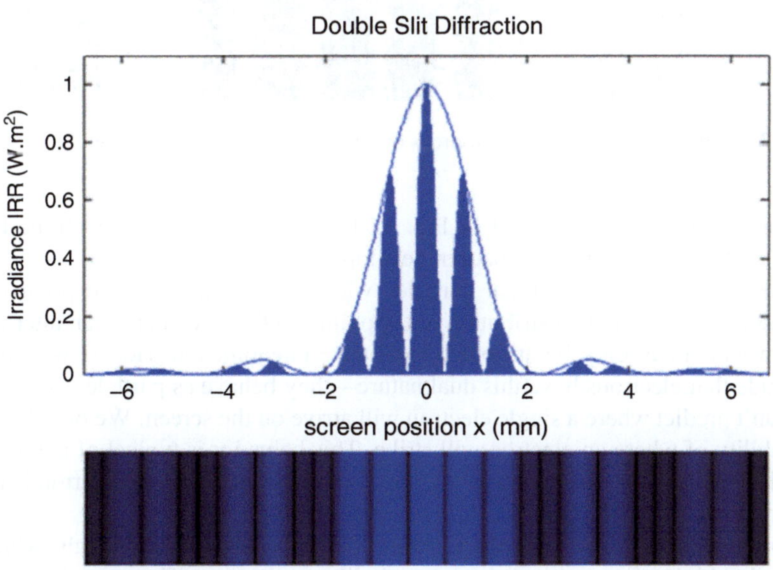

Fig. 1.69 Fraunhofer diffraction from a double slit

Fig. 1.70 Pattern formed by 2000 electrons on passing through the equivalent of a double slit

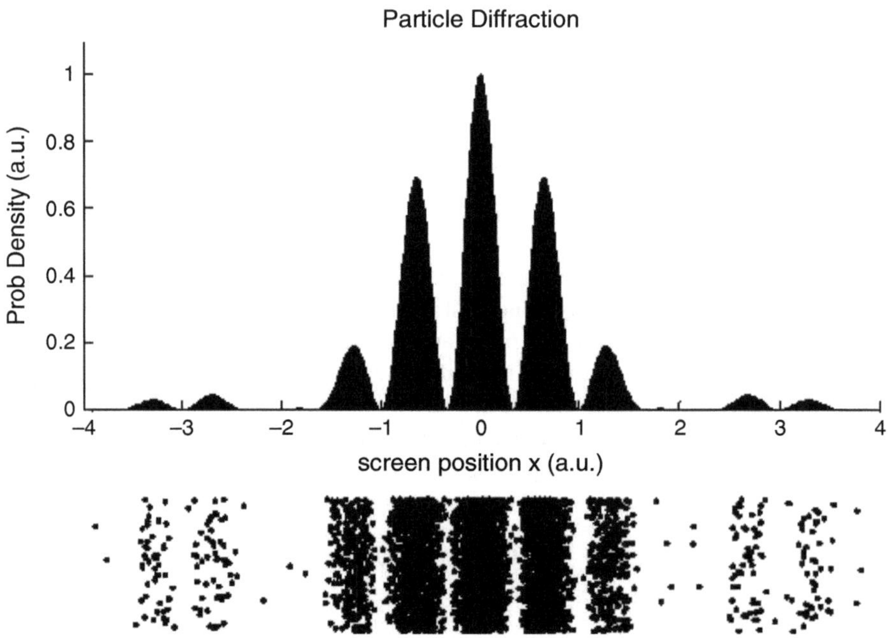

Fig. 1.71 Pattern formed by 80,000 electrons on passing through the equivalent of a double slit

Considering what is plotted in Fig. 1.71, as more and more electrons hit the screen, a two-slit interference pattern develops.

The electrons are individual particles when they strike a single point on the detection screen, but the distribution of the points on the screen gives an interference pattern which can only be attributed to a wave phenomenon. Hence, we can only conclude that electrons have this dual nature—they behave as particles or as waves. We can't predict where a single electron will arrive on the screen. We only know the probability of where an electron will strike. This behavior is typical of the quantum world and is a good example of the interplay between indeterminism and determinism.

So particles exhibit wave characteristics, but we also find that light which we normally think of as a wave has particle-like properties. The particle nature of electromagnetic waves is observed in the photoelectric effect—when light of a sufficient frequency strikes a metal surface, electrons are emitted from the surface. To account for the emission of the electrons from the surface, the light is modeled as a stream of particles called photons. The energy of each photon is $E = hf$.

Particles—particle and wave properties
Waves—wave and particle properties

As a final note for this section, bear in your mind that *mechanical waves*, unlike *electromagnetic waves*, require the presence of a material medium in order to transport their energy from one location to another. However, electromagnetic waves are waves which can travel through the vacuum of outer space. Mechanical waves, unlike electromagnetic waves, require the presence of a material medium in order to transport their energy from one location to another. Sound waves are examples of mechanical waves while light waves are examples of electromagnetic waves.

1.10 Radar Range Equation

In a typical radar behavior, assume that the power P_r is returning to the receiving antenna and it is given by the Equation 1.19 in the form of:

$$P_r = \frac{P_t G_t A_r \sigma F^4}{(4\pi)^2 R_t^2 R_r^2} \tag{1.19}$$

where:

P_t = transmitter power
G_t = gain of the transmitting antenna
A_r = effective aperture (area) of the receiving antenna; this can also be expressed as $G_r\lambda^2/4\pi$, where:
λ = transmitted wavelength
G_r = gain of receiving antenna
σ = radar cross section, or scattering coefficient, of the target
F = pattern propagation factor
R_t = distance from the transmitter to the target
R_r = distance from the target to the receiver

In the common case where the transmitter and the receiver are at the same location, $R_t = R_r$, and the term $R_t^2 R_r^2$ can be replaced by R^4, where R is the range. This would yield as:

$$P_r = \frac{P_t G_t A_r \sigma F^4}{(4\pi)^2 R^4} \tag{1.20a}$$

The relationship in Equations 1.20a and 1.20b shows that the received power declines as the fourth power of the range, which means that the received power from distant targets is relatively very small.

Additional filtering and pulse integration modifies the radar equation slightly for pulse Doppler radar performance, which can be used to increase detection range and reduce transmit power.

The equation above with $F = 1$ is a simplification for transmission in a vacuum without interference. The propagation factor accounts for the effects of multipath and shadowing and depends on the details of the environment. In a real-world situation, path loss effects should also be considered.

In summary, the radar range equation (RRE) provided the most useful mathematical relationship available to the engineer in assessing both the need for and the resulting effectiveness of efforts to reduce radar target cross section. In its complete form, the radar equation accounts for [34]:

- Radar system parameters
- Target parameters
- Background effects such as clutter, noise, interference, and jamming
- Propagation effects such as reflection, refraction, and diffraction
- Propagation medium such as absorption and scatter

When fully implemented, the radar equation can be used to estimate radar system performance, and the bottom line for any radar cross section (RCS) control task is its effect on radar performance.

The radar cross section of a complex target, such as an aircraft or ship, varies considerably with change in aspect or change in frequency so that a single number cannot adequately describe the radar cross section of a target. Nevertheless, Table 1.5 lists "example" values for various targets at microwave frequencies. These are for illustrative purposes to show the relative "sizes" of common targets as "seen" by radar.

Therefore, a thorough knowledge of the radar equation and its implications are vitally necessary in the area of radar cross section reduction (RCSR). Luckily, the fundamental form of the equation is based on very simple geometric principles, as shown before.

Furthermore, the radar range equation is important not only for predicting the range performance of a radar but to act as a focus for radar design and for better understanding the factors that affect radar performance. The simple form of the radar range equation is given by Equation 1.20a.

However, if an antenna is used for both transmitting and receiving, as is usually the case and using Equation 1.20a, then we can assume $G_t = G = 4\pi A/\lambda^2$ where λ is the radar wavelength in meters. Then, we have a new form of equation as Equation 1.20b:

$$P_r = \frac{P_t G^2 \lambda^2 \sigma}{\left[(4\pi)^3 R^4\right]} = \frac{P_t A_e^2 \sigma}{\left[4\pi \lambda^2 R^4\right]} \tag{1.20b}$$

Table 1.5 "Example" values of radar cross section (RCS) [35]

Target	σ (square meters)
Conventional unmanned winged missile	0.1
Small single-engine aircraft	1
Small fighter or four-passenger jet	2
Large fighter	6
Medium bomber or medium jet airline	20
Large bomber or large jet airliner	40
Jumbo jet	100
Helicopter	3
Small open boat	0.02
Small pleasure boat	2
Cabin cruiser	10
Ship, grazing angle greater than zero	Displacement tonnage expressed in m^2
Pickup truck	200
Automobile	100
Bicycle	2
Man	1
Large bird	10^{-2}
Medium bird	10^{-3}
Large insect (locust)	10^{-1}
Small insect (fly)	10^{-5}

where:

P_r = received signal power in watts
G = antenna gain
σ = radar cross section of target in square meters
λ = radar wavelength in meters
P_t = peak power in watts
A_e = antenna effective aperture in square meters
R = range in meters

The maximum range R_{max} of a radar occurs when the received signal $P_r = S_{min}$ the minimum detectable signal. The minimum detectable signal is a statistical quantity limited by receiver noise. It can be written as:

$$S_{min} = kT_0BF_n(S/N)_1 \tag{1.21}$$

where:

k = Boltzmann's constant
T_0 = standard temperature (290 K)
$kT_0 = 4 \times 10^{-21}$ W/Hz
B = receiver bandwidth in hertz

F_n = receiver noise figure
$(S/N)_1$ = minimum signal-to-noise ratio required for reliable detection

The received echo signal power can be increased by integrating (adding) a number of echo signal pulses n. This can be incorporated into the radar equation by dividing $S_{\min}$ by $nE_i(n)$, where $E_i(n)$ is the efficiency with which the n pulses can be integrated. Since the average power P_{avg} is more indicative of radar capability than is the peak power, it is introduced via the relation:

$$P_{\text{avg}} = P_t \tau f_p \tag{1.22}$$

where:

τ = pulse width in seconds
f_p = pulse repetition frequency in hertz

With the above, the form of the radar equation suitable for calculating the range is:

$$R_{\max} = \left[\frac{P_{\text{avg}} G^2 \lambda^2 \sigma n E_i(n)}{(4\pi)^3 k T_0 F_n (B\tau) f_p (S/N)_1 L_s} \right]^{1/4} \tag{1.23}$$

The radar system losses L_s (number greater than 1) have been included. For most radars designed with a matched filter receiver (a filter that maximizes the output signal-to-noise ratio), the product $B\tau = 1$. In Equation 1.22, $(S/N)/nE_i(n)$ is the required signal-to-noise ratio per pulse $(S/N)_n$.

Figure 1.72 shows the relationship of the required signal-to-noise ratio $(S/N)_1$ to the probability of detection and the probability of false alarm. The probability of detection is usually taken as 0.90, but sometimes it is quoted as 0.5 or 0.8. Its choice is usually the prerogative of the customer. The probability of a false alarm P_{fa} is given here as [35]:

$$P_{\text{fa}} = \frac{1}{BT_{\text{fa}}} \tag{1.24}$$

where:

B = receiver bandwidth in hertz
T_{fa} = average time between false alarms

The reciprocal of P_{fa} is ns, the false alarm number. The false alarm time T_{fa} is usually specified for radar performance rather than the probability of false alarm or the false alarm number.

Figure 1.73 is a plot of the integration-improvement factor $nE_i(n)$ as a function of n. The number of pulses returned from a target when an antenna of beam width θ_B

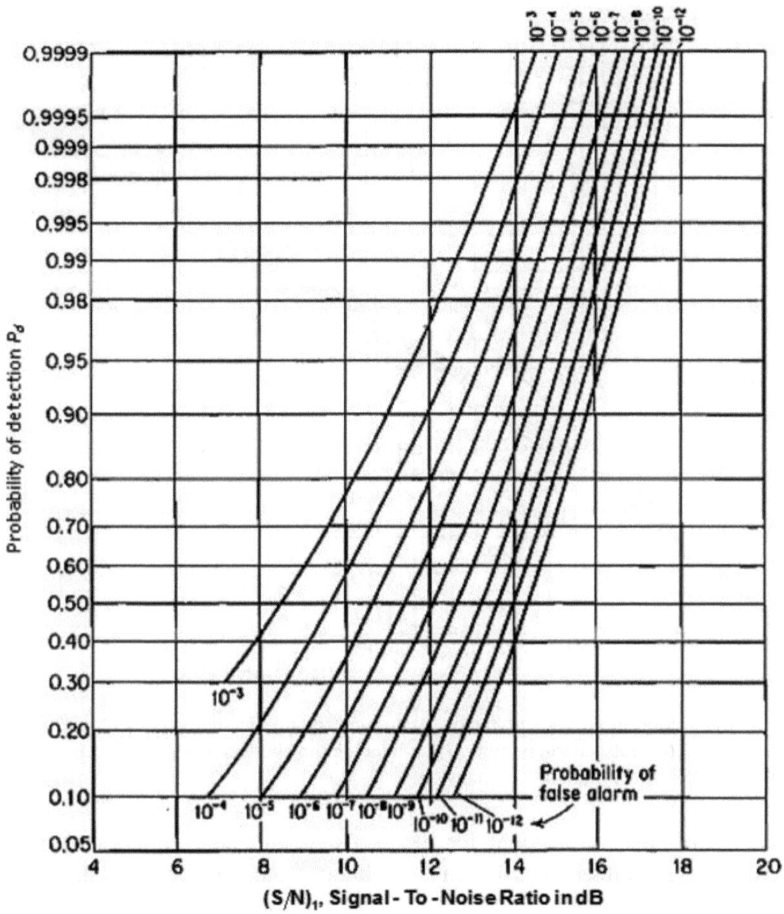

Fig. 1.72 Probability of detection for a sine wave in noise as a function of the signal-to-noise (power) ratio and the probability of false alarm [35]. (Courtesy of McGraw-Hill Book Company)

degrees rotates at a rate of ω_{nt} revolutions per minute with a pulse repetition rate of f_p Hz is given as:

$$n = \frac{\theta_B f_p}{6\omega_{nt}} \tag{1.25}$$

Failure to include the many factors that contribute to the system losses L can result in considerable difference between the calculated range and the actual range. Losses include:

- Loss in the transmission line connecting the antenna to the transmitter and receiver
- Loss in the duplexer, rotary joint, and other microwave components

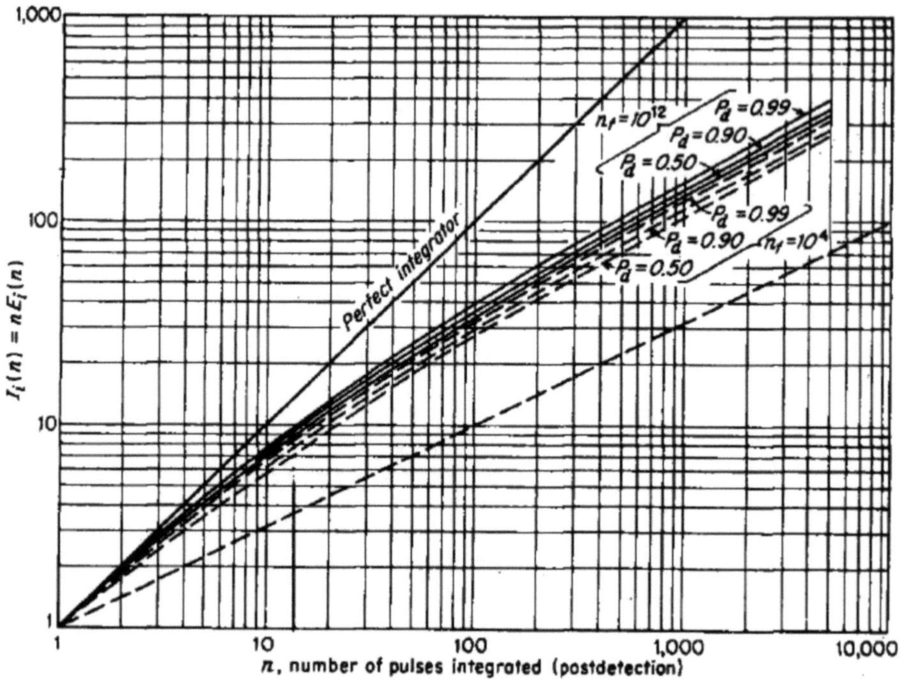

Fig. 1.73 Integration-improvement factor, assuming square-law detector, P_d = probability of detection, $n_f = T_\mathrm{fa}B$ = false alarm number, T_fa = average time between false alarms, B = bandwidth. (Courtesy of McGraw-Hill Book Company)

- Beam shape loss, to account for the fact that the radar equation employs the maximum gain rather than a gain that changes pulse to pulse as the antenna is scanned past the target
- Signal processing losses, which can sometimes be surprisingly large
- Loss due to degradation of transmitter power and receiver noise figure

The system losses from all factors might be from 10 to 20 dB or even greater. (A loss of *16* dB reduces the radar range by a factor of 2) [35].

1.11 Other Forms of Radar Equation

The radar equation is used for the calculation of range; but it is also used as a basis for assessing the trade-offs in radar design. The simple forms of the radar equation given above (Equations 1.19, 1.20a and 1.20b) are seldom sufficiently complete, however, and they must be extended. Each specific radar application has some particular requirements or constraints that can result in a slightly different form of the radar equation. Examples are presented below.

1.11.1 Surveillance Radar Equation

Surveillance radar equation is given by:

$$R_{\max}^{4} = \frac{P_{\mathrm{av}} A_{\mathrm{e}} \sigma E_i(n)}{4\pi k T_0 F (S/N)_1 L_{\mathrm{s}}} \cdot \frac{t_{\mathrm{s}}}{\Omega} \tag{1.26}$$

where:

$R_{\max}$ = maximum radar range in meter
P_{av} = average power in watts
A_{e} = antenna effective aperture in square meters
σ = radar cross section of target in square meters
$E_i(n)$ = efficiency in integrating n pulses
t_{s} = scan time or revisit time in second
k = Boltzmann's constant = 1.38×10^{-23} J/K
T_0 = standard temperature = 290 K
$F_n = F$ = receiver noise figure per pulse (i.e., single hit $n = 1$)
$(S/N)_1$ = minimum signal-to-clutter ratio necessary to detect a target with a specified probability of detection and probability of false alarm for a single pulse
L_{s} = system losses
Ω = solid angular region (steradians) of radar coverage

Equation 1.21 applies to a radar that must observe all targets within an angular region of solid angle Ω steradians once every t_{s} seconds.

When the surveillance radar utilizes a conventional rotating fan beam whose elevation beam width is θ_{e}, the solid angle Ω equals $2\pi \sin \theta_{\mathrm{e}}$, and t_{s} is the azimuth rotation period or revisit time.

1.11.2 Tracking Radar Equation

Equations 1.20a and 1.20b is basically the tracking radar equation, where $n/f_{\mathrm{p}} = t_0$ is the signal integration time. It has also been called the searchlight equation.

1.11.3 Surface Clutter Range Equation

The equation for this application is given by:

$$R_{\max} = \frac{\sigma n_{\mathrm{e}}}{\left[(S/C)_0 \sigma^0 \theta_{\mathrm{a}} (c\tau/2) \sec \psi\right]} \tag{1.27}$$

where:

n_e = effective number of pulses integrated
σ^0 = radar cross section of surface clutter per unit area
$(S/C)_0$ = minimum signal-to-clutter ratio necessary to detect a target with a specified probability of detection and probability of false alarm, for a single pulse
θ_a = azimuth beam width in radians
ψ = grazing angle or glancing angle
τ = pulse width in seconds

Note that a great deal of processing is used to help detect targets in clutter. Nowadays, much of this takes place in the digital domain, but the important things to understand are the fundamental concepts in general. In Sect. 1.12 we describe some of them such as "Sea Clutter Suppression" and Sect. 1.13 on "Rain Clutter" as well.

1.11.4 Volume Clutter Radar Equation

The equation for this form of radar is given by:

$$R_{\max}^2 = \frac{\sigma G n_e}{\left[(S/C)_0 \eta (\pi^3/4)(c\tau/2)\right]} \tag{1.28}$$

where:

η = volume clutter of reflectivity, or radar cross section of clutter per unit volume, in meters^{-1}

As we defined the reflectivity η is the radar cross section of the clutter per unit volume.

1.11.5 Noise Jamming Radar Equation (Surveillance)

This equation is given by:

$$R_{\max}^2 = \frac{P_{\text{avg}} E_i(n)}{G_{\text{SL}} L_{\text{S}}} \cdot \frac{\sigma}{(S/N)_1} \cdot \frac{t_s}{\Omega} \cdot \frac{B_j}{P_j G_j} \tag{1.29}$$

where:

B_j = jammer bandwidth in hertz
P_j = jammer power in watts

G_j = jammer antenna gain
G_{SL} = antenna side-lobe gain
L_S = system loss
Ω = solid angular region (steradians) of radar coverage
P_{avg} = average power in watts
$E_i(n)$ = efficiency in integrating n pulses
$(S/N)_1$ = minimum signal-to-clutter ratio necessary to detect a target with a specified probability of detection and probability of false alarm for a single pulse
t_s = scan time, or revisit time, in seconds

This equation assumes that the jamming noise enters the antenna side-lobes whose gain is G,. When the jamming enters the main beam, $G_{SL} = G$. The jammer power P_j is spread over a bandwidth B_j and is radiated by an antenna whose gain is G_j.

1.11.6 Noise Jamming Radar Equation (Tracking)

The equation for this application is given by:

$$R_{max}^2 = \frac{P_{avg}G^2E_i(n)t_0}{4\pi G_{SL}} \cdot \frac{\sigma}{(S/N)_1} \cdot \frac{B_j}{P_jG_j} \tag{1.30}$$

All the quantities in Equation 1.30 have been define as before, and $t_0 = n/f$ is presenting signal integration time in seconds. When the jamming noise enters the radar via the main beam, $G_{SL} = G$. For more details on jamming-to-signal (J/S) ratio, see Sect. 1.14 of this chapter in this book.

1.11.7 Self-Screening Range Equation

This is the range at which the radar echo signal S received from a target exceeds the received jamming noise power J by the amount S/J (i.e., jamming-to-signal ratio). It is also called the crossover range. The self-screening range is found from either Equation 1.29 or Equation 1.30 (depending on the application) by setting $G_{SL} = G$, setting $(S/N)_i = S/J$, and calling R_{max} the self-screening range R_{ss}. The value of required S/J is often taken to be the same as $(S/N)_i$ found for receiver noise.

1.11.8 Weather Radar Equation

The equation for this application is given by:

$$\overline{P}_{\mathrm{r}} = \frac{2.4 P_{\mathrm{r}} G \tau r^{1.6}}{\left[R^2 \lambda^2 L_{\mathrm{s}}\right]} \tag{1.31}$$

relate the average echo signal power $\overline{P}_{\mathrm{r}}$, to the rainfall rate r (mm/h). It assumes that rain uniformly fills the radar resolution cell. P_{r} is presenting the average received signal power in watts.

1.11.9 Synthetic-Aperture Radar Equation

The equation for this application is given by:

$$\frac{S}{N} = \frac{2 P_{\mathrm{avg}} \rho_{\mathrm{a}}^2 \sigma^0 \delta_{\mathrm{cr}} \delta_{\mathrm{r}}}{\pi f k T_0 F_{\mathrm{n}} R S_{\mathrm{w}} L_{\mathrm{s}} \sin^2 \psi} \tag{1.32}$$

where:

δ_{r} = range resolution in meters
δ_{cr} = cross-range resolution in meters
F_{n} = receiver noise figure
S_{w} = swath width in meters
T_0 = standard temperature = 290 K
L_{s} = system losses
σ^0 = radar cross section of surface clutter per unit area
f = radar frequency in hertz
k = Boltzmann's constant = 1.38×10^{-23} J/K
P_{avg} = average power in watts
ρ_{a} = antenna efficiency
R = range in meters
ψ = grazing angle or glancing angle

This equation relates the signal-to-noise ratio of a resolution cell (sometimes called a pixel) with range resolution δ_{r} and cross-range resolution δ_{cr}, located within a swath S centered at a range R. The above takes account of the combined restriction on cross-range resolution and swath necessary to avoid ambiguities in either range or cross range.

1.11.10 HF Over-the-Horizon Radar Equation

The equation for this application is given by:

$$R_{\max}^2 = \frac{P_{\text{avg}} G_t G_r \lambda^2 \sigma F_p^2 T_c}{(4\pi)^3 N_0 (S/N)_1 L_s} \tag{1.33}$$

where:

P_{avg} = average power in watts
G_r = radar receiving antenna gain
G_t = transmitting antenna gain
λ = wavelength in meters
σ = radar cross section of target in square meters
F_p = propagation factor
T_c = coherent processing time
N_0 = noise power per unit bandwidth
$(S/N)_1$ = minimum signal-to-noise ratio necessary to detect a target with a specified probability of detection and probability of false alarm, for a single pulse
L_s = system losses

The transmitting antenna gain G_t and the receiving antenna gain G_r are shown separately since two different antennas are often used for transmit and receive. The propagation loss is accounted for by F_ν (number less than unity), and T_c is the coherent processing time. The noise power per unit bandwidth N_0 (W/Hz) at the receiver is determined by external noise.

1.12 Sea Clutter Suppression

As we stated, a great deal of processing is used to help detect targets in clutter. Nowadays, much of this takes place in the digital domain, but the important things to understand are the fundamental concepts, as described in this section.

The amelioration of sea clutter effects is intimately tied up with sensitivity time control (STC). Bear in your mind that the range effect is countered by a technique known as sensitivity time control (STC), which adjusts the radar receiver gain to being very low immediately after the pulse is transmitted and then climbing at a rate approximately proportional to R^4 until it reaches its maximum gain. STC is also sometimes known as swept time constant, a term that accentuates the time-varying nature of the gain. The changing amplifier gain (also called sensitivity) can be applied at any amplification stage—RF, IF, and video—and also after digitization of the signal. There are various advantages and disadvantages as to where it is best applied. In fact, it is often optimized by being separately applied at several different stages of the receiver.

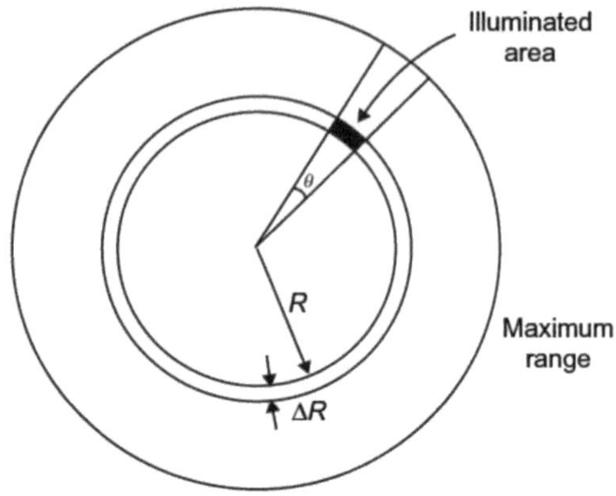

Fig. 1.74 Area of range/azimuth cell

For normal targets, as underlined by the radar equation, we have seen that the returned signal power varies according to $1/R^4$. However, the sea does not act like a normal target, as its effective radar cross section (RCS) varies according to range. If at first we imagine a sea area on a totally flat Earth, the effective illuminated area of sea at any one instant would be bounded by the beam width of the radar antenna and the length of the range cell; see Fig. 1.74.

Mathematics readily shows that the area A_s of the sea being "illuminated" by the radar is given by:

$$A_s = K \times \theta \times R \times \Delta R \tag{1.34}$$

where:

$K =$ a constant irrelative to this discussion
$\theta =$ the beam width of the antenna
$R =$ the range of the range cell
$\Delta R =$ the range cell increment

The important fact is that the area of the sea being illuminated is proportional to the range, R, resulting in the equivalent RCS of the sea also being proportional to R. The sea's RCS increases with range, unlike regular targets, which have a range independent of RCS.

However, it is also worth recognizing that the radar cross section (RCS) of the other major component of clutter (precipitation, such as rain) also tends to increase with range. This time, because the volume of precipitation illuminated also increases with range _ and hence with the increase proportional to the square of the range, assuming that the precipitation continues to occupy all the vertical angle of the radar antenna, which in reality is unlikely over any appreciable distance.

For sea clutter, and assuming a flat Earth, the fact that its RCS is proportional to range effectively means that the R^4 term in the radar equation reduces to R^3, simply because the RCS (σ) should now be represented by $R \times \sigma_S$, where σ_s is a fixed quantity, effectively representing the RCS of the particular sea state at $R = 1$.

Mathematically we have from the representation of the radar equation:

$$P_r = \frac{kP_tG^2\sigma}{R^4} = \frac{kP_tG^2R\sigma_s}{R^4} = \frac{kP_tG^2\sigma_s}{R^3} \tag{1.35}$$

All the quantities parameters are as defined in before.

If we applied STC following the standard R^4 law on such sea clutter, it would mean that the clutter would become increasingly dominant with range. Therefore, the applied STC should vary according to R^3 in order to compensate for this, whenever sea clutter is dominant. When we switch on the sea clutter control for a marine radar, this is effectively what happens. However, one particular practicality has to be taken into account, which is that we are not operating on a flat Earth. At distances beyond the radar horizon, there is no sea clutter as the radar beam does not illuminate the sea at such distances, and, therefore the STC needs to change to R^4 at this point. The actual horizon is dependent on the height of the radar antenna and so this is generally set by the engineer when the equipment is installed.

The standard STC curve when sea clutter is switched on is illustrated in Fig. 1.75. The manual sea clutter control adjusts the exact shape of this curve, according to how the equipment has been specifically designed. Very low settings would approximate to an R^4 curve and higher settings to an R^3 curve.

Switching to automatic sea clutter reduction would normally introduce additional processing that effectively altered the shape of the curve according to the actual level of returns.

The user may also retain some manual control as to how intensely the automatic system is effectively "allowed" to distort the curve from the standard R^3 through to R^4 settings.

The use of scan-to-scan correlation, also called rotation-to-rotation correlation, is an additional way of reducing sea clutter effects. Peaks in sea clutter are unlikely to occur at the same range and bearing for consecutive scans and so a digital process that attempts to eliminate the display of reflected signals that are only visible during one scan can significantly reduce such clutter.

Unfortunately, weak wanted signals can also be eliminated by such a process.

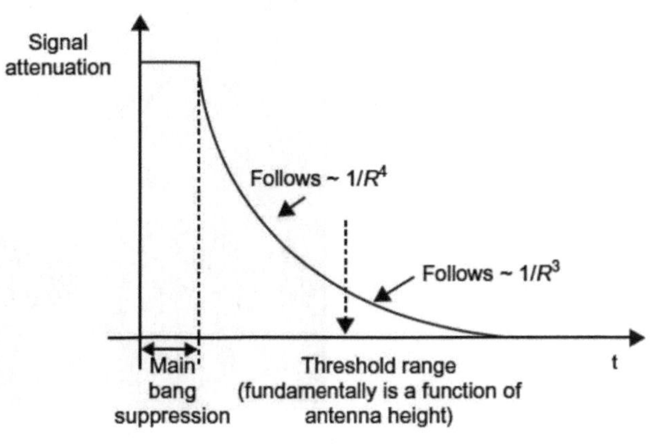

Fig. 1.75 Sensitivity time control (STC)

1.13 Rain Clutter

Rain clutter, and other types of precipitation clutter such as from hail and snow, is typified by having a continuous return over a long range and at wide angles. Unlike the returns from sea clutter, which tend to be very "spiky"—the spikes resulting from particular instantaneous sea waves—rain clutter has a very smooth overall response. It is a problem for the user of the radar because the generally increased levels of the total radar return caused by precipitation clutter can mask other targets, as shown in Fig. 1.76.

On a large area of rain clutter, falling from a well-defined rain clod, for instance, the reflected signal would rise suddenly and then remain high over a large range, until it would fall suddenly. The effects of such clutter can be mitigated by ensuring that the gain thresholds in such affected areas are appropriately reduced.

Before digital signal processing, this was typically performed by having an analogue circuit that performed a differential process. By differentiating the signal with respect to time (i.e., range), the resultant signal will be large at the start and end of the rain area, where the signal changes amplitude suddenly, and near zero where the signal is virtually constant at ranges where the rain was falling.

Since echoes of wanted targets rise and fall sharply with range, the differentiating process keeps these very visible, but now with much reduced contribution from the rain. This is illustrated in Fig. 1.77.

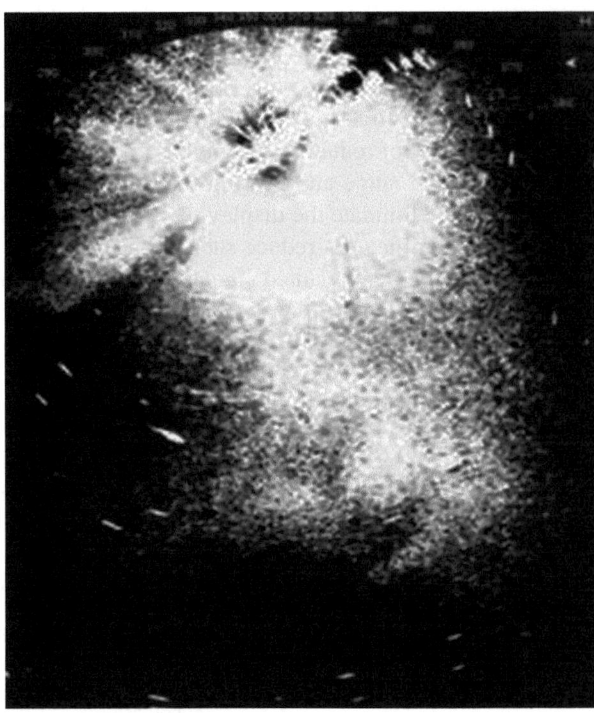

Fig. 1.76 Precipitation clutter with no fast time constant (FTC)

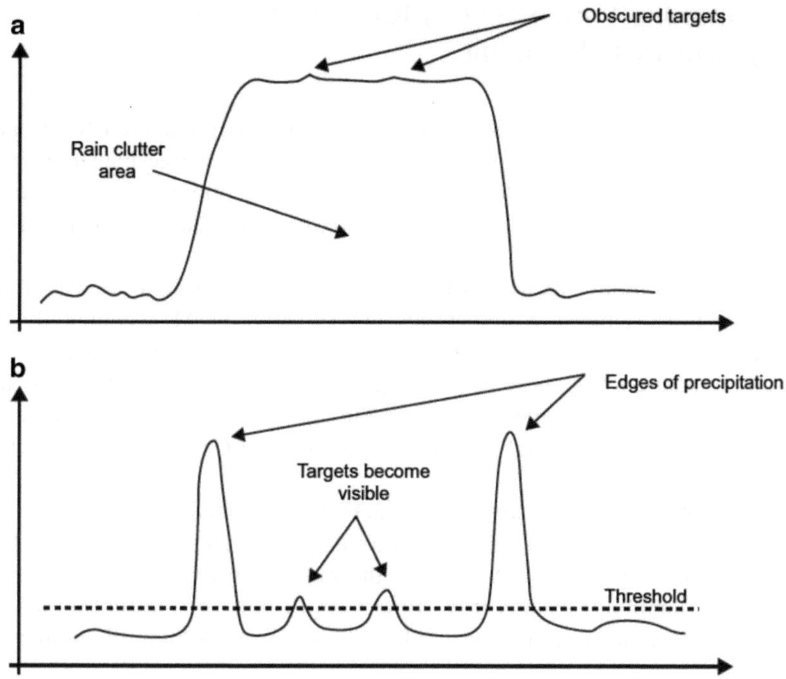

Fig. 1.77 Differential (FTC) processing, (**a**) signal before differentiation and (**b**) signal after differentiation

Today, modern radars can use various digital methods to optimally set the threshold to be able to see small targets in rain, including those based on differentiation. The user's rain control adjusts the level of the resultant threshold up and down, assisted by manufacturer-specific algorithms aimed at getting the best performance. Reflecting the original analogue processing techniques used for differentiation, the rain clutter control process is sometimes known as the fast time constant (FTC) control.

This reflects the fact that the fast changing elements of the signal in time, for instance, the edges of rain clutter and normal target reflections, create a larger processed signal than the slower changing elements, such as that from a large area of rain.

STC is sometimes understood to mean slow time constant, as the effective gain applied to the received signal is moving relatively slowly in time—at least compared to that implied by the use of FTC. The safe and successful operation of the main manual controls of a radar, gain, rain (FTC), and sea (STC) as we have described them very briefly in above. More details can be found in book by Alan Bole et al. [36].

1.14 Jamming-to-Signal (*J*/*S*) Ratio: Constant Power [Saturated] Jamming

This section derives the *J*/*S* ratio from the one-way range equation for *J* and the two-way range equation for *S* and deals exclusively with active transmitting electronic countermeasure (ECM) devices or systems. Furthermore, the only purpose of the ECM considered is to prevent, delay, or confuse the radar processing of target information.

By official definition, ECM can be either jamming or deception. This may be somewhat confusing because almost any type of active ECM is commonly called "jamming," and the calculations of ECM signal in the radar compared to the target signal in the radar commonly refer to the "jamming-to-signal" ratio ("*J*-to-*S*" ratio). Therefore, this section uses the common jargon and the term "jammer" refers to any ECM transmitter, and the term "jamming" refers to any ECM transmission, whether deception or concealment.

Table 1.6 contains a summary of the equations developed in this section.

"Official" jamming should more aptly be called concealment or masking. Essentially, concealment uses ECM to swamp the radar receiver and hide the targets. Concealment (jamming) usually uses some form of noise as the transmitted ECM signal. In this section, concealment will be called "noise" or "noise jamming."

1.14.1 *Jamming*

Deception might be better called forgery. Deception uses ECM to forge false target signals that the radar receiver accepts and processes as real targets.

"*J*" designates the ECM signal strength whether it originates from a noise jammer or from a deception ECM system.

Basically, there are two different methods of employing active ECM against hostile radars:

- Self-protection ECM
- Support ECM

For most practical purposes, self-protection ECM is usually deception, and support ECM is usually noise jamming. As the name implies, self-protection ECM is ECM that is used to protect the platform that it is on. Self-protection ECM is often called "self-screening jamming," and also "DECM," which is an acronym for either "defensive ECM (DECM)" or "deception ECM."

The top half of Fig. 1.78 shows self-screening jamming defensive electronic countermeasure (DECM).

Table 1.6 Jamming-to-Signal (*J*/*S*) ratio monostatic

JAMMING TO SIGNAL (*J*/*S*) RATIO (MONOSTATIC)	*Keep R and σ in same units		
$J/S = (P_j G_{ja} 4\pi R^2)/(P_t G_t \sigma)$ (ratio form)* or:	Target gain factor, (in dB)		
	$G_\sigma = 10 \log \sigma + 20 \log f_1 + K_2$		
$10 \log J/S = 10 \log P_j + 10 \log G_{ja} - 10 \log P_t - 10 \log G_t - 10 \log \sigma^* + 10.99 \text{ dB} + 20 \log R^*$	K_2 Values (dB):		
Note (1): Neither f nor λ terms are part of these equations	RCS (σ)	f_1 in MHz	f_1 in GHz
	(units)	$K_2 =$	$K_2 =$
If simplified radar equations developed in previous sections are used:	m^2	−38.54	21.46
$10 \log J/S = 10 \log P_j + 10 \log G_{ja} - 10 \log P_t - 10 \log G_t - G_\sigma + \alpha_1$ (in dB)	ft^2	−48.86	11.14
Note (2): the $20 \log f_1$ term in $-G_\sigma$ cancels the $20 \log f_1$ term in α_1			
JAMMING TO SIGNAL (*J*/*S*) RATIO (BISTATIC)	One-way free space loss (dB)		
R_{Tx} is the range from the radar transmitter to the target. See note (1)	α_1 or $\alpha_{Tx} = 20 \log (f_1 R) + K_1$		
	K_1 Values (dB):		
$J/S = \left(P_j G_{ja} 4\pi R_{Tx}^2\right)/(P_t G_t \sigma)$ (ratio form)* or:	Range	f_1 in MHz	f_1 in GHz
	(units)	$K_1 =$	$K_1 =$
$10 \log J/S = 10 \log P_j + 10 \log G_{ja} - 10 \log P_t - 10 \log G_t - 10 \log \sigma^* + 10.99 \text{ dB} + 20 \log R_{Tx}^*$	NM	37.8	97.8
	km	32.45	92.45
If simplified radar equations developed in previous sections are used: see note (2)	m	−27.55	32.45
$10 \log J/S = 10 \log P_j + 10 \log G_{ja} - 10 \log P_t - 10 \log G_t - G_\sigma + \alpha_{Tx}$ (in dB)	ft	−37.87	22.13

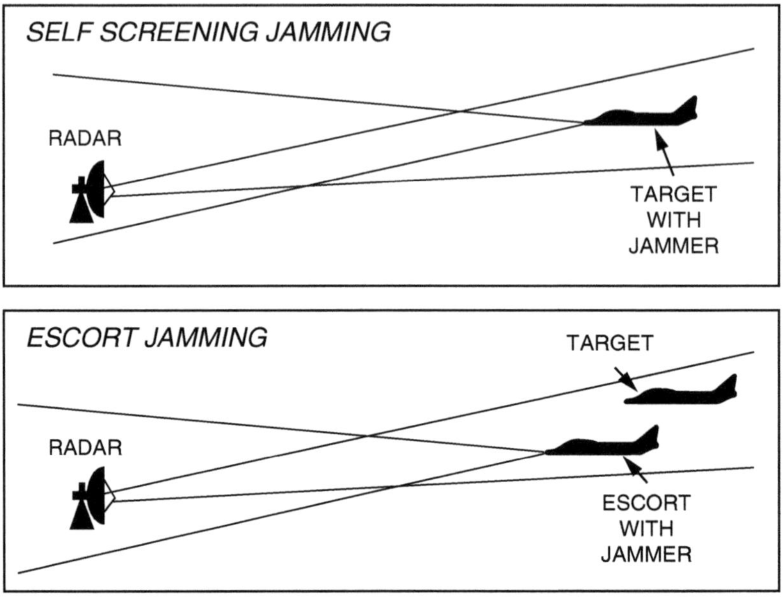

Fig. 1.78 Self-protection and escort jamming [37]. (Courtesy of Naval Air Warfare Center Weapon Division, US Navy)

The bottom half of Fig. 1.78 illustrates escort jamming which is a special case of support jamming. If the escort platform is sufficiently close to the target, the J-to-S calculations are the same as for DECM.

Support ECM is ECM radiated from one platform and is used to protect other platforms. Figure 1.79 illustrates two cases of support jamming—standoff jamming (SOJ) and stand-in jamming (SIJ).

For standoff jamming (SOJ), the support jamming platform is maintaining an orbit at a long range from the radar—usually beyond weapons range. For stand-in jamming (SIJ), a remotely piloted vehicle is orbiting very close to the victim radar. Obviously, the jamming power required for the SOJ to screen a target is much greater than the jamming power required for the SIJ to screen the same target.

When factoring ECM into the radar equation, the quantities of greatest interest are "J-to-S" and burn-through range.

"J-to-S" is the ratio of the signal strength of the ECM signal (J) to the signal strength of the target return signal (S). It is expressed as "J/S" and, in this section, is always in dB. J usually (but not always) must exceed S by some amount to be effective; therefore, the desired result of a J/S calculation in dB is a positive number. Burn-through range is the radar to target range where the target return signal can first be detected through the ECM and is usually slightly farther than crossover range where $J = S$. It is usually the range where the J/S just equals the minimum effective J/S.

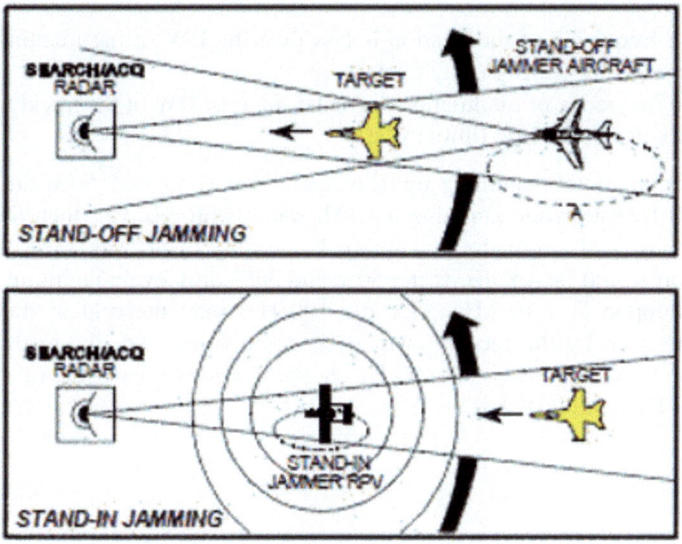

Fig. 1.79 Support jamming [37]. (Courtesy of Naval Air Warfare Center Weapon Division, US Navy)

The significance of "*J*-to-*S*" is sometimes misunderstood. The effectiveness of ECM is not a direct mathematical function of "*J*-to-*S*." The magnitude of the "*J*-to-*S*" required for effectiveness is a function of the particular ECM technique and of the radar it is being used against. Different ECM techniques may very well require different "*J*-to-*S*" ratios against the same radar. When there is sufficient "*J*-to-*S*" for effectiveness, increasing it will rarely increase the effectiveness at a given range. Because modern radars can have sophisticated signal processing and/or ECCM capabilities, in certain radars too much "*J*-to-*S*" could cause the signal processor to ignore the jamming or activate special anti-jamming modes. Increasing "*J*-to-*S*" (or the jammer power) does, however, allow the target aircraft to get much closer to the threat radar before burn-through occurs, which essentially means more power is better if it can be controlled when desired [37].

Important Note

If the signal *S* is continuous wave (CW) or pulse Doppler (PD) and the jamming *J* is amplitude modulated, then the *J* used in the formula has to be reduced from the peak value (due to sin *x*/*x* frequency distribution). The amount of reduction is dependent upon how much of the bandwidth is covered by the jamming signal. To get an exact value, integrals would have to be taken over the bandwidth. As a rule of thumb however:

(continued)

- If the frequency of modulation is less than the BW of the tracking radar, reduce *J/S* by 10 Log (duty cycle).
- If the frequency of modulation is greater than the BW of the tracking radar, reduce *J/S* by 20 Log (duty cycle).

For example, if your jamming signal is square wave chopped (50% duty cycle) at a 100 Hz rate while jamming a 1 kHz bandwidth receiver, then the *J/S* is reduced by 3 dB from the maximum. If the duty cycle was 33%, then the reduction would be 4.8 dB. If the 50% and 33% duty cycle jamming signals were chopped at a 10 kHz (vice the 100 Hz) rate, the rule of thumb for jamming seen by the receiver would be down 6 and 9.6 dB, respectively, from the maximum since the 10 kHz chopping rate is greater than the 1 kHz receiver bandwidth (BW).

1.14.2 J/S Self-Protection Electronic Attack (EA) Versus Monostatic Radar

Figure 1.80 is radar jamming visualized. The physical concept of Fig. 1.80 shows a monostatic radar that is the same as Fig. 1.78 in above and a jammer (transmitter) to radar (receiver) that is the same as Fig. 1.80 here. In other words, Fig. 1.80 is simply the combination of the previous two visual concepts where there is only one receiver (the radar's).

The equivalent circuit shown in Fig. 1.81 applies to jamming monostatic radars with either self-protect or support electronic attack (EA), which is very similar to older term for electronic countermeasure (ECM). For self-protect (or escort) vs. a monostatic radar, the jammer is on the target, and the radar receiving and transmitting antennas are collocated so the three ranges and three space loss factors (αs) are the same.

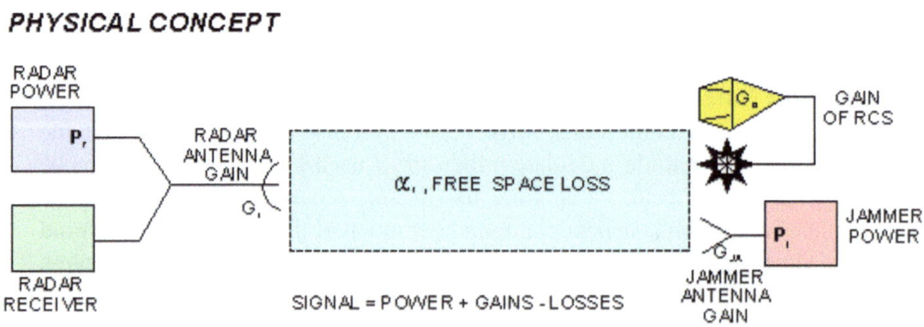

Fig. 1.80 Radar jamming visualized [37]. (Courtesy of Naval Air Warfare Center Weapon Division, US Navy)

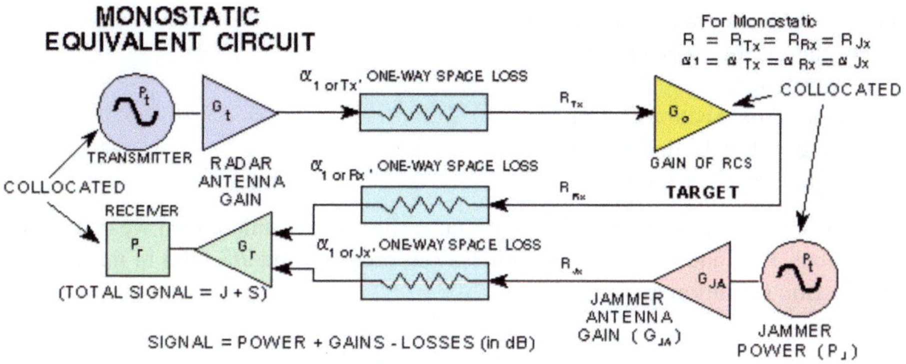

Fig. 1.81 Monostatic radar electronic attack equivalent circuit [37]. (Courtesy of Naval Air Warfare Center Weapon Division, US Navy)

1.15 Jamming-to-Signal Ratio (Monostatic)

The ratio of the power received (P_{r1} or J) from the jamming signal transmitted from the target to the power received (P_{r2} or S) from the radar skin return from the target equals J/S.

From the one-way range equation as before, we can write:

$$P_{r1} \text{ or } J = \frac{P_j G_{ja} G_r \lambda^2}{(4\pi R)^2} \tag{1.36}$$

where:

P_{r1} = power received at point 1
P_j = power of a jammer transmitter
G_{ja} = gain of the jammer antenna
G_r = gain of the receiver antenna
λ = wavelength
R = range (straight line distance)
J = jamming signal (receiver input)

From a two-way range equation, we can write:

$$P_{r2} \text{ or } S = \frac{P_t G_t G_r \lambda^2 \sigma}{(4\pi)^3 R^4} \tag{1.37}$$

where:

P_{r2} = power received at point 2
P_t = power of a transmitter

G_t = gain of the transmitter antenna
G_r = gain of the receiver antenna
λ = wavelength
σ = radar cross section (RCS)
R = range (straight line distance)
S = signal (receiver input)

Moreover, if we keep R and σ in the same dimension (units), then we can divide Equations 1.35 and 1.36 by Equation 1.37 in order to find J/S ratio in the form of monostatic condition as:

$$\frac{J}{S} = \frac{P_j G_{ja} G_r \lambda^2 (4\pi)^3 R^4}{P_t G_t G_r \lambda^2 \sigma (4\pi R)^2} = \frac{P_j G_{ja} 4\pi R^2}{P_t G_t \sigma} \quad \text{Ratio form} \tag{1.38}$$

On reducing the above equation to log form, we have:

$$\begin{aligned} 10 \log J/S = {} & 10 \log P_j + 10 \log G_{ja} - 10 \log P_t - 10 \log G_t \\ & - 10 \log \sigma + 10 \log 4\pi + 20 \log R \end{aligned} \tag{1.39}$$

or

$$\begin{aligned} 10 \log J/S = {} & 10 \log P_j + 10 \log G_{ja} - 10 \log P_t - 10 \log G_t \\ & - 10 \log \sigma + 10.99\ \text{dB} + 20 \log R \end{aligned} \tag{1.40}$$

Note: Neither f nor λ terms are part of the final form of Equations 1.38 and 1.40.

1.16 *J/S* Calculations (Monostatic) Using a One-Way Free-Space Loss

The simplified radar equations developed in previous sections can be used to express *J/S*.

From the one-way range equation:

$$10 \log (P_{r1} \text{ or } J) = 10 \log P_j + 10 \log G_{ja} + 10 \log G_r - \alpha_1 \text{ (in dB)} \tag{1.41}$$

From the two-way range equation:

$$\begin{aligned} 10 \log (P_{r2} \text{ or } S) = {} & 10 \log P_t + 10 \log G_t + 10 \log G_r + G_\sigma \\ & - 2\alpha_1 \text{ (in dB)} \end{aligned} \tag{1.42}$$

Table 1.7 List of target gain factor and one-way free-space loss

Target gain factor, $G_\sigma = 10 \log \sigma + 20 \log f_1 + K_2$ (in dB)				One-way free space loss, $\alpha_1 = 20 \log (f_1 R) + K_1$ (in dB)			
K_2 values (dB)	RCS (σ)	f_1 in MHz	f_1 in GHz	K_1 Values (dB)	Range	f_1 in MHz	f_1 in GHz
	(units)	$K_2 =$	$K_2 =$		(units)	$K_1 =$	$K_1 =$
					NM	37.8	97.8
	m^2	−38.54	21.46		km	32.45	92.45
	ft^2	−48.86	11.14		m	−27.55	32.45
					yd	−28.33	31.67
					ft	−37.87	22.13

$$10 \log (J/S) = 10 \log P_j + 10 \log G_{ja} - 10 \log P_t - 10 \log G_t - G_\sigma + \alpha_1 \text{ (in dB)} \tag{1.43}$$

$$10 \log (P_{r2} \text{ or } S) = 10 \log P_t + 10 \log G_t + 10 \log G_r + G_\sigma - 2\alpha_1 \text{ (in dB)} \tag{1.44}$$

$$10 \log (J/S) = 10 \log P_j + 10 \log G_{ja} - 10 \log P_t - 10 \log G_t - G_\sigma + \alpha_1 \text{ (in dB)} \tag{1.45}$$

Note: To avoid having to include additional terms for these calculations, always combine any transmission line loss with antenna gain. The 20 log f_1 term in $-G_\sigma$ cancels the 20 log f_1 term in $-G_\sigma$; see Table 1.7.

1.17 *J/S* for Self-Protection Electronic Attack (EA) Versus Bistatic Radar

The semi-active missile illustrated in Fig. 1.82 is the typical bistatic radar which would require the target to have self-protection electronic attack (EA) to survive. In this case, the jammer is on the target, and the target to missile receiver range is the same as the jammer to receiver range, but the radar to target range is different.

The following equations as:

α_{Tx} = The one-way space loss from the radar transmitter to the target for range R_{Tx} (1.46)

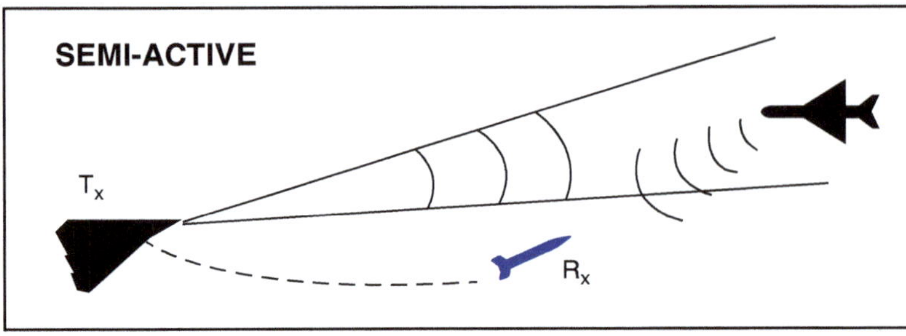

Fig. 1.82 Bistatic radar

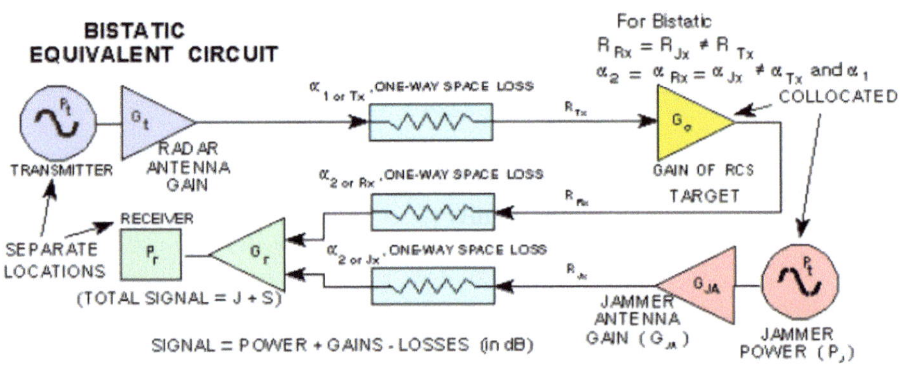

Fig. 1.83 Bistatic radar electronic attack equivalent circuit [37]. (Courtesy of Naval Air Warfare Center Weapon Division, US Navy)

α_{Rx} = The one-way space loss from the target to the missile receiver for range R_{Rx} (1.47)

Like the monostatic radar, the bistatic jamming and reflected target signals travel the same path from the target and enter the receiver (missile in this case) via the same antenna. In both monostatic and bistatic *J/S* equations, this common range cancels, so both *J/S* equations are left with an R_{Tx} or 20 log R_{Tx} term.

Therefore, only two of the ranges and two of the αs (Fig. 1.83) are the same.

Since in the monostatic case $R_{\mathrm{Tx}} = R_{\mathrm{Rx}}$ and $\alpha_{\mathrm{Tx}} = \alpha_{\mathrm{Rx}}$, only R or α_1 is used in the equations.

Therefore, the bistatic *J/S* Equations 1.50, 1.52, or 1.53 and 1.54 will work for monostatic *J/S* calculations, but the opposite is only true if bistatic R_{Tx} and α_{Tx} terms are used for R or α_1 terms in monostatic Equations 1.38, 1.40, and 1.45.

The equivalent circuit shown in Fig. 1.83 applies to jamming bistatic radar. For self-protect (or escort) vs. a bistatic radar, the jammer is on the target, and the radar receiving and transmitting antennas are at separate locations so only two of the three ranges and two of the three space loss factors (αs) are the same.

1.18 *J*-to-*S* Ratio (Bistatic)

When the radar's transmitting antenna is located remotely from the receiving antenna (Fig. 1.83), the ratio of the power received (P_{r1} or J) from the jamming signal transmitted from the target to the power received (P_{r2} or S) from the radar skin return from the target equals J/S. For jammer effectiveness J normally has to be greater than S.

From the one-way range equation, we can write the following equation as:

$$P_{r1} \text{ or } J = \frac{P_j G_{ja} G_r \lambda^2}{(4\pi R_{Rx})^2} \quad R_{Jx} = R_{Rx} \tag{1.48}$$

From the two-way range equation, we can write the following equations as:

$$P_{r2} \text{ or } S = \frac{P_t G_t G_r \lambda^2 \sigma}{(4\pi)^3 R_{Tx}^2 R_{Rx}^2} \quad \text{Ratio form} \tag{1.49}$$

Thus, J/S will be found by dividing Equation 1.48 by Equation 1.49 as follows with assumption that R and σ have the same dimensional units:

$$\frac{J}{S} = \frac{P_j G_{ja} G_r \lambda^2 (4\pi)^3 R_{Tx}^2 R_{Rx}^2}{P_t G_r \lambda^2 \sigma (4\pi R_{Rx})^2} = \frac{P_j G_{ja} 4\pi R_{Tx}^2}{P_t G_t \sigma} \tag{1.50}$$

On reducing the above equation to log form, we have:

$$\begin{aligned} 10 \log J/S = {} & 10 \log P_j + 10 \log G_{ja} - 10 \log P_t - 10 \log G_t \\ & - 10 \log \sigma + 10 \log 4\pi + 20 \log R_{Tx} \end{aligned} \tag{1.51}$$

or

$$\begin{aligned} 10 \log J/S = {} & 10 \log P_j + 10 \log G_{ja} - 10 \log P_t - 10 \log G_t \\ & - 10 \log \sigma + 10.99 \text{ dB} + 20 \log R_{Tx} \end{aligned} \tag{1.52}$$

Note: To avoid having to include additional terms for these calculations, always combine any transmission line loss with antenna gain. Neither f nor λ terms are part of the final form of Equations 1.50 and 1.52.

1.19 Bistatic *J*/*S* Calculations (Bistatic) Using a One-Way Free-Space Loss

The simplified radar equations developed in previous sections can be used to express *J*/*S*.

From the one-way range equation, we can write the following relation as:

$$10\ \log\ (P_{r1}\text{ or }J) = 10\ \log\ P_j + 10\ \log\ G_{ja} + 10\ \log\ G_r - \alpha_{Rx}\ (\text{all factors dB}) \tag{1.53}$$

From the two-way range equation in, we can write the following relation as:

$$10\ \log\ (P_{r2}\text{ or }S) = 10\ \log\ P_t + 10\ \log\ G_t + 10\ \log\ G_r + G_\sigma - \alpha_{Tx} - \alpha_{Rx}\ (\text{all factors dB}) \tag{1.54}$$

$$10\ \log\ (J/S) = 10\ \log\ P_j + 10\ \log\ G_{ja} - 10\ \log\ P_t - 10\ \log\ G_t - G_\sigma + \alpha_{Tx}\ (\text{all factors dB}) \tag{1.55}$$

Note: To avoid having to include additional terms for these calculations, always combine any transmission line loss with antenna gain. The 20 log f_1 term in $-G_\sigma$ cancels the 20 log f_1 term in α_1. See Table 1.8.

1.20 Standard *J*/*S* (Monostatic) Example (Constant Power Jamming)

Assume that a 5 GHz radar has a 70 dBm signal fed through a 5 dB loss transmission line to an antenna that has 45 dB gain. An aircraft is flying 31 km from the radar. The aft EW antenna has −1 dB gain and a 5 dB line loss to the EW receiver (there is an

Table 1.8 Target gain factor and one-way free-space loss

Target gain factor, $G_\alpha = 10\log\sigma + 20\log f_1 + K_2$ (in dB)				One-way free space loss $\alpha_{Tx\text{ or }Rx} = 20\log f_1 R_{Tx\text{ or }Rx} + K_1$ (in dB)			
K_2 Values (dB)	RCS (σ)	f_1 in MHz	f_1 in GHz	K_1 Values (dB)	Range	f_1 in MHz	f_1 in GHz
	(units)	$K_2 =$	$K_2 =$		(units)	$K_1 =$	$K_1 =$
					NM	37.8	97.8
	m^2	−38.54	21.46		km	32.45	92.45
	ft^2	−48.86	11.14		m	−27.55	32.45
					yd	−28.33	31.67
					ft	−37.87	22.13

additional loss due to any antenna polarization mismatch, but that loss will not be addressed in this problem). The aircraft has a jammer that provides 30 dBm saturated output if the received signal is above −35 dBm. The jammer feeds a 10 dB loss transmission line which is connected to an antenna with a 5 dB gain. If the RCS of the aircraft is 9 m^2, what is the *J*/*S* level received by the tracking radar?

1.21 Millimeter-Wave Radar Equation

The signal transmitted by a radar and reflected from a target (or targets) is well-characterized in the literature to follow the radar Equation 1.29. To improve the accuracy of the channel model, it is common to include additional factor to account for losses such as atmospheric absorption as discussed in Sect. 1.11.5 as a distance-dependent factor η. Thus, the modified radar equation takes the form as:

$$S = \underbrace{\frac{P_0 \eta G_t}{4\pi R^2}}_{\text{Incident Signal}} \times \underbrace{\frac{\sigma_c}{4\pi R^2}}_{\text{Reflected Signal}} A_e = \gamma_1 \gamma_2 \frac{P_0 \eta}{R^4} \tag{1.56}$$

which models the returned signal power, where P_0 is the radar transmit power; R is the target range, that is, the distance to the target; G_t and A_e are the antenna gain and its effective area, respectively; and σ_c is the radar cross section (RCS) area of the target. The parameters γ_1 and γ_2 are given as:

$$\begin{cases} \gamma_1 = \dfrac{G_t A_e}{4\pi} = G_t^2 \left(\dfrac{c}{4\pi f}\right)^2 \\ \gamma_2 = \dfrac{\sigma_c}{4\pi} \end{cases} \tag{1.57}$$

where f is the operating frequency. We illustrate the parameters affecting a radar signal in Fig. 1.84, for millimeter-wave radar in the typical automotive scenario. See reference by Rama Chellappa and Sergios Theodoridis [38] for further discussions on the radar equation.

Furthermore, millimeter waves are electromagnetic (radio) waves typically defined to lie within the frequency range of 30–300 GHz. The microwave band is just below the millimeter-wave band and is typically defined to cover the 3–30 GHz range. The terahertz band is just above the millimeter-wave band and is typically defined to cover the 300 GHz to 3 + THz range. The wavelength of electromagnetic radiation is given by $\lambda = c/f$, where $c = 3 \times 10^8$ m/s is the speed of light and f is the frequency (in Hz). The millimeter-wave band thus corresponds to a wavelength range of 10 mm at 30 GHz decreasing to 1 mm at 300 GHz [39].

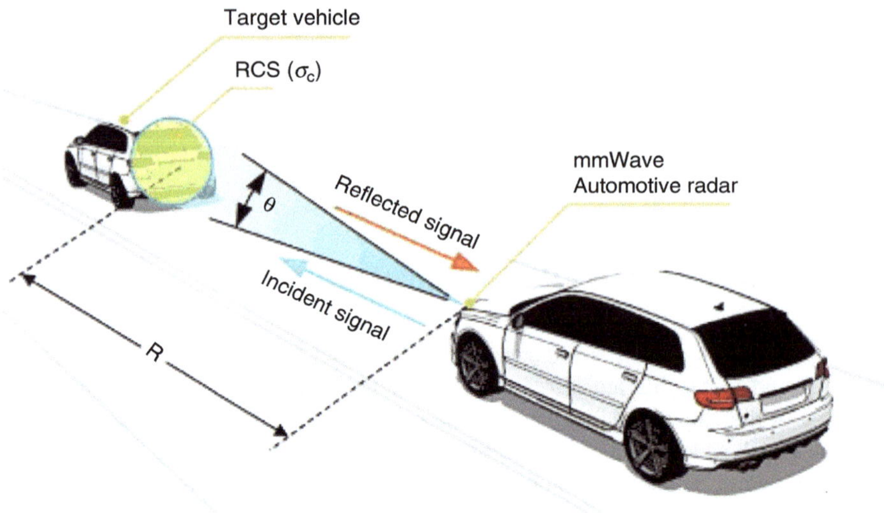

Fig. 1.84 Illustration of some of the factors that affect the strength of the returned signal in typical automotive scenario [38]. (Courtesy of Academic Press)

Millimeter waves are effective for explosive detection on personnel because the waves readily pass through common clothing materials and reflect from the body and any concealed items. These reflected wavefronts can be focused by an imaging system that will reveal the size, shape, and orientation of the concealed object. Diffraction generally limits resolution to spot sizes of $\lambda/2$ or larger, so resolution spot sizes of <10 mm are readily achievable at millimeter wavelengths.

There is little spectral (frequency) variation in the reflection or emission of millimeter waves from most bulk materials, including the human body and most concealed objects. This means that millimeter-wave imaging systems cannot uniquely identify specific materials, such as explosives. They can, however, form high-resolution images that will reveal discrepancies from the expected image of a person and reveal the shape and position of the concealed items, which enables the development of high-performance and versatile concealed weapon detection imaging systems.

Millimeter waves can be used for both active and passive imaging systems. Active imaging systems primarily image the reflectivity of the person/scene including the effect of the object's shape and orientation. Passive systems measure the thermal (black-body) emission from the scene, which will include thermal emission from the environment that is reflected by objects in the scene (including the person) [39].

For both active and passive personnel screening systems to be effective, it is necessary that most clothing be relatively transparent at the frequency of operation of the system, so that concealed items will be detected. Fabrics can be considered to be a thin layer of dielectric material. The thickness of most materials will be much less than the wavelength throughout most of the millimeter-wave band. Additionally, most materials have relatively low attenuation losses over the millimeter-wave band. The combination of thinness (thickness $\ll\lambda$) and low loss means that fabrics will cause only slight absorption and reflection losses to the millimeter-wave signals. Bjarnason et al. [40] have published a number of fabric attenuation measurements covering the millimeter-wave, terahertz, and infrared (IR) frequency bands. These measurements confirm the relative transparency of most materials in the frequency range below 300 GHz.

In contrast to most fabrics, the human body can be considered a good conductor and strongly reflects and absorbs waves in the millimeter-wave range. Concealed objects can generally be classified as dielectrics with unknown shape and dielectric properties. Metals can be considered to be a limiting case of a highly conductive dielectric. Dielectric objects including metals, the human body, and concealed items will all produce reflections based on the Fresnel reflection at each air-dielectric or dielectric-dielectric interface [41]. Additionally, these reflections will be altered by the shape, texture, and orientation of the surfaces. This complexity renders it difficult to directly measure dielectric properties of concealed items. However, it does create significant variation in the reflectivity which provides significant contrast in active imaging systems.

Passive systems exploit the natural thermal emission of radiation that emanates from all warm objects (above absolute zero). For objects or bodies near room temperature, these emission spectra peak near wavelengths of 10 μm, which is in the long-wave IR region of the spectrum [42]. IR imaging cameras typically operate near this wavelength or at shorter wavelengths, closer to visible light. For longer wavelengths, such as in the millimeter-wave band, this radiation is at much lower intensity but is still present and can be used to form passive millimeter-wave imaging systems. These systems are analogous to IR imaging camera systems but are tuned to take advantage of the unique properties of millimeter waves, which includes effective clothing penetration to detect concealed objects. Owing to the significantly reduced signal levels available in the millimeter wave, it is considerably more difficult to develop sensitive imaging systems. Sensitive receivers employing advanced integrated low-noise amplifiers have allowed the development of effective systems.

Passive systems form an image of the emitted millimeter-wave radiation that is the sum of energy directly emitted from the target or scene and energy that originates elsewhere and is reflected by the target or scene. This emission increases directly with temperature; therefore, imaging systems frequently display their imaging results calibrated to an effective temperature scale, with contrast represented as differential temperature. Noise in the image is also characterized as a noise-effective differential temperature.

The emission of millimeter waves from concealed objects is complicated somewhat by the environment in which they are employed. Targets within the image, including the human body and any concealed items, emit millimeter waves based on both their temperature and their emissivity. Objects with high emissivity radiate at close to the black-body limit, whereas objects with low emissivity radiate proportionally less. Metals and other good reflectors have low emissivity, whereas good absorbers have relatively high emissivity. The human body has both moderate emissivity and reflectivity—so it is easily visible in both active and passive systems. These differences in target/scene emissivity provide contrast in images even if the temperatures of different components of the image are all close to the same value.

Objects in passive images that have moderate to high reflectivity will typically contain signals due to both thermal emission and reflected radiation from the background. Outside, the sky represents a relatively cold background, whereas indoors the background is relatively warm. These factors can significantly reduce the thermal contrast available in passive imagery, particularly for systems operated inside. Passive systems rely on effective temperature contrast in the images, which is altered by the environment in which the systems are used. Active systems essentially measure reflectivity and are not significantly affected by the environment.

Atmospheric attenuation properties of millimeter waves can be important, especially in specific bands. Electromagnetic waves effectively pass through the atmosphere without significant losses over much of the spectrum, including many portions of the microwave, millimeter-wave, IR, and optical bands. However, significant absorption because of water vapor or other atmospheric constituents does occur over several narrow frequency bands in the millimeter-wave band and is extremely significant over much of the terahertz band.

References

1. *Translation Bureau (2013). "Radar definition". Public Works and Government Services Canada. Retrieved 8 November 2013.*
2. McGraw-Hill dictionary of scientific and technical terms/Daniel N. Lapedes, editor in chief. Lapedes, Daniel N. New York ; Montreal : McGraw-Hill, 1976. [xv], 1634, A26 p.
3. *Nees, Michael A. (September 2016). "Acceptance of Self-driving Cars: An Examination of Idealized versus Realistic Portrayals with a Self- driving Car Acceptance Scale". Proceedings of the Human Factors and Ergonomics Society Annual Meeting.* ***60*** *(1): 1449–1453. doi:*https://doi.org/10.1177/1541931213601332. *ISSN 1541-9312.*
4. *Fakhrul Razi Ahmad, Zakuan; et al. (2018). "Performance Assessment of an Integrated Radar Architecture for Multi-Types Frontal Object Detection for Autonomous Vehicle". 2018 IEEE International Conference on Automatic Control and Intelligent Systems (I2CACIS). Retrieved 9 January 2019.*
5. *LIDAR—Light Detection and Ranging—is a remote sensing method used to examine the surface of the Earth".* NOAA. Archived from the original on May 30, 2013. Retrieved June 4, 2013.
6. Oxford English Dictionary. *2013. p. Entry for "lidar".*
7. James Ring, "The Laser in Astronomy." pp. 672–73, *New Scientist* June 20, 1963.
8. Carter, Jamie; Keil Schmid; Kirk Waters; Lindy Betzhold; Brian Hadley; Rebecca Mataosky; Jennifer Halleran (2012). "Lidar 101: An Introduction to Lidar Technology, Data, and Applica tions." (NOAA) Coastal Services Center" (PDF). Coast.noaaa.gov. p. 14. Retrieved 2017-02-11.
9. Kostenko, A.A., A.I. Nosich, and I.A. Tishchenko, "Radar Prehistory, Soviet Side," *Proc. of IEEE APS International Symposium 2001,* vol. 4. p. 44, 2003.

10. Christian Huelsmeyer, the inventor". radarworld.org.
11. Patent DE165546; Verfahren, um metallische Gegenstände mittels elektrischer Wellen einem Beobachter zu melden.
12. Verfahren zur Bestimmung der Entfernung von metallischen Gegenständen (Schiffen o. dgl.), deren Gegenwart durch das Verfahren nach Patent 16556 festgestellt weird.
13. GB 13170 Telemobiloscope
14. "gdr_zeichnungpatent.jpg". Retrieved 24 February 2015.
15. "Making waves: Robert Watson-Watt, the pioneer of radar". BBC. 16 February 2017. Hyland, L.A, A.H. Taylor, and L.C. Young; "System for detecting objects by radio," U.S. Patent No. 1981884, granted 27 Nov. 1934
16. Howeth, Linwood S.; "Radar," Ch. XXXVIII in History of Communications -Electronics in the United States Navy, 1963; Radar
17. Watson, Raymond C., Jr. (25 November 2009). Radar Origins Worldwide: History of Its Evolution in 13 Nations Through World War II. Trafford Publishing. ISBN 978-1-4269-2111-7.
18. Mark A. Richards, Fundamentals of Radar Signal Processing", The McGraw-Hill Companies, Inc. 2005.
19. Bonnier Corporation (December 1941). Popular Science. Bonnier Corporation. p.56.
20. ICAO: Global Air Navigation Plan for CNS/ATM Systems, Second Edition — 2002, Chapter 7 Surveillance Systems.
21. ICAO: Annex 10 - Aeronautical Communications, Volume I, Chapter 3, Item 3.2.3: The precision approach radar element (PAR), page 3-25 (PDF-page 33)
22. ICAO: Annex 6 - Operation of Aircraft, Part I, Chapter 1, Definitions, page 1-1 (PDF-page 25)
23. ICAO: NON-PRECISION INSTRUMENT APPROACH, in Advisory Circular for Air Operators, November 2012, AC No: 008A-CDFA, page 3.
24. http://www.radartutorial.eu/06.antennas/Feeding%20Systems.en.html
25. http://www.radartutorial.eu/06.antennas/Angle%20of%20the%20Irradiation.en.html
26. *Radar Modulator". radartutorial.eu*; http://www.radartutorial.eu//08.transmitters/Radar%20Modulator.en.html
27. http://www.radartutorial.eu/11.coherent/co07.en.html
28. http://www.radartutorial.eu//08.transmitters/Crossed-Field%20Amplifier%20%28Amplitron%29.en.html
29. *"Fully Coherent Radar". radartutorial.eu*: .http://www.radartutorial.eu//08.transmitters/Fully%20Coherent%20Radar.en.html
30. http://www.radartutorial.eu/08.transmitters/Traveling%20Wave%20Tube.en.html
31. http://www.radartutorial.eu/01.basics/Time-dependences%20in%20Radar.en.html
32. http://www.radartutorial.eu/01.basics/Doppler%20Dilemma.en.html
33. http://www.radartutorial.eu/19.kartei/13.labs/karte007.en.html
34. Blake, L. V., "Radar Range Performance Analysis", Artech House, Norwood, MA, 1986.
35. Reference Data for Engineers: Radio, Electronics, Computer, and Communications, Ninth Edition, Wendy M. Middleton, Editor-in-Chief, Ninth Edition, Newnes Publishing Company, Boston USA, Chapter 36, written by Merrill I. Skolnik.
36. Alan Bole, Alan Wall and Andy Norris, "Radar and ARPA Manual, Radar, AIS and Target Tracking for Marine Radar Users" Third Edition, Published by Elsevier, Butterworth-Heinemann is an imprint of Elsevier, New York, NY, 2008.
37. Electronic Warfare and Radar System, Engineering handbook 4th Edition, Naval Air Warfare Center Weapons Division, Point Mugu, California Approved for public release, October 2003 published by Avionics Department, US Navy.
38. Rama Chellappa and Sergios Theodoridis, "Academic Press Library in Signal Processing , Volume 7) 1st edition, Published by Academic Press; 1 edition (December 15, 2017), New Your New York.
39. Jehuda Yinon, "Counterterrorist Detection Techniques of Explosives" 1st Edition, Published by Elsevier Science 3rd July 2007

40. J. E. Bjarnason, T. L. J. Chan, A. W. M. Lee, M. A. Celis, and E. R. Brown, "Millimeter-wave, terahertz, and mid-infrared transmission through common clothing," *Applied Physics Letters*, vol. 85, pp. 519, 2004.
41. J. A. Kong, *Electromagnetic Wave Theory*. New York: John Wiley and Sons, 1986.
42. R. W. Boyd, *Radiometry and the Detection of Optical Radiation*. New York: John Wiley and Sons, 1983.

Chapter 2
Electronic Countermeasure and Electronic Counter-Countermeasure

The evolution of electronic warfare has been driven by the competition between electronic countermeasures (ECM) and electronic counter-countermeasures (ECCM). Electronic warfare involves not only harnessing the electromagnetic spectrum but defending against enemy use of the spectrum, and, if possible, denying their ability to use it in the first place. Since the earliest attempts at jamming radio communications, techniques have been developed to counteract enemy ECM. Today we'll discuss how ECCM continues to evolve and shape the nature of modern warfare.

2.1 Introduction

Electronic warfare (EW) involves denying an enemy the use of the electromagnetic spectrum (EMS) or gathering intelligence of an enemy's intended actions or capabilities through analysis of transmitted electromagnetic (EM) signals. Electronic warfare uses electromagnetic spectrum for offense, attack, and mission support. From air, land, and sea, it can target forces, communication, radar, and other assets (military and civilian).

Dealing with electronic warfare technology, we are in need of delivering integrated measurement tools that help military and government personnel accurately recreate the physical and electromagnetic (EM) environment exposed during field operations and electronic warfare such as real-time spectrum analyzer (RTSA) instrument or technology that can be used to capture, visualize, and trigger on threats and electronic countermeasure response.

Notably, that spectrum analyzer measures the amplitude of an input signal versus frequency within the full frequency range of the instrument. The primary use is to measure the power of the spectrum of known and unknown signals [1].

Moreover, the electronic warfare includes signals intelligence (SIGINT) in its heart. Intelligence-driven decision-making is at the heart of daily operations and

B. Zohuri, *Radar Energy Warfare and the Challenges of Stealth Technology*,
https://doi.org/10.1007/978-3-030-40619-6_2

strategic planning for modern militaries and intelligence agencies, and signals intelligence (SIGINT) is a big part of what makes it possible.

Here in this chapter also, we will discuss how SIGINT works and why it is so important, especially as it applies to electronic warfare applications.

Modern deception electronic countermeasure (ECM) techniques are enhanced by ways of accurate replication and reproduction of the radar signal through the digital radio-frequency memory (DRFM), which attempts to deceive the radar systems and therefore make it hard to discriminate true and false targets as illustrated in Fig. 2.1. This kind of jamming signals can be fully coherently processed by the radar receiver, which means that it can be processed similar to the real targets. A variety of ECM heuristic approaches have been proposed such as range false targets (RFT) that are signals transmitted by the DRFM jammer that reasonably look like the target return but that appear in different ranges (negative or positive range offsets) from the target return, whereas range-velocity gate stealers(R-VGS) aim to mislead the radar in tracking mode, assuming that the target is in track, and therefore this track must be annihilated.

Thus, the tracker range or velocity gate must be pulled off from the target return. In view of both RFT and R-VGS, therefore, it is crucial to propose an effective ECCM method to suppress these jamming threats. Irrespective of the ECCM technique employed, the radar must guarantee the continuity of its normal work in good performance under these ECM conditions.

2.2 Explanation of Signals Intelligence (SIGINT)

Signals intelligence (SIGINT) is the interception of signals for the purpose of gathering intelligence. It is divided into three sub-disciplines:

- **Communications intelligence (COMINT)** which is the interception of communication between people and groups
- **Electronic intelligence (ELINT)** which is the intercepting of electronic signals which are not specifically used for communication
- **Foreign instrumentation signals intelligence (FISINT)**, which is the collection of signals created by the testing and use of foreign weapons systems [2]

SIGINT is collected in a variety of ways depending on the type of signal targeted. National Security Agency (NSA) collects the raw SIGINT, and then NSA translators, cryptologists, analysts, and other technical experts turn the raw data into something that an all-source analyst can use.

Once the NSA has collected, processed, and analyzed SIGINT, it is passed on to CIA and Intelligence Community analysts who use it to complement information from other sources to produce finished intelligence.

The volume and variety of today's signals add challenges to the timely production of finished intelligence for policymakers. It is a lot of work to track and analyze all the SIGINT collected.

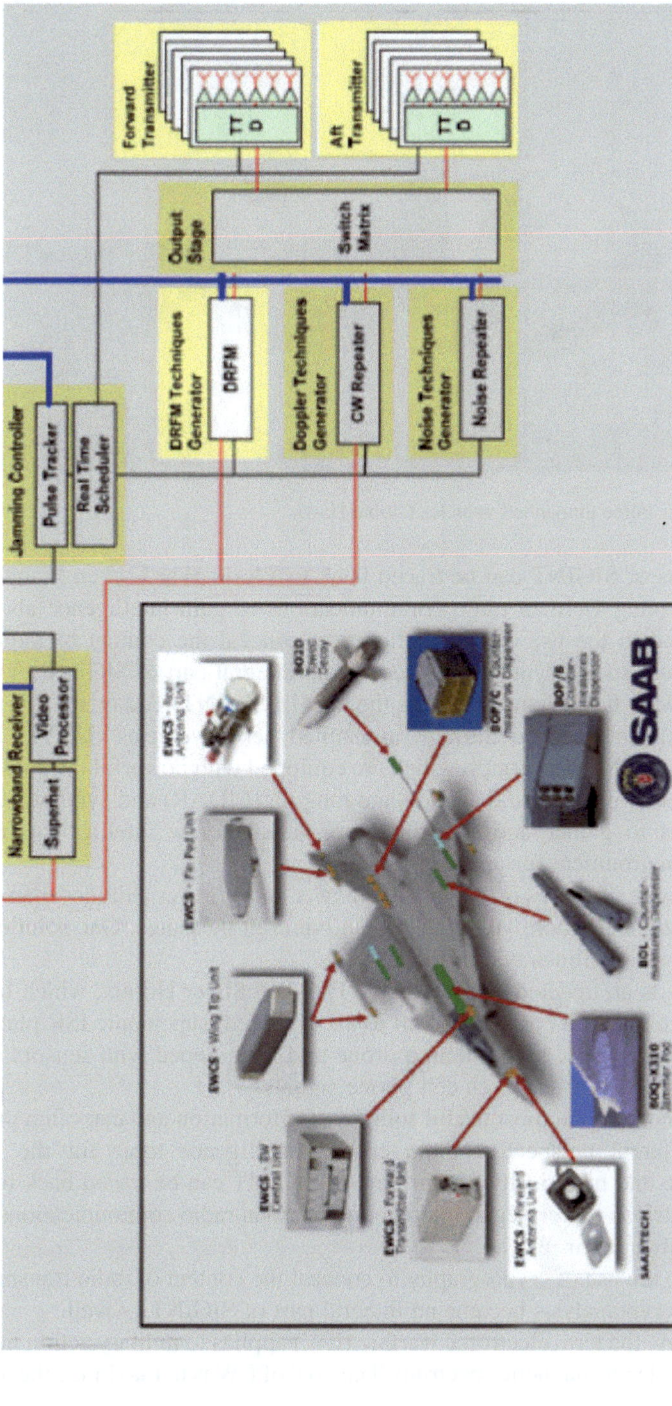

Fig. 2.1 Global illustration of digital radio-frequency memory

Fig. 2.2 US Air Force unmanned vehicles Global Hawk

The origins of SIGINT can be traced back to World War I when British forces began intercepting German radio communications to gain intelligence about their plans. This led to the use of cryptography to conceal the content of radio transmissions, and as such, cryptanalysis became an integral part of SIGINT as well.

As technology has advanced, so has the field of SIGINT. Today, the US Military gathers signals intelligence through unmanned aerial vehicles (UAVs) like the Global Hawk and Reaper drones, which are equipped with powerful infrared sensors and cameras, as well as light detection and ranging (LIDAR) and synthetic-aperture radar systems to gather and transmit back valuable raw intelligence from the operational environment for analysis (Fig. 2.2).

One downside of UAVs is that they fly slower and at lower altitudes than manned aircraft, leaving them more vulnerable to anti-aircraft measures. One solution is the EA-18G Growler as illustrated in Fig. 2.3.

This plane is an updated version of the F/A-18F Super Hornet, which has been repurposed from a pure combat aircraft to an advanced, supersonic ISR platform. It can fly much faster and higher than a drone and is equipped with sensors that can detect enemy RADAR and even cell phone signals.

SIGINT is one of the most useful sources of information and can often provide a new and different perspective on a critical intelligence topic for the nation's policymakers, and historically, the origins of SIGINT can be traced back to World War I when British forces began intercepting German radio communications to gain intelligence about their plans.

This led to the use of cryptography to conceal the content of radio transmissions, and as such, cryptanalysis became an integral part of SIGINT as well.

In summary, the term electronic warfare (EW) applies to military action involving the use of the electromagnetic spectrum. The goal of EW is to maximize the ability of

Fig. 2.3 EA-18G Growler image

friendly forces to access and exploit the spectrum while disrupting and denying the enemy's ability to do the same. It also encompasses the use of technology to defend against attacks on spectral capabilities and the use of offensive directed-energy weapons. Examples of EW include radar jamming, communication jamming, and electronic masking, as well as countermeasures against such techniques.

As with SIGINT, EW can be divided into three sub-disciplines. These include:

- **Electronic attack (EA)**, which includes offensive use of directed energy against the enemy
- **Electronic protection (EP)**, which is defensive, like the electronic warfare self-protection (EWSP) suite built into fighter jets
- **Electronic warfare support (EWS)**, the practice of locating and identifying the sources of electromagnetic energy signals for the purpose of supporting decision-making

It is in this third category of EWS that we see the overlap of electronic warfare and SIGINT because the systems and equipment used for ES can simultaneously collect intelligence. While ES is more focused on immediate threats in the operational environment, much of the data obtained can be used to enhance raw signals intelligence and SIGINT decision-making.

Electronic warfare system (EWS) can detect the source of an electromagnetic signal, the type of equipment generating that signal, and relevant data like frequency, modulation, etc. For example, EWS personnel can detect an unknown radar signal emanating from somewhere in the battlespace. They can analyze the signal and determine the type of radar that is being used and compare their findings with

Fig. 2.4 Holistic and artistic depiction of electronic warfare

countries known to use this type of radar and what vehicles, ships, aircraft, etc. it is typically used with. They can then ascertain the nature of the radar source and make intelligent predictions on what the unknown actor's intentions are.

These are exciting times to be working in the military aerospace and aviation industries. Electronic warfare is slated to become a significant area of investment and R&D within the defense sector, and as technology becomes more advanced, the value of SIGINT will only increase (Fig. 2.4) [3].

Furthermore, note that the electronic warfare (EW) functionality and capability as depicted in Fig. 2.5 via electronic support measure (ESM), electronic countermeasure (ECM), and electronic counter-countermeasure (ECCM) combined together do play a big role in the black world of stealth technology and research and development.

Figure 2.5 is drawn as a conclusion of electronic warfare based on a radar environment that is presented in Fig. 2.6, holistically

Bear in mind that this is a very generic view of electronic warfare scenario within an artistic battlefield environment.

Electronic warfare (EW) is not rigorously "electronic," i.e., it is not conducted utilizing electrons; rather it is electromagnetic and uses the entire range of the *electromagnetic spectrum* as it was explained in Chap. 1 of this book and depicted here in Fig. 2.7.

Because of this, some people will call it electromagnetic warfare (EW).

The rudimentary concept of EW is to exploit the enemy's electromagnetic emissions in all components of the electromagnetic spectrum in order to provide

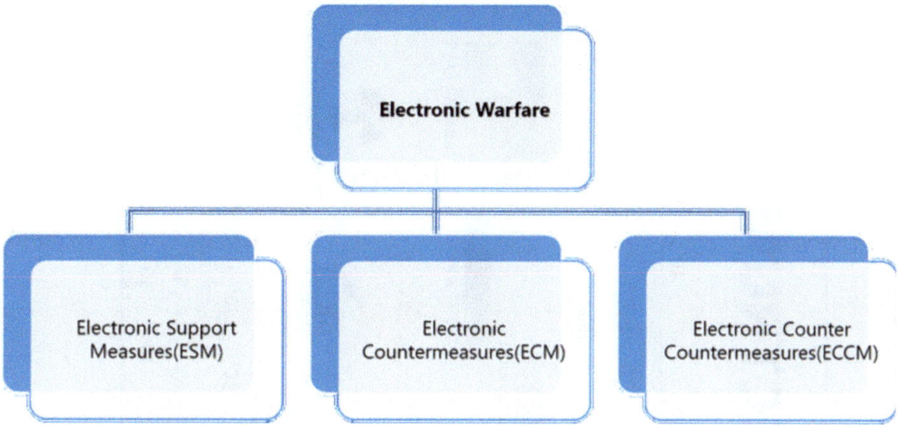

Fig. 2.5 Top view of electronic warfare chart

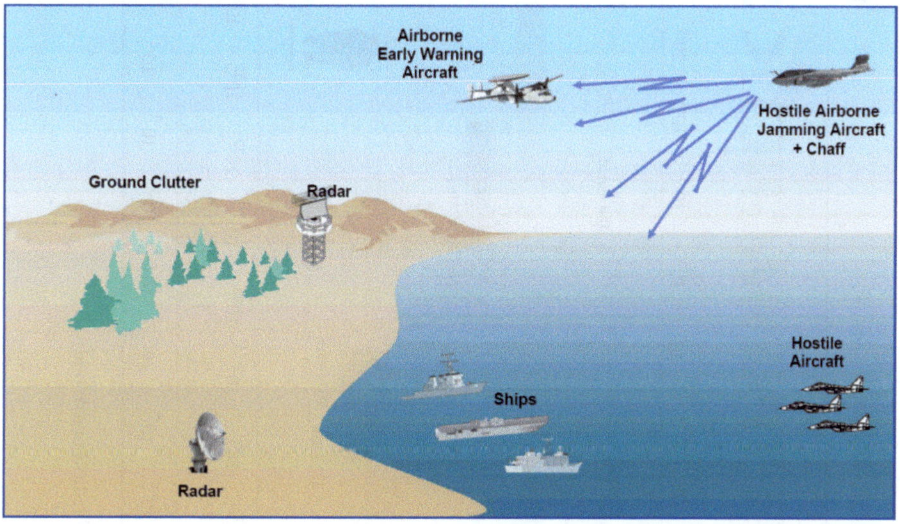

Fig. 2.6 Holistic view of radar environment. (Courtesy: IEEE New Hampshire Section)

perspicacity on the enemy's order of battle, intentions, and capabilities and to use countermeasures to gainsay efficacious use of communications and weapons systems while protecting one's own efficacious use of the same spectrum.

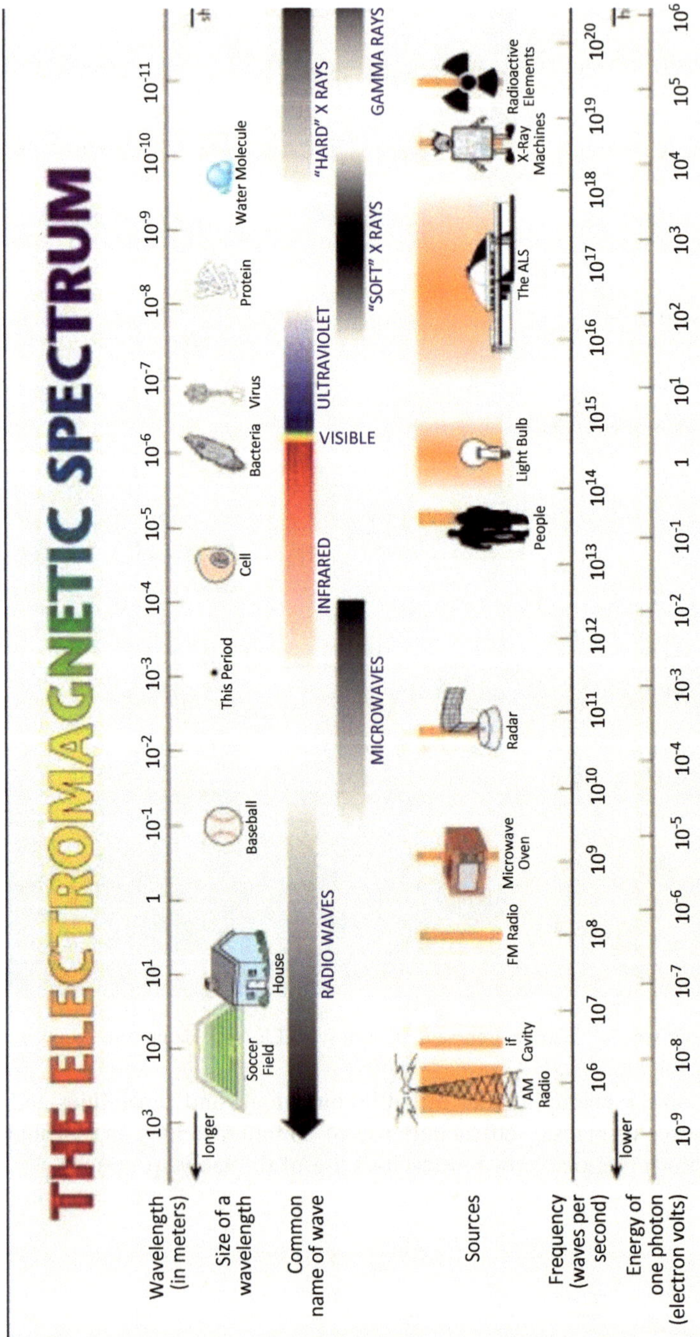

Fig. 2.7 The electromagnetic spectrum. (Credit: MicroWorld)

2.3 Electronic Support Measure (ESM)

In military telecommunications, the terms electronic support (ES) or electronic support measures (ESM) (Fig. 2.8) describe the division of electronic warfare (EW) involving actions taken under direct control of an operational commander to detect, intercept, identify, locate, record, and/or analyze sources of radiated electromagnetic energy for the purposes of immediate threat recognition (such as warning that fire-control *RADAR* has locked on a combat vehicle, ship, or aircraft) or longer-term operational planning [4]. Thus, electronic support provides a source of information required for decisions involving electronic protection (EP), electronic attack (EA), avoidance, targeting, and other tactical employment of forces. Electronic support data can be used to produce signals intelligence (SIGINT), communications intelligence (COMINT), and electronics intelligence (ELINT) [5].

Electronic support measure (ESM) involves actions taken to probe for, intercept, locate, record, and analyze radiated electromagnetic energy, for the purpose of exploiting such radiations to fortify military operations. Thus, ESM is a paramount source of EW information to carry out electronic countermeasures and electronic, counter-countermeasures. ESM involves, in general, accumulation of EW information through electronic perspicacity such as electronic intelligent (ELINT), communications perspicacity (COMINT), and ESM.

In summary, electronic support measures (ESM) are designed for electronic warfare techniques involving actions to detect, intercept, identify, locate, record, and/or analyze sources of radiated electromagnetic energy for the purposes of immediate threat recognition.

The systems have been specially designed and built to suit the needs of military services. The systems are engineered and built to provide services even in the most rugged, tactical, and extreme environments.

Fig. 2.8 A typical military electronic support measurement system

Fig. 2.9 US Navy P-3C Orion

As bottom line, electronic support measures gather intelligence through passive "listening" to electromagnetic radiations of military interest [4]. Electronic support measures can provide:

1. Initial detection or knowledge of foreign systems
2. A library of technical and operational data on foreign systems
3. Tactical combat information utilizing that library

ESM collection platforms can remain electronically silent and detect and analyze RADAR transmissions beyond the RADAR detection range because of the greater power of the transmitted electromagnetic pulse (EMP) with respect to a reflected echo of that pulse [5]. United States-airborne ESM receivers are designated in the AN/ALR series, which is a maritime patrol ESM system that enhances an aircraft's survivability by detecting, identifying, and locating hostile radar signal i.e., ALR-97 (V) such as US Navy P-3 Orion as illustrated in Fig. 2.9.

This information enables the aircrew to effectively respond to threats.

ALR-97(V) Maritime Patrol Aircraft ESM System applications include:

- Maritime domain awareness
- Sovereignty patrol
- Long-range surveillance
- Monitoring the economic exclusion zone
  - Illegal immigration and human trafficking prevention
  - Smuggling prevention
  - Protection of environmental resources
  - Fisheries enforcement

Note: The Lockheed P-3 Orion is a four-engine turboprop anti-submarine and maritime surveillance aircraft developed for the United States Navy and introduced in the 1960s.

Desirable characteristics for electromagnetic surveillance and collection equipment include:

1. Wide-spectrum or bandwidth capability because foreign frequencies are initially unknown
2. Wide dynamic range because signal strength is initially unknown
3. Narrow bandpass to discriminate the signal of interest from other electromagnetic radiation on nearby frequencies
4. Good angle-of-arrival measurement for bearings to locate the transmitter [5]

The frequency spectrum of interest ranges from 30 MHz to 50 GHz [5]. Multiple receivers are typically required for surveillance of the entire spectrum [5], but tactical receivers may be functional within a specific signal strength threshold of a smaller frequency range.

2.4 Electronic Countermeasure (ECM)

In Chap. 1 under Sect. 1.14, we touched upon the electronic countermeasure (ECM); however in this section, we describe and explain the ECM further. The ECM also was touched at the introductory of this chapter as well.

Electronic countermeasures are the actions taken to avert or reduce the enemy's efficacious utilization of the electromagnetic spectrum as presented in Fig. 2.7.

The second major division of electronic warfare is ECM, and of the three divisions, it is probably the best known. Partly this is because ECM tends to be visualized as "black boxes" that display a visible realization of electronic warfare. Often it appears that if one understands the black boxes, then one has an understanding of ECM, but such an attitude is very narrow because it ignores the two types of ECM: jamming and deception (Fig. 2.10).

Thus, the approach in this section will be more general; an attempt will be made to lay down the framework within which the black boxes function.

Of the two types of electromagnetic radiating systems against which ECM may be employed—either sensors or communications systems—enemy sensors receive by far the greatest attention. The primary reasons for this fact are:

1. The enemy sensor system produces an immediate threat, whereas the communications system does not.
2. The sensor system is usually specifically directed toward the friendly forces, and communications are not.

The emphasis of ECM employment in this section will be against sensor systems. However, some mention of the theory and practice of employing ECM against communications systems is considered appropriate, particularly in the contemporary

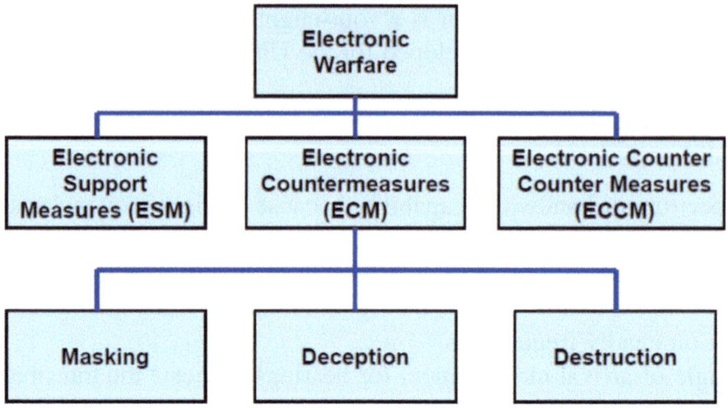

Fig. 2.10 Top view of electronic warfare chart. (Courtesy: IEEE New Hampshire Section)

Navy, which is so heavily dependent upon communications—including the various computer data links that provide the backbone to the fleet-wide command and control efforts.

From a strategic point of view, using ECM against an enemy communications system is questionable, for by doing so, the opportunity to gain valuable information by eavesdropping is lost. Tactically, however, it may be very advantageous to jam the enemy communications system in order to cause a breakdown in his battle plan. This was vividly illustrated during the 1973 Middle East War when the Egyptians successfully jammed the Israeli UHF/VHF radio frequencies, which resulted in a complete disruption of the Israelis' air-to-ground communications and consequently significantly reduced the effectiveness of their close air support.

Typical electronic sensors against which ECM might be used include long-range passive detectors; radar warning picket ships; airborne radar patrols (AWACS); long-range early-warning radar sets; ground-controlled intercept radar sets; fighter intercept radar; missiles guided by radar or infrared; radio and radar navigation equipment; electronic bombing equipment; electronic identification equipment such as identification, friend or foe(IFF) (that is illustrated in Fig. 2.11); terrain-following radar; anti-aircraft artillery (AAA); fire-control radar; surface-to-air (SAM) control radar, etc. The particular method used will depend upon the tactical situation.

Note: Identification, friend or foe (IFF) is a radar-based identification system designed for command and control. It uses a transponder that listens for an *interrogation* signal and then sends a *response* that identifies the broadcaster. It enables military and civilian air traffic control interrogation systems to identify aircraft, vehicles, or forces as friendly and to determine their bearing and range from the interrogator. IFF may be used by both military and civilian aircraft. IFF was first developed during World War II, with the arrival of radar, and several friendly fire incidents.

Figure 2.11 is holistic artistic depiction of an air traffic control (ATC) radar in the loop with other radars and how they complement each other. The way an ATC with a

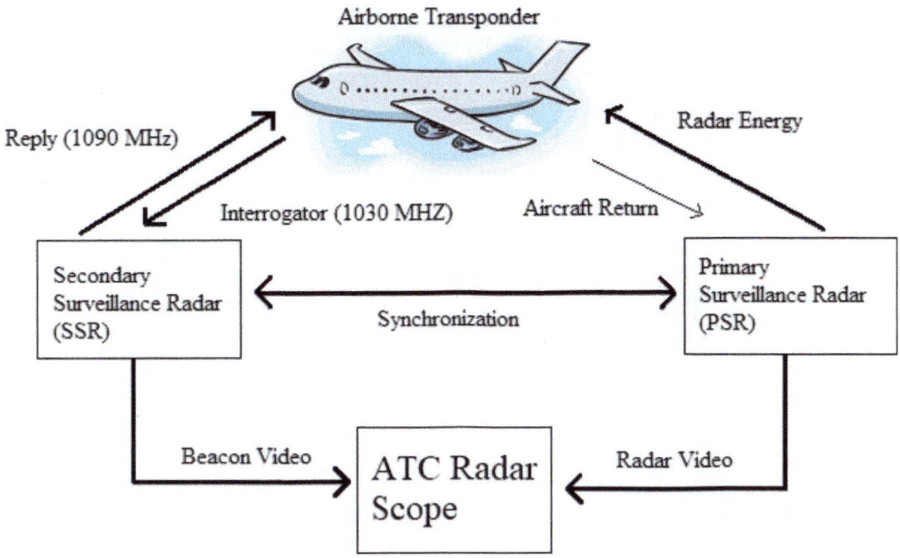

Fig. 2.11 Artistic depiction of identification, friend or foe

Fig. 2.12 Typical air control tower (www.wikipedia.com)

Fig. 2.13 The antenna system of a typical air traffic control radar beacon, the ASR-9 (www.wikipedia.com)

typical air control tower (ACT) (i.e., Fig. 2.12), works is that the ground station of this radar consists of two radar systems and their associated support components.

A typical antenna for an air traffic control radar is illustrated in Fig. 2.13.

The transponder emits a signal when it is interrogated by the secondary radars. In a transponder-based system, signals drop off as the inverse square of the distance to the target, instead of the fourth power in primary radars.

The air traffic control radar beacon system (ATCRBS) (i.e., Fig. 2.14) is a system used in air traffic control (ATC) to enhance surveillance radar monitoring and separation of air traffic. It consists of a rotating ground antenna and transponders in aircraft. The ground antenna sweeps a narrow vertical beam of microwaves around the airspace. When the beam strikes an aircraft, the transponder transmits a return signal back giving information such as altitude and the squawk code, a four-digit code assigned to each aircraft that enters a region. Information about this aircraft is then entered into the system and subsequently added to the controller's screen to display this information when queried.

This information can include flight number designation and altitude of the aircraft. ATCRBS assists air traffic control (ATC) surveillance radars by acquiring information about the aircraft being monitored and providing this information to the radar controllers. The controllers can use the information to identify radar returns from aircraft (known as *targets*) and to distinguish those returns from ground clutter.

In summary, the basic principles of ECM effectiveness can be stated as follows. The basic purpose of ECM is to interfere with the operation of the sensors of the air/surface defense system and through them to interfere with the operation of the

Fig. 2.14 A typical air traffic control radar beacon system (www.wikipedia.com)

system itself. Briefly, ECM attempts to make the defense more uncertain as to the threat it faces. The greater the defense uncertainty, the more effective the ECM. To state this principle in another way, ECM attempts to reduce the information content of the signals the defense receives with its sensors.

The objective of ECM, then, is to force the air/surface defense system to make mistakes or errors. An artistic global depiction of such system is presented in Fig. 2.15.

One should always keep in mind that ECM does not have to prevent tracking completely to be effective. In an age where rapid reaction is critical to survival, delaying the establishment of a solid track on a target, causing a moment's confusion, or forcing the decision-maker to wait just those few more seconds to be sure of the proper response can enable weapons to penetrate an adversary's defenses.

There exist three classes of electronic countermeasures (ECMs), and given that we want to interfere with an enemy air/surface defense radar, how may we go about it?

1. Radiate active signals to interfere with the radar.
2. Change the electrical properties of the medium between the aircraft/ship and the radar.
3. Change the reflective properties of the aircraft or ship itself.

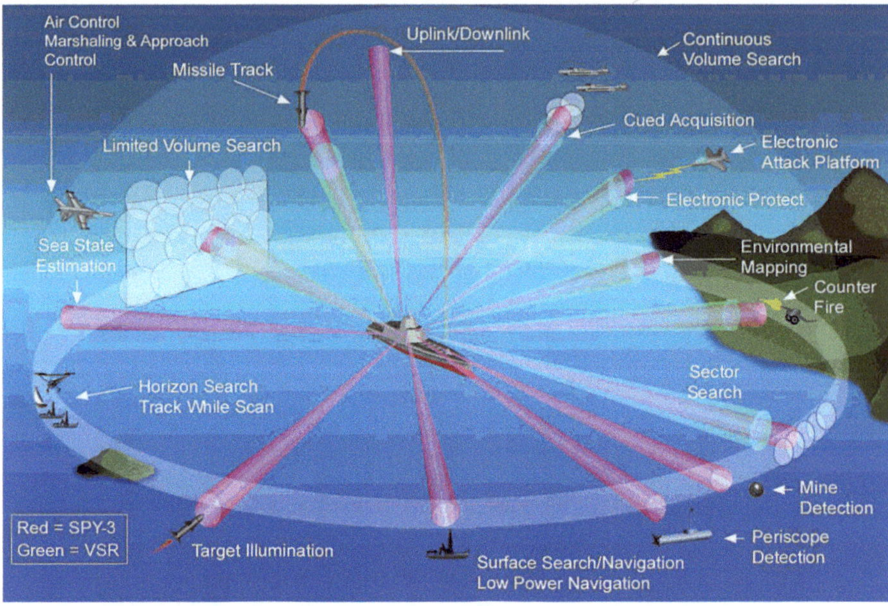

Fig. 2.15 Overall artistic picture of ECM system (www.wikipedia.com)

In general, there are four fundamental ways as listed below and illustrated in Fig. 2.16.

1. Jamming
2. Deception
3. Manipulative
4. Imitation

Each of the above four cases has been defined in their own box as they are written and may be implemented in many ways.

The three classifications above are established based on techniques by class and type as shown in Table 2.1.

We may possibly add a couple more points to the fundamental aspect of ECM in addition to the four ones that are mentioned above including deception, and they are listed as below and depicted in Fig. 2.17:

5. Masking
6. Deception
7. Destruction

In case of masking electronic countermeasure, Fig. 2.18 shows the basic flowchart that gives some idea of what is behind the masking.

As illustrated in Fig. 2.18 by red-dashed line, attributes of chaff can be listed as:

- Large number of resonant dipoles (i.e., metallic or metallic coated)

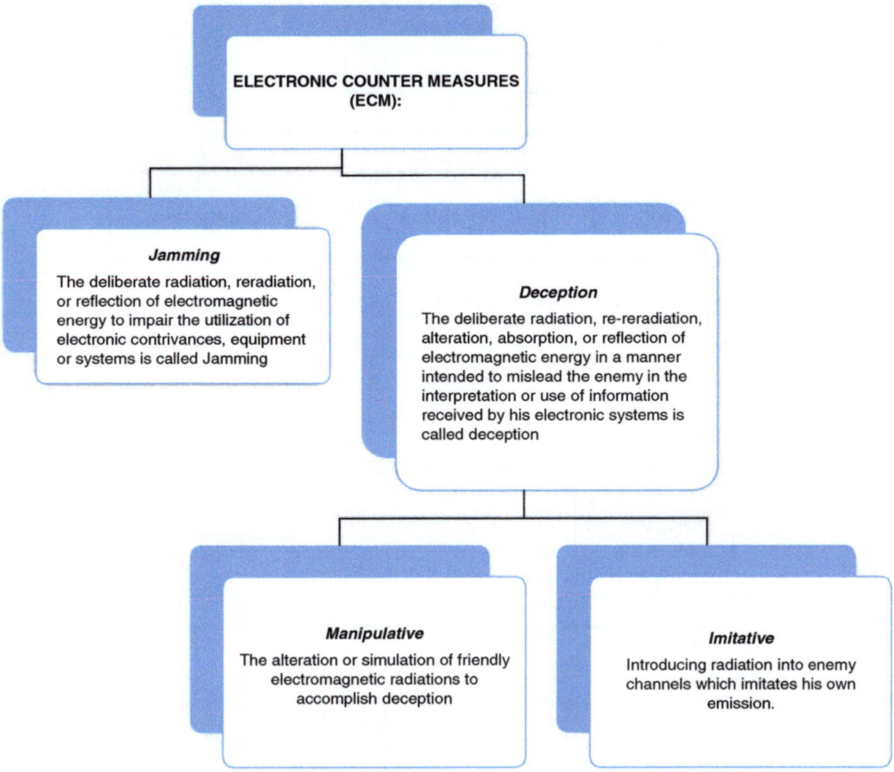

Fig. 2.16 Four fundamental aspects of ECM

Table 2.1 ECM techniques by class and type

1	Active radiators Noise Spot False target generators radiation barrage
	• Track breakers
	• Swept
2	Medium modifiers Chaff corridors Random chaff
	• Chaff bursts
	• Vehicle design
3	Reflectivity modifiers RAM (radar-absorbent materials)
	• Echo enhancers
	• Corner reflectors

- High reflectivity per pound
- Optimum length ½ of radar wavelength
- Movers horizontally with the wind

- Uses of chaff
  - Masking

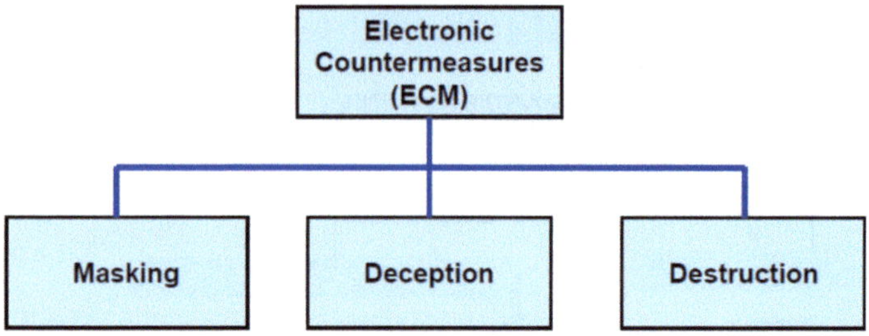

Fig. 2.17 Additional fundamental aspect of ECM

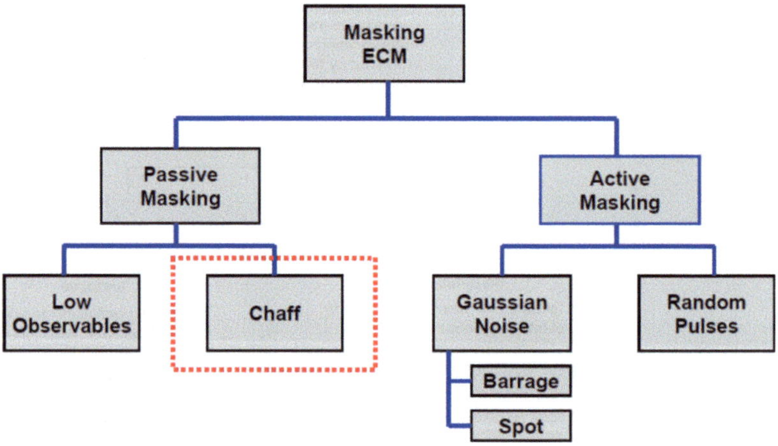

Fig. 2.18 Basic flowchart aspect of ECM masking

Large cloud can shield aircraft or missiles in or near the cloud.
Diffuse clutter similar in characteristics to rain.

– Deception

Chaff "puff" can emulate a missile or aircraft and cause false detections.
Packets of chaff seeded in a row can cause radar tracker to track the chaff rather than the aircraft being tracked.

Chaff reflectivity and density can be listed as:

- Resonant dipoles (metallic)
  - $\sigma = 0.86\lambda^2$ (in m^2) (maximum cross section per dipole)
  - $\lambda =$ wavelength in meters
- Random orientation of a large number of dipoles
  - $\sigma = 0.18\lambda^2$ (in m^2) (average cross section per dipole)

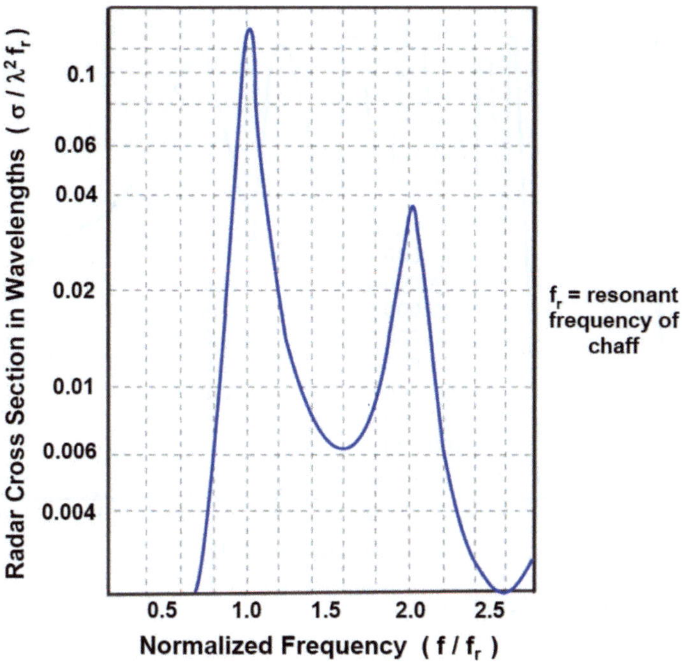

Fig. 2.19 Frequency response of resonant chaff

- Aluminum foil dipoles (0.001 in. thick, 0.01 in. wide, $\lambda/2$ long)
  - $\sigma = 3000$ W/f (in m^2)
  - $W =$ weight in lb
  - $f =$ frequency in GHz
- At S-band, 400 lb yields = 400,000 m^2 or 56 dBsm
  as chaff properties, we can write:
- Bandwidth 10–15% of center frequency
- Wideband chaff 1–10 GHz
  - $\sigma = 60$ m^2/lb
  - Variable length dipoles in a single package
- Fall rates of chaff 0.5–3 m/s
  - Nylon (coated) ~0.6 m/s
  - Aluminum ~1.0 m/s
  - Copper ~3.0 m/s

Frequency response of resonant chaff is presented in Fig. 2.19 as:

Bear in mind that chaff finds its main applications in electromagnetic countermeasures. To meet the increasing need of ballistic missile countermeasure and to

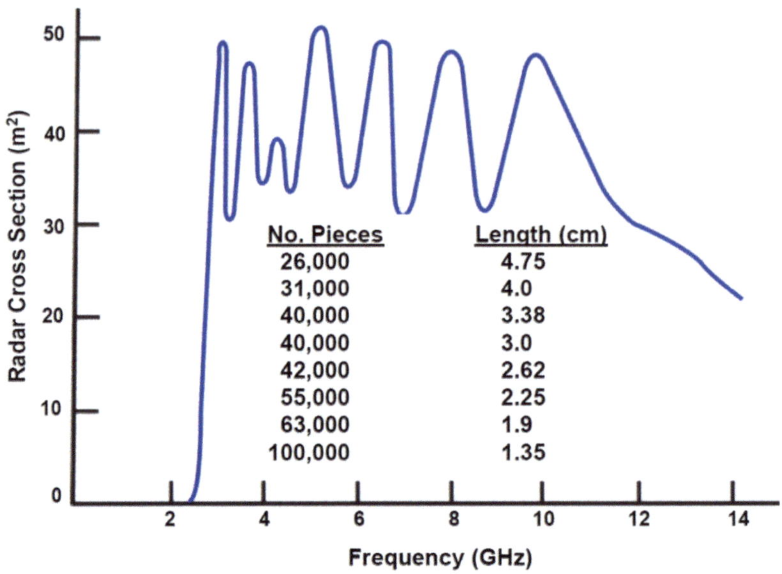

Fig. 2.20 RCS of multiband chaff package

maximize the chaff performance, it has become highly important to model the radar cross section (RCS) behavior of chaff cloud.

Furthermore, chaff constitutes a passive electronic countermeasure device for military equipments against radar sensors. It refers to the large numbers of microwave reflecting material elements deployed into the atmosphere as a countermeasure against hostile electronic systems. Most chaff payloads comprise metal strips or wires called as chaff dipoles or elements, dispensed in space so as to produce radar echoes of desirable type. It is deployed as a decoy for masking an offensive attack. In decoys the chaff is dispensed in the close vicinity of targets, i.e., tank or aircrafts, permitting the target to move away, leaving the missile locked on to the chaff cloud echo. The projectile or cartridge which deploys the chaff in a large cloud is fired in a selected position to give a strong radar return within the view and range of the missile's seeker. To optimize chaff performance, it becomes important to model the RCS behavior of chaff cloud.

Moreover, radar cross section (RCS) of multiband chaff package is presented in Fig. 2.20.

However, efficient use of chaff for the protection of aircraft in tactical situations demands knowledge of the radar scattering characteristics of small, dense clouds of dipoles. In extreme cases, just after ejection from the vehicle into the wind stream, dipole densities of thousands per cubic wavelength are encountered. Moments later, lower densities on the order of a hundred dipoles per cubic wavelength are encountered.

Bottom line is that the importance of electronic warfare in the area of defense is becoming increasingly emphasized. More than the well-known "stealth" concept, a thorough understanding of available techniques in this discipline helps to increase the survival rate on the battlefield. One such technique is the chaff cloud, a system where thousands of small printed dipoles are thrown from military vehicles to create a false radar signature, making the correct identification of the target by the enemy more complicated.

2.5 Electronic Counter-Countermeasure (ECCM)

During the last decades, many electronic counter-countermeasure (ECCM) techniques have been described to suppress the deception jamming. The ECCM techniques are utilized against active deception jamming electronic countermeasure (ECM).

Firstly, these schemes are classified into two groups according to the ECM threat: techniques which are used to counter the range false target (RFT) and techniques to counter the range-velocity gate stealers (R-VGS) deception jamming. Secondly, the pros and cons of these schemes are highlighted and compared under different viewpoints.

In this section we briefly review the ECCM schemes; specifically, these schemes are classified into two as:

1. ***Techniques that aim to suppress the range false target (RFT)***
   Although there are other ways to suppress the RFT, pulse diversity is commonly used. It should be noted that this technique is mostly used in synthetic-aperture radar (SAR) [6]

   In view of the fact that the repeat jammer lags at least one pulse behind the radar and benefiting from orthogonal pulse block, which was first used by Alamouti in wireless communication [7], Akhtar proposed schemes [8–10] to combat the range false target via the orthogonal pulse block design set in slow-time domain in which the process is assumed to be stationary (no remarkable changes in the received signals position) known as coherent processing interval (CPI). Thus, the jamming signals can be easily suppressed in the output of the matched filter. These techniques require integration over several pulses in order to separate the false target. However, it can also be based on the transmission of pulses which comprise of two, to decrease the integration over several pulses. It is worth remarking that most of the pulse diversity methods assumed pulse block with four pulses.

   Note that for a coherent radar, the total time to be sampled is referred to as the coherent processing interval (CPI). Only a perfect coherence of the radar guarantees that the number of samples per row is constant with the time [11].
2. ***Techniques that aim to eliminate the R-VGS***
   Generally, these techniques use different approaches to preserve the radar tracking. However, it should be selected properly, since the jammer can modulate information to delay amplitude, frequency, and phase on the basis of received

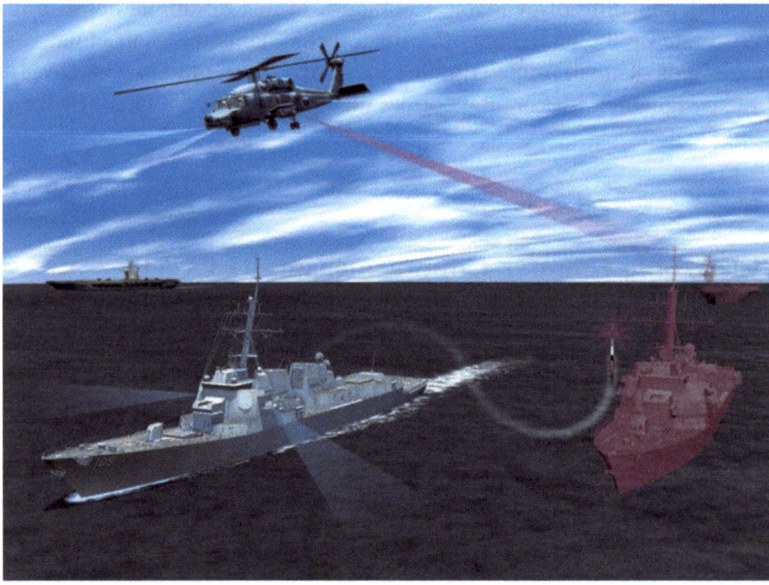

Fig. 2.21 Artistic depiction of Lockheed Martin's Advanced Off-Board Electronic Warfare (AOEW) Active Mission Payload (AMP) system. (Courtesy of Lockheed Martin Corporation)

radar signal to realize coherent interference. These techniques make the radar system focus on the target incessantly or constantly, without any interruption [6].

Furthermore, the actions taken to ascertain cordial, efficacious utilization of the electromagnetic spectrum despite the enemy's utilization of electronic warfare (EW) are termed as electronic counter-countermeasure (ECCM). In fact, a radar's ability to change frequency within its operating band is usually on a pulse-to-pulse basis. This is an ECCM technique employed to avoid spot jamming and to force the jammer to go into a less effective barrage mode.

The battlefield of electronic warfare is ecumenical, and its intensity varies according to different national interests and perceptions of potential threats. In fact, electronic warfare is toward the maintenance of regional and ecumenical balances which deter the outbreak of armed conflict. The mere possession of a certain number of electronic support measure (ESM) or electronic countermeasure (ECM) contrivances is not enough to ascertain prosperity in war. In EW what works today may not work tomorrow, and the developments in EW systems must always proximately and opportunely follow developments in the threat. With the illimitable evolution of applied military technology, electronically guided weapons are coming more proximate and more proximate to perfection, and thus constant updating and refinement of electronic warfare (EW) equipment are required.

Figure 2.21 is an artistic depiction of Lockheed Martin's Advanced Off-Board Electronic Warfare (AOEW) Active Mission Payload (AMP) system, a pod hosted on an MH-60R, which will enhance the way the US Navy detect and response to anti-ship missile threats as part of ECM and ECCM.

Electronic counter-countermeasures are the art of reducing the effectiveness of an EW threat with the objective of making the cost of effective EW prohibitive for the enemy. As in ECM, ECCM includes both radar design and operator training. The radar ECCM designer must understand the various forms of ECM that his radar is likely to encounter; hence he is very interested in intelligence about the ECM threat. Likewise, the radar operator would like to know what ECM he will be facing. But in both cases, detailed intelligence will probably be lacking. Therefore, the designer must provide a variety of options to be used against the expected threats. And the operator must be trained both to recognize the various countermeasures that might be used against him and to select the appropriate combination of options against each of them. The most effective measure to combat ECM is an up-to-date piece of equipment operated by a well-trained operator. Radar design for ECCM can be broken down into three areas: radar parameter management, signal processing techniques, and design philosophy.

Another aspect of ECCM design philosophy is the relationship between automatic equipment and the human operator. The trained radar operator fulfills a useful and necessary role in a countermeasure environment and cannot be completely replaced by automatic detection and data processors. An automatic processor can be designed to operate only against those interfering or jamming signals known beforehand; that is, any capability against such signals must be programmed into the equipment beforehand. New jamming situations not designed into the data processor might not be readily handled. On the other hand, a human being has the ability to adapt to new and varied situations and is more likely to be able to cope with, and properly interpret, a strange new form of interference than a machine can. Therefore, a skilled operator is the most important counter-countermeasure for maintaining radar operation in the presence of deliberate and clever countermeasures.

2.6 Electronic Countermeasure (ECM) vs. Electronic Counter-Countermeasure (ECCM)

The difference between ECM and ECCM can be simply described as the fact that the electronic countermeasures have two primary focuses, and they are:

1. Countering the effectiveness of enemy radar
2. Interrupting enemy communications

ECM can be directed at aircraft, ships, sensors, or weapon systems like radar-guided missiles. Examples of ECM include radar jamming or releasing chaff (clouds of small metal strips released by aircraft) to confuse the returning radar systems. Another example is the use of radar-absorbent coating on aircraft to weaken the returning signal or altering the shape of an aircraft in such a way as to deflect

Fig. 2.22 Example of ECCM techniques on C-130 platform

incoming radar waves. This passive form of ECM is called "target modification" and is a core component of stealth technology.

ECCM on the other hand is about defending against ECM techniques and rendering them ineffective. ECCM can be traced back to World War II when the British disrupted German radio communications using jamming techniques. To counter this, the German military increased the transmission power of their radio signals to overpower the jamming. Figure 2.22 is an example of modern electronic warfare augmenting new modern technology as part of ECCM techniques on board of a C-130 airplane.

The simplest ECCM technique against jamming is to just increase radio transmission power to "burn" through the enemy's jamming attempt, as described above. But this is only the most basic example of what ECCM can do. Let's explore some more sophisticated methods for overcoming ECM.

1. **ECM Detection and Radiation Homing Weapons**
   This includes the use of sensors which can recognize attempts to deceive enemy radars (like chaff) and disregard them. This can be taken even further with "radiation homing" weapon systems like missiles which can detect and target radio emissions themselves, known as "anti-radiation missiles" (ARMs). Some radiation homing weapons are designed to redirect toward the source of the enemy jamming signal if the interference makes it impossible to hit their original target. Others are specifically designed to seek out the location where the enemy signal is emanating from. This is almost like applying the principles of Judo to electronic warfare, turning the enemy's ECM attack against them; if the enemy uses their jamming capabilities, they are also giving away their position and opening themselves up to a kinetic counterattack.

Fig. 2.23 New radar AESA Bolsters F-15 fleet

2. **Frequency-Hopping Spread Spectrum (FHSS)**
   Frequency-hopping spread spectrum (FHSS) is a spread-spectrum modulation technique, meaning that the transmitter rapidly switches the frequency of the carrier wave in an apparently random fashion across a wide spectrum. This makes it much more difficult for an enemy force to jam the signal, since it is always changing, and it is almost impossible to predict what frequency it will jump to next. FHSS is usually used along with encryption for an added layer of radio communication security.
3. **Pulse Compression**
   Pulse compression is a radar ECCM technique which consists of modulating the pulse of the radar signal transmission and then cross-correlating it upon reception. There are different methods of pulse compression which are suited to different purposes, but for ECCM, it is done by linear frequency modulation, a practice also known as "chirping." Using this technique, the frequency of a radar signal is changed within individual pulses of the carrier, like how the sound of a grasshopper or cricket can change within an individual chirp—hence the name. This form of pulse compression is highly resistant to jamming and is used in active electronically scanned arrays (AESA) radar systems as illustrated in Fig. 2.23.

Active electronically scanned arrays are considered a phased array **system**, which consists of an array of antennas which form a beam of radio waves that can be aimed in different directions without physically moving the antennae themselves. The primary use of AESA technology is in radar systems.

Electronic warfare capabilities will only become more important as time goes on. As both ECM and ECCM continue to advance, the advantage will go to those militaries who partner with the best minds in radio-frequency engineering. The experts at Bliley have been pushing the envelope in RF technology for decades, and we can't wait to put our skills to the test and help our armed forces succeed in the twenty-first century.

2.7 How ECCM Techniques Take Electronic Warfare to the Next Level

The evolution of electronic warfare has been driven by the competition between electronic countermeasure (ECM) and electronic counter-countermeasure (ECCM). Electronic warfare involves not only harnessing the electromagnetic spectrum but defending against enemy use of the spectrum, and, if possible, denying their ability to use it in the first place. Since the earliest attempts at jamming radio communications, techniques have been developed to counteract enemy ECM. Today we'll discuss how ECCM continues to evolve and shape the nature of modern warfare.

Figure 2.24 is conceptual image that is presenting ECCM techniques that take electronic warfare to the next level.

From the very beginning of radar, attempts have been made to disrupt its use through various forms of electronic and nonelectronic countermeasures and associated techniques. These countermeasures include active jamming, or the attempt to introduce extraneous electronic signals into the radar receiver and processor; passive techniques such as chaff, decoys, and so on; intercept equipment and techniques such as direction finding (DF), radar warning receivers, and electronic intelligence (ELINT) receivers; and radar homing missiles or anti-radiation missiles (ARM). In

Fig. 2.24 Conceptual illustration of ECCM techniques

addition, target evasive actions, maneuvers, and flight plans can be developed as countermeasures against radar.

Techniques included in the radar or as part of the radar's general operational philosophy primarily to counter these countermeasures are appropriately designated *radar counter-countermeasures.* Even though not all of the techniques are electronic, the general term *electronic counter-countermeasures (ECCM)* is normally used to refer to the collection of both passive or nonelectronic and electronic techniques used to counter or reduce the effectiveness of radar countermeasures used by the enemy or opposing forces in today's modern warfare, where engagement between friendly force and foe is basically at speed of light [12].

Radar electronic counter-countermeasure (ECCM) is a very broad subject. Several books are available which deal with radar ECCM either exclusively [13, 14] or in conjunction with discussions of electronic warfare and countermeasures [15–18].

The subject is much too large to cover in detail in this chapter and consequently this section as well. Only a generalized summary or overview of radar counter-countermeasures is presented. This treatment deals with nomenclature, definitions, and semantics rather than with specific technical descriptions and equipment details. However, keep in mind that survivability enhancement with respect to electronic countermeasure (ECM) is generally referred to as electronic counter-countermeasures (ECCM).

Furthermore, to make it easy to understand the concept of electronic warfare (EW), we reach out to the United States Field Manual FM-1005, Operation [19] as depicted in Fig. 2.25 to define certain terms, which relate to radar ECCM and more generally to electronic warfare.

Figure 2.25 gives the US Department of Defense's accepted definitions for electronic warfare and associated components [7]. As noted, electronic warfare includes all actions required to prevent hostile use of the electromagnetic (EM) spectrum and retain friendly use.

Sub-elements include electronic support measures (ESM), generally passive electronic eavesdropping and location techniques, electronic countermeasures (ECM), active approaches to prevent or reduce the use of the EM spectrum by the enemy, and ECCM, actions taken to retain friendly use of the spectrum.

In contrast to the presentation in Fig. 2.25 and electronic warfare, where military action involves the use of electromagnetic energy to engage with hostile situation, we can present the electronic attack (EA) technology as depicted in Fig. 2.26.

Over the years, numerous ECCM techniques have been developed. Table 2.2 lists more than 150 types of radar ECCM [20]. Many of them are discussed in reference to E. K. Ready [12].

In order to come up with some technological approach as we have defined so far, several basic considerations or objectives dictate radar ECCM strategy. The primary objective is always to negate the effects of the enemy's ECM on the radar. However, as previously indicated, *counter-countermeasures* is a generic term which includes *anything or any action* resulting in the degradation of enemy ECM activities. It is certainly not limited to electronic techniques or approaches but can include tactics, deployment, operational doctrines, and so on.

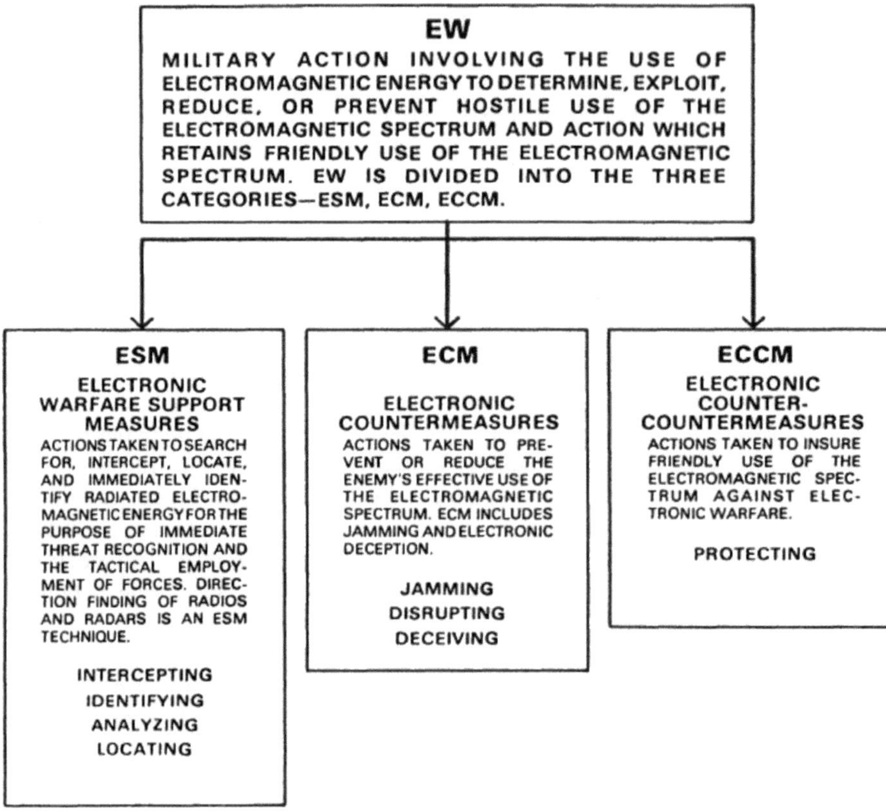

Fig. 2.25 Electronic warfare definition. (Source: The U.S. Army Field Manual FM-1005, Operations) [19]

Although it is sometimes not obvious, radar ECCM is equivalent in a hostile EM environment to considerations of EM compatibility which involve techniques and approaches associated with reduction of the susceptibility of electronic equipment to interference—either man-made or natural. Another consideration sometimes overlooked is that natural ECM (clouds, inclement weather, ground returns, and other clutter) requires what can be thought of as ECCM.

In this case, ECCM takes the form of clutter rejection processing such as moving target indicator (MTI) or constant false alarm rate (CFAR) processing.

It is noteworthy to state that moving target indication (MTI) is a mode of operation of a radar to discriminate a target against the clutter. It describes a variety of techniques used to find moving objects, like an aircraft, and filter out unmoving ones, like hills or trees. Furthermore, the MTI radar uses low pulse repetition frequency (PRF) to avoid range ambiguities.

Moving target indicator (MTI) begins with sampling two successive pulses. Sampling begins immediately after the radar transmits pulse ends. The sampling continues until the next transmit pulse begins [21].

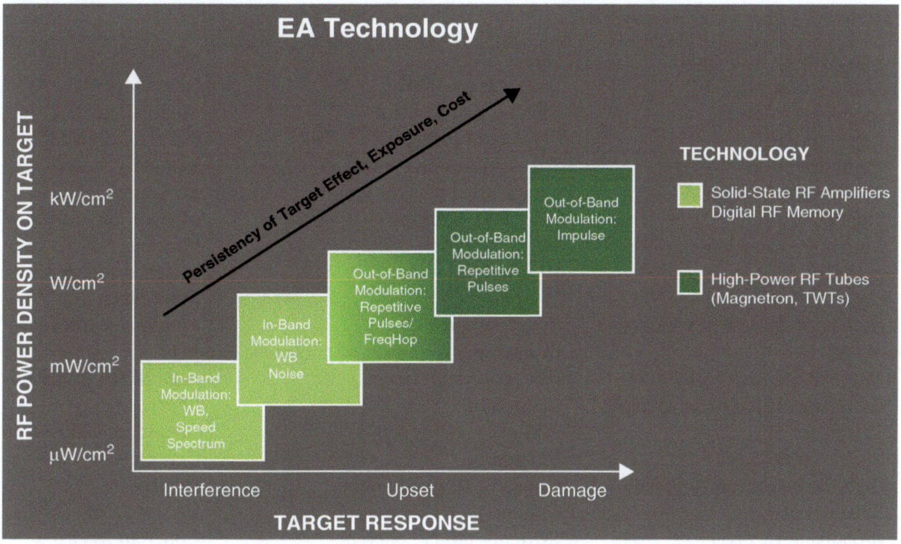

Fig. 2.26 Electronic act technologies chart

Sampling (i.e., Fig. 2.27) is repeated in the same location for the next transmit pulse, and the sample taken (at the same distance) with the first pulse is rotated 180° and added to the second sample. This is called destructive interference.

If an object is moving in the location corresponding to both samples, then the signal reflected from the object will survive this process because of constructive interference. If all objects are stationary, the two samples will cancel out and very little signal will remain.

It is pointed out that electronic counter-countermeasures (ECCM) capability is one of the most important functions of a radar. There is, however, no simple and generalized method for measuring it. Johnston [20] has suggested that the radar ECCM effectiveness should be expressed by ECCM improvement factors (EIF). A disadvantage of EIF is related to the fact that it cannot measure the ECCM capability of the whole radar system. The present investigation is concerned with a new method for measuring radar ECCM capability.

Attention is given to a study of basic radar and ECCMs against type of ECM, the use of output target signal and ECM signal power ratio, supplementary factors, the ECCM capability for a multichannel radar system, and calculation examples involving four air-defense multichannel surveillance radar systems.

By definition, the ECCM improvement factor (EIF) is the ratio of the:

$$\frac{\text{ECM signal require to produce a given output at the radar with ECCM}}{\text{ECM signal required to produce the same radar output without ECCM}}$$

Table 2.2 An ECCM lexicon [20]

Acceleration limitation	Gated FAGC	Main lobe cancel-
Angle sector blanking	Instantaneous AGC	ation (MLC)
Angular resolution	Manual gain control	Monopulse MLC
Audio limiter	Pulse gain control	Polarization MLC
Aural detection	Sensitivity-time control	Manually aided
Autocorrelation signal	Guard-band blanker	tracking
Processing	High PRF tracking	Manual rate-aided
Automatic cancelation of	High-resolution radar	tracking
extended	IF diversity	Matched filtering
targets (ACET)	IF limiter	Monopinch
Automatic threshold variation	Image suppressor	Monopulse tracker
(ATV)	Instantaneous frequency correlator	MTI
Automatic tuner (SNIFFER)	(IFC-CRAFT)	Area MTI (velocity
Automatic video noise leveling	Integration	filter)
(AVNL)	AM video delay line	Cascaded feedback
Back-bias receiver	Integration	Canceler
Baseline-break (on A-scope)	Coherent IF integration	(MTI)
Bistatic radar	Coherent (IF) integration	Clutter gating (MTI)
Broadband receiver	(Moving target coherent	Coherent MTI
Coded waveform modulation	(IF) integration or stationary target)	Noncoherent MTI
Coherent long pulse discrimina-	Display integration	Pulse Doppler
tion	FM delay line integration	Pseudo-coherent
Compressive IF amplifier	Noncoherent (video) Integration	MTI
Constant false alarm rate (CFAR)	Pulse integration	Single-delay line
Cross-gated CFAR	Video delay-line integration	(MTIC
Dispersion fix (CFAR)	Inter-pulse coding (PPM)	canceler)
IF Dicke-fix CFAR (Dicke-fix)	Jamming cancelation Receiver	Re-entrant data pro-
MTI CFAR	jittered PRF	cessor
Unipolar video CFAR	Kirba Fix	Three-pulse canceler
Video Dicke-fix CFAR (Dicke-	Least voltage coincidence Detector	Two-pulse canceler
fix)	Linear intrapulse FM (CHIRP)	(single delay)
Zero-crossing CFAR	Lin-log IF	Line MTI cancel-
Contiguous filter-limiter	Lin-log receiver	ation
Cross-correlation signal	Lobe-on-receive-only (LORO) also	Multifrequency radar
processing	(SORO)	Multi-visual antenna
CW jamming canceler	Log fix (also, log FTC)	Phased array radar
Detector back-bias (DBB) (same	Logarithmic receiver	Polarization diversity
as	Logical ECCM processing	Polarization selector
detector balanced bias)		Post canceler log
Dicke-fix		FTC
Clark Dicke-fix (cascaded)		PRF discrimination
Dicke-fix		Pulse burst mode
Coherent MTI Dicke-fix		Pulse coding and
Craft receiver		correlation
Dicke log fix		Pulse compression,
IF canceler MTI Dicke-fix		stretching
IF Dicke-fix CFAR (zero-cross-		(CHIRP)
ings)		Pulse edge tracking
Dicke-fix CFAR		Pulse interference
Instantaneous frequency Dicke-		elimination
fix		(PIE)

(continued)

Table 2.2 (continued)

Noncoherent MTI Dicke-fix Video Dicke-fix CFAR Diplexing Doppler range rate Comparison Double threshold detection Electronic implementation of baseline-break technique Fast manual frequency shift Fast time constant (FTC) Fine frequency Frequency agility Frequency diversity Frequency preselection (narrow bandwidth) Frequency shift Gain control Automatic gain control (AGC) Dual gated AGC Fast AGC		Pulse shape discrimination Pulse-to-pulse frequency shift (RAINBOW) Pulse width discrimination (PWD) Pulse length discrimination (PLD) Random-pulse blanker Random-pulse discrimination (RPD) Range angle rate memory Range gating Range rate memory Scan-rate amplitude Modulation Short pulse radar Side-lobe blanker Side-lobe canceler Side-lobe reduction Side-lobe suppression (SLS) Side-lobe suppression by absorbing Material Staggered PRF Transmitter power Two-pulse autocorrelation Variable bandwidth receiver Variable PRF Variable scan rate Velocity tracker Video correlator Wide-bandwidth radar Zero-crossing counter

Thus, it helps in quantifying the ECCM efficiency in a system-oriented evaluation. The radar's ECCM effect can be conveniently viewed under the following generalized grouping as:

- Functional sensor-wise
- Response to specific ECM

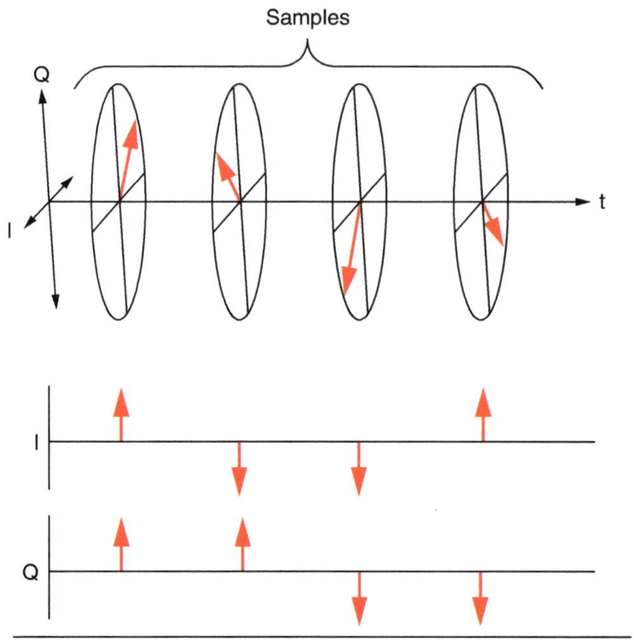

Fig. 2.27 Moving target indicator signal sampling process. (Source: www.wikipedia.com) [21]

- Deployment type in respect to field/environment
- Total weapon system efficiency

In expanding the above models with appropriate examples, the ECCM implementation and its evaluation can be examined:

- As pertaining to search, track, or weapon guidance radar
- According to the ECM-ECCM matrix well defined in all literature
- In relation to the vulnerability experienced in a deployment pattern
- By determining the effectiveness and survivability as a total weapon system against and ECM attack such as a missile-site radar complex with its sensors, weapons, and inter/intra communication equipment is concern.

The last category is what the EIF specifies as a total figure of merit. However, the diverse subsystems and their operations from sensor to weapon point of view make it a difficult task to evaluate the result. The complexity due to the nature and spread of technologies involved in such a system makes this evaluation a complex one.

Moreover, we can also elaborate on constant false alarm rate (CFAR) processing. CFAR detection refers to a common form of adaptive algorithm used in radar systems to detect target returns against a background of noise, clutter, and interference, which is related to the statistical signal processing, and this is briefly described below [22].

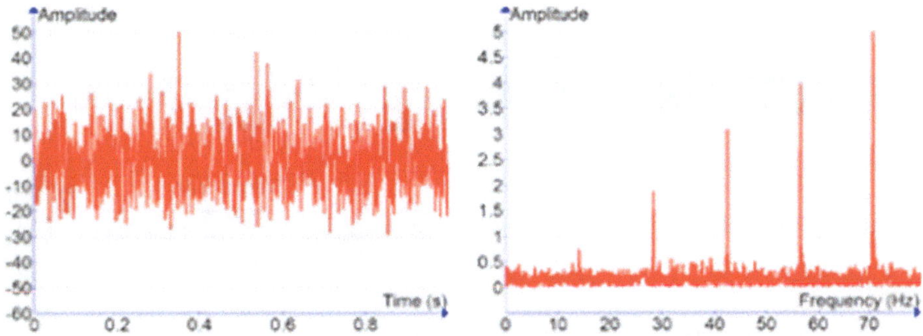

Fig. 2.28 Statistical signal processing illustration. (Source: www.wikipedia.com)

In the radar receiver, the returning echoes are typically received by the antenna, amplified, down-converted, and then passed through detector circuitry that extracts the envelope of the signal (known as the *video signal*). This video signal is proportional to the power of the received echo and comprises the wanted echo signal and the unwanted power from internal receiver noise and external clutter and interference.

The role of the constant false alarm rate circuitry is to determine the power threshold above which any return can be considered to probably originate from a target. If this threshold is too low, then more targets will be detected at the expense of increased numbers of false alarms. Conversely, if the threshold is too high, then fewer targets will be detected, but the number of false alarms will also be low. In most radar detectors, the threshold is set in order to achieve a required probability of false alarm (or equivalently, false alarm rate or time between false alarms).

If the background against which targets are to be detected is constant with time and space, then a fixed threshold level can be chosen that provides a specified probability of false alarm, governed by the probability density function of the noise, which is usually assumed to be Gaussian. The probability of detection is then a function of the signal-to-noise ratio of the target return. However, in most fielded systems, unwanted clutter and interference sources mean that the noise level changes both spatially and temporally. In this case, a changing threshold can be used, where the threshold level is raised and lowered to maintain a constant probability of false alarm. This is known as constant false alarm rate (CFAR) detection.

Figure 2.28 is an indication of a statistical signal processing, which is an electrical engineering subfield that focuses on analyzing, modifying, and synthesizing signals such as sound, images, and biological measurements [22].

Signal processing techniques can be used to improve transmission, storage efficiency, and subjective quality and to also emphasize or detect components of interest in a measured signal.

In Fig. 2.28, the signal on the left looks like noise, but the signal processing technique known as the Fourier transform (right) shows that it contains five well-defined frequency components.

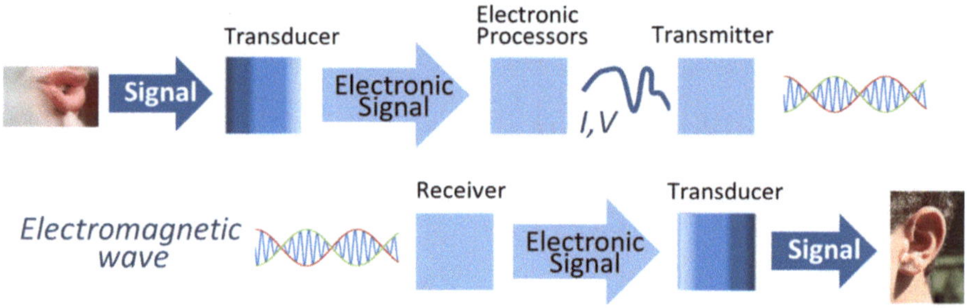

Fig. 2.29 Signal transmission process. (Source: www.wikipedia.com)

Note that signal transmission uses electronic signal processing as indicated in Fig. 2.29.

Transducers convert signals from other physical waveforms to electric current or voltage waveforms, which then are processed, transmitted as electromagnetic waves, and received and converted by another transducer to final form.

In conclusion, the topics of this chapter are current interest of all the military brasses and companies involved with development of such technology and subject of this book here.

References

1. https://www.tek.com/spectrum-analyzer
2. https://www.cia.gov/news-information/featured-story-archive/2010-featured-story-archive/intelligence-signals-intelligence-1.html
3. https://defensesystems.com/articles/2017/09/06/dod-electronic-warfare.aspx
4. Polmar, Norman "The U. S. Navy Electronic Warfare (Part 1)" *United States Naval Institute Proceedings* October 1979 p.137.
5. This article incorporates public domain material from the General Services Administration document "Federal Standard 1037C
6. Ahmed Abdalla Ali, Zhao Yuan, Sowah Nii Longdon, Joyce Chelangate Bore, Tang Bin, "A study of ECCM techniques and their performance", Conference Paper · September 2015, https://www.researchgate.net/publication/308848636
7. Alamouti, S. M. "A simple transmit diversity technique for wireless communications," IEEE Journal Selected Areas in Communications, 1998. Vol. 16, No. 8, 1451-1458.
8. J. Akhtar. "An ECCM Signaling Approach for Deep Fading of Jamming Reflectors" 978-0-86341-848-8 IET 2007.
9. J. Akhtar. "Orthogonal block coded ECCM schemes against repeat radar jammers". IEEE Transactions on Aerospace and Electronic Systems, vol.45, no.3, pp.1218-1226, 2009.
10. J. Akhtar. "An ECCM scheme for orthogonal independent range-focusing of real and false targets", In ICR '2007 Proceedings, Massachusetts, USA, pp. 846-849, (2007).
11. J.-M. Muñoz-Ferreras, R. Gómez-García and C. Li, "Human-aware localization using linear-frequency modulated continuous-wave radars" Chapter 5, Page 191, "Principles and Applications of RF/Microwave in Healthcare and Biosensing" book edited by Changzhi Li Mohammad-Reza Tofighi Dominique Schreurs Tzyy-Sheng Jason Horng, published by Academic Press, 2017, New York, NY.

12. Edward K. Ready, “Radar ECCM Considerations and Techniques” Principles of Modern Radar pp 681-699, Editors Jerry L. Eaves Edward K. Reedy, Chapman & Hall, New Your, NY, and International Thomson Publishing, 1987 Singapore.
13. S. L. Johnston, *Radar Electronic Counter-Countermeasures.* Artech House, Dedham, Mass., 1979.
14. M. V. Maksimov et al., *Radar Anti-Jamming Techniques,* Artech House, Dedham, Mass., 1979.
15. L. B. Van Brunt, *Applied ECM,* EW Engineering, Inc., Dunn Loring, Va., 1978.
16. 4. J. A. Boyd et al., *Electronic Countermeasures,* Peninsula Publishing, Los Altos, Calif., 1978.
17. H. F. Eustace, The *International Countermeasures Handbook, 1979-1980,* EW Communications, Palo Alto, Calif., 1980.
18. P. Tsipouras, et aI., "ECM Technique Generation," *Microwave Journal,* vol. 27, no. 9, September 1984, pp. 38-74.
19. U.S. Army Field Manual, FM 100-5, *Operations,* Headquarters, Department of the Army, July 1976.
20. S. L. Johnston, "ECCM Improvement Factors (ElFs)," *Electronic Warfare,* vol. 6, no. 3, May--June 1974, pp. 41-45.
21. https://en.wikipedia.org/wiki/Moving_target_indication
22. *Scharf, Louis L. Statistical Signal Processing: Detection, Estimation, and Time Series Analysis. Addison Wesley, NY. ISBN 0-201-19038-9*

Chapter 3
Radar-Absorbent Material and Radar Cross Section

Radiation-absorbent material, usually known as RAM, is a material which has been specially designed and shaped to absorb incident RF radiation, as effectively as possible, from as many incident directions as possible. The more effective the RAM, the lower the resulting level of reflected radio-frequency (RF) radiation. Many measurements in electromagnetic compatibility (EMC) and antenna radiation patterns require that spurious signals arising from the test setup, including reflections, are negligible to avoid the risk of causing measurement errors and ambiguities. Radar cross section (RCS) of a target is also subject of this chapter, where this phenomenon is the ratio of the radar power scattered by the target in the direction of the radar receive antenna to the power density incident on the target.

3.1 Introduction

Radiation-absorbent material (RAM) is a material which has been specially designed and shaped to absorb incident radio-frequency (RF) radiation also known as non-ionizing radiation, as effectively as possible, from as many incident directions as possible. The more effective the RAM, the lower the resulting level of reflected RF radiation. Many measurements in electromagnetic compatibility (EMC) and antenna radiation patterns require that spurious signals arising from the test setup, including reflections, are negligible to avoid the risk of causing measurement errors and ambiguities.

One of the most effective types of RAM comprises arrays of pyramid-shaped pieces as illustrated in Fig. 3.1, each of which is constructed from a suitably lossy material, which is a material that dissipates energy of electromagnetic or acoustic energy passing through it.

Note the gray color in the image of Fig. 3.1. The gray paint helps to protect the delicate radiation-absorbent material (RAM).

B. Zohuri, *Radar Energy Warfare and the Challenges of Stealth Technology*,
https://doi.org/10.1007/978-3-030-40619-6_3

Fig. 3.1 Pyramid radiation-absorbent material

The physical property of material has dielectrics that exhibit electromagnetic loss at microwave frequencies. In these lossy dielectrics, absorbed microwave energy is converted into heat, which must be removed from the circuit; consequently, for high average power applications, the thermal conductivity of the lossy material is critical. Furthermore, dielectric loss quantifies a dielectric material's inherent dissipation of electromagnetic energy (e.g., heat). It can be parameterized in terms of either the loss angle δ or the corresponding loss tangent tanδ. Both refer to the phasor in the complex plane whose real and imaginary parts are the resistive (lossy) component of an electromagnetic field and its reactive (lossless) counterpart.

Radar cross section (RCS) reduction can be achieved by controlling the reflections from the surface of the structure. Plasma envelope is one of the ways to control the reflections and scattering from the surface. RCS of a target is the measure of its detectability. It is related to the scattering characteristics of the target, depending on the incident wave frequency, target shape, material, and orientation of the target with respect to the incident wave. The target having metallic components is easy to be detected by the RADAR due to the high reflections. Thus, it becomes important to design the targets based on low observable principle for reducing the reflections from the conductive surface. This may be achieved using radar-absorbent structures (RAS), radar-absorbent materials (RAMs), or plasma apart from the shaping methods.

3.2 Types of Radar-Absorbent Material (RAM)

One of the most commonly known types of RAM is iron ball paint. It contains tiny spheres coated with carbonyl iron or ferrite. Radar waves induce molecular oscillations from the alternating magnetic field in this paint, which leads to conversion of the radar energy into heat. The heat is then transferred to the aircraft and dissipated.

Fig. 3.2 Lockheed F-117 Nighthawk

The iron particles in the paint are obtained by decomposition of iron pentacarbonyl and may contain traces of carbon, oxygen, and nitrogen.

One technique used in the F-117A Nighthawk (Fig. 3.2) and other such stealth aircraft is to use electrically isolated carbonyl iron balls of specific dimensions suspended in a two-part epoxy paint. Each of these microscopic spheres is coated in silicon dioxide as an insulator through a proprietary process. Then, during the fabrication panel process, while the paint is still liquid, a magnetic field is applied with a specific Gauss strength and at a specific distance to create a magnetic field patterns in the carbonyl iron balls within the liquid paint ferrofluid.

The paint then cures (hardens), while the magnetic field holds the particles in suspension, locking the balls into their magnetic pattern. Some experimentation has been done applying opposing north-south magnetic fields to opposing sides of the painted panels causing the carbonyl iron particles to align (standing up on end so they are three-dimensionally parallel to the magnetic field). The carbonyl iron ball paint is most effective when the balls are evenly dispersed, are electrically isolated, and present a gradient of progressively greater density to the incoming radar waves.

A related type of RAM consists of neoprene polymer sheets with ferrite grains or conductive carbon black particles (containing about 0.30% of crystalline graphite by cured weight) embedded in the polymer matrix. The tiles were used on early versions of the F-117A Nighthawk, although more recent models use painted RAM. The painting of the F-117 is done by industrial robots so the paint can be applied consistently in specific layer thicknesses and densities. The plane is covered in tiles "glued" to the fuselage, and the remaining gaps are filled with iron ball "glue."

The US Air Force introduced a radar-absorbent paint made from both ferrofluidic and nonmagnetic substances. By reducing the reflection of electromagnetic waves, this material helps to reduce the visibility of RAM-painted aircraft on radar. The Israeli firm Nanoflight has also made a radar-absorbent paint that uses nanoparticles [1].

Nanomaterials that are used as part of stealth airframe structure are listed below here as: [2]

1. *Carbon nanotube (CNT)-based polymer composites* having a wide range of Young's modulus, high specific strength, crash resistance, and thermal performance, and these properties can provide conventional composites and lightweight metals.
2. *Nano-clay-reinforced polymer composites* having thermal and flame-retardant properties.
3. *Metal nanoparticle-incorporated composites*: The extraordinary electrostatic discharge and electromagnetic interference (EMI) shielding properties of these composites make them the probable futuristic solution for making the structure which is resistant to lightning strikes.
4. *Nano-coatings for aeroengine parts*: SiC nanoparticles in SiC-particle reinforced alumina yttria-stabilized nano-zirconia can facilitate crack healing, resulting in improved high temperature and strength and creep resistance as compared to monolithic ceramics. TiN nano-crystallites embedded in amorphous Si3N4 are used for wear-resistant coatings. The nano-composite coatings made of crystalline carbide, diamond like carbide and metal dichalcogenide, and TiN are used for low-friction and wear-resistant applications of aircraft. Nanotube and nanoparticles (nano-graphite, nano-aluminum) containing polymer coating are used for electrostatic discharge, EMI shielding, and low-friction applications of aircraft surfaces.
5. *Nanomaterials for aircraft electrocommunication components*: Magnetic nanoparticles (iron oxide nanoparticles, i.e., Fe_2O_3 and Fe_3O_4)-incorporated polymer films and composites can be used in various data storage media. Ceramic nanoparticles like barium titanate and barium strontium titanate are used for making super-capacitors. MEMS (microelectromechanical systems) and NEMS (nanoelectromechanical systems) offer the possibility of developing a standard fuel management unit which controls the fuel control in aeroengines.

Furthermore, properties of nanomaterials are described as follows. Nanomaterials have the structural features in between of those of atoms and the bulk materials. While most microstructured materials have similar properties to the corresponding bulk materials, the properties of materials with nanometer dimensions are significantly different from those of atoms and bulk materials. This is mainly due to the nanometer size of the materials which render them: [2]

- Large fraction of surface atoms
- High surface energy
- Spatial confinement
- Reduced imperfections, which do not exist in the corresponding bulk materials

Due to their small dimensions, nanomaterials have extremely large surface area-to-volume ratio, which makes a large to be the surface or interfacial atoms, resulting in more "surface"-dependent material properties. Especially when the sizes of nanomaterials are comparable to length, the entire material will be affected by the surface properties of nanomaterials [2].

Moreover, the stealth material used in stealth technology is identified as follows.

The modern aviation design requirements like faster, miniature, highly maneuverable, self-healing (i.e., known as memory material as well), intelligence-guided, smart, eco-friendly, lightweight, and stealth systems warrant for materials with extraordinary mechanical and multifunctional properties:

1. *Carbon nanotube (CNT)-based polymer composites*
   Properties of CNT-based polymer composites are their wide range of Young's modulus, high specific strength, crash resistance, and thermal performance, and these properties can provide conventional composites and lightweight metals. Some CNT-based composites which can be used for airframe structure are CNT/epoxy, CNT/polyimide, and CNT/PP.
2. *Nano-clay-reinforced polymer composites*
   Properties of these composites are barrier and thermal and flame-retardant properties.
3. *Metal nanoparticle-incorporated composites*
   The extraordinary electrostatic discharge and electromagnetic interference (EMI) shielding properties of these composites make them the probable futuristic solution for making the structure which is resistant to lightning strikes.

Many modern military aircraft incorporate some type of surface treatment that provides radar cross section reduction to thereby transform these aircraft into "low observable" or "stealth" airplanes. Generally, these treatments employ materials that absorb or conduct incident radar energy and typically include adhesive bonding or spray-paint-like processes for material adherence. Electromagnetic radiation-absorbent/electromagnetic radiation-shielding materials and structures are well-known.

Such electromagnetic radiation-absorbent/electromagnetic radiation-shielding materials and structures are commonly used in electromagnetic capability/electromagnetic interference (EMC/EMI) test cells to eliminate reflection and interference during testing. Electromagnetic radiation-absorbent materials and structures are also utilized in electromagnetic anechoic chambers for testing high-frequency radar, in antennas, and in low observable (LO) structures.

Radar-absorbent material (RAM) reduces the radar cross section making the object appear smaller. These materials are both very heavy and very costly, two key limitations to their adoption for many applications. The materials which come under RAM are as follows:

- Iron ball paint
- Foam absorber
- Jaumann absorber

The above details show the potential of nanomaterials with stealth technology in aviation (defense) sector. Using nanotechnology with stealth technology in aviation gives the low observability with light weight, high strength, high toughness, corrosion resistance, easy reparability and reusability, less maintenance and durability

Fig. 3.3 Two stealth F-35 taxing

with increase in carrying pay load hence it becomes cheaper, safer and used for protecting to be the target than the conventional. However, as electronic sensors have replaced the eyes of pilots as the primary means of tracking other aircraft, more intricate means of defense were needed. This paper also concluded the potential of nanomaterials with stealth technology in aviation (defense) sector. Using Nano-technology with Stealth Technology in aviation gives the Low Observability with Light Weight, High Strength, High Toughness, Corrosion Resistance, Less Maintenance & Durability with increase in carrying Pay load hence it becomes cheaper, safer and used for protecting to be the target than the conventional. These technologies have some drawback also, but due to the above reason, it can be ignored.

The Republic of China (Taiwan)'s military has also successfully developed radar-absorbent paint which is currently used on Taiwanese stealth warships and the Taiwanese-built stealth jet fighter which is currently in development in response to the development of stealth technology by their rival, the mainland People's Republic of China, which is known to have displayed both stealth warships and planes to the public.

However, an article published on Sept. 29, 2019, by C4ISRNET [3] tells the interesting story of German radar maker Hensoldt that claims to have tracked two F-35A jets attending Berlin ILA airshow back in 2018. So, this story reported by C4ISRNET puts the F-22, F-35, and even F-117 warplanes in different spectrum than what US Air Force was and is lauding.

The 35 stealth fighters are lauded by the US Air Force as almost invisible to radar—which is why it has spent $100 million on each of the jets (Fig. 3.3).

Furthermore, as this article indicates, an experimental German radar "tracked two US F-35s for 100 mi" when they were ready to take off from Berlin Air Show, where this German radar maker claims to have tracked two of the jets from a horse or pong farm for nearly 100 mi using an emerging generation of sensors and processors. See next chapter for more details.

Moreover, the German radar maker used a new "passive tracking radar (PTR) system" (i.e., see previous chapters for more information about (PTR)) that analyzes how a civilian communication—such as radio and television broadcasts' and mobile phone stations' electromagnetic (EM) radiation—bounce off an airborne object such as Lockheed Stealth F-35 Lightning that is supposed to be 100% stealthy, which seems to be not the case. In fact, it turns out the aircraft were flying with radar reflectors and ADS-B transponder that could have made the task easier. See the following section for introduction to an automatic dependent surveillance-broadcast (ADS-B) radar. See Sect. 3.3 of this chapter for further description of ADS-B radar.

3.3 Introduction to ADS-B Radar

Automatic dependent surveillance-broadcast or ADS-B is the latest technological leap in airspace surveillance as Fig. 3.4 shows the basic operating scheme of a passive radar system (PSR).

ADS-B uses a Trig transponder, typically combined with a global positioning satellite (GPS), to transmit highly accurate positional information to ground controllers and also directly to other aircraft. This transmission is known as ADS-B Out and its accuracy is greater than using conventional radar surveillance. This gives air traffic controllers the potential to reduce the required separation distance between aircraft that are ADS-B equipped.

ADS-B is seen as being vital to maintaining future efficient airspace management in busy airspace. It also provides advantages in remote "non-radar" areas too—here

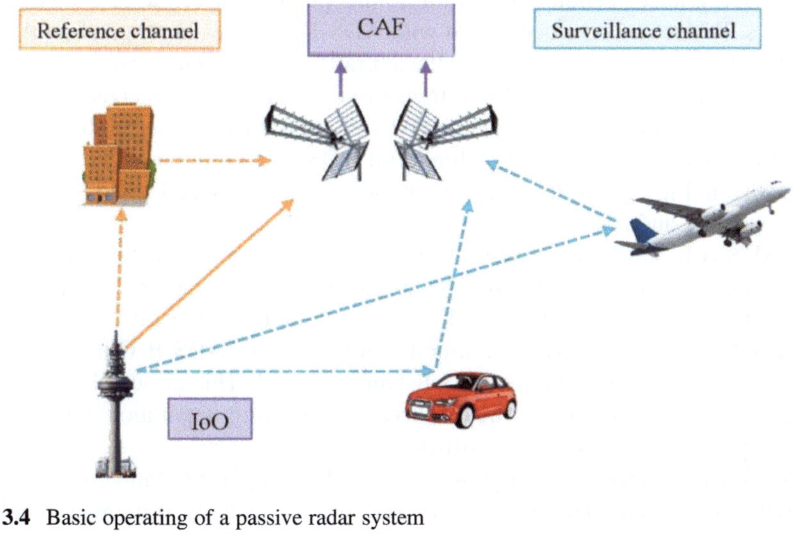

Fig. 3.4 Basic operating of a passive radar system

suitably equipped aircraft, with a traffic receiver connected to a display, can see other aircraft without conventional radar coverage. This enhances aircraft visibility and reduces the risk of air-to-air collision.

ADS-B uses satellite and transponder technology to provide the following benefits in an ADS-B environment:

- More aircraft can operate safely in the same airspace, so congestion is reduced.
- ADS-B technology enables more direct aircraft routing—this can generate significant time and fuel savings.
- With appropriate equipment it's possible to have a live "traffic picture" in the cockpit.
- ADS-B enhances flight safety and collision avoidance.

Aircraft with ADS-B Out provide air traffic controllers with "pinpoint" positional and flight information data. Trig supports 1090ES, the International Civil Aviation Organization (ICAO) international standard for ADS-B that can be used throughout the world.

From an infrastructure point of view, one can say that ADS-B is just in time, and in a given time, the ADS-B will supersede existing surface-based radar technology. Today air traffic controllers have to place significant separation between aircraft to ensure safe flight operations. Existing surface radar technology often has limitations: the slow speed of the radar's return beam, the impact local geography can hide, or mask returns and finally limitations in a radar's range and power. The cost of installing and maintaining radar coverage is expensive. These costs are a challenge for many governments. ADS-B is an attractive alternative technology for national air traffic requirements. An ADS-B infrastructure consists of a network of ADS-B ground stations. These typically consist of an ADS-B tower and mast (similar to a mobile phone mast). This provides a cost-effective national airspace solution, providing accurate data, with fewer gaps or blind spots.

The driving factor behind ADS-B traffic-independent surveillance technology is as follows. When an aircraft equipped with ADS-B Out flies in range of an ADS-B ground station, the station will receive the aircraft's ADS-B Out transmission. At the same time, the aircraft's ADS-B Out transmission can also be received directly by other aircraft that are in range and equipped with an ADS-B In Traffic Receiver. See Fig. 3.5 for an aircraft operating within a full ADS-B environment.

Note that in Fig. 3.5, ground services are only available in the United States (these include weather and traffic services).

An ADS-B In Traffic Receiver translates local airspace information, and using a suitable cockpit traffic display provides a real-time "traffic picture" which significantly enhances a pilot's situational awareness. In remote areas where ADS-B ground stations may not exist, an aircraft equipped with ADS-B Out and ADS-B In can see other aircraft that are ADS-B Out equipped. This allows the aircraft to operate independent of air traffic services, maintaining separation without any reliance upon a ground-based infrastructure (Fig. 3.6).

ADS-B surveillance technology that does not rely upon ground controllers was first trialed in Alaska. This region was selected as an early proving ground for

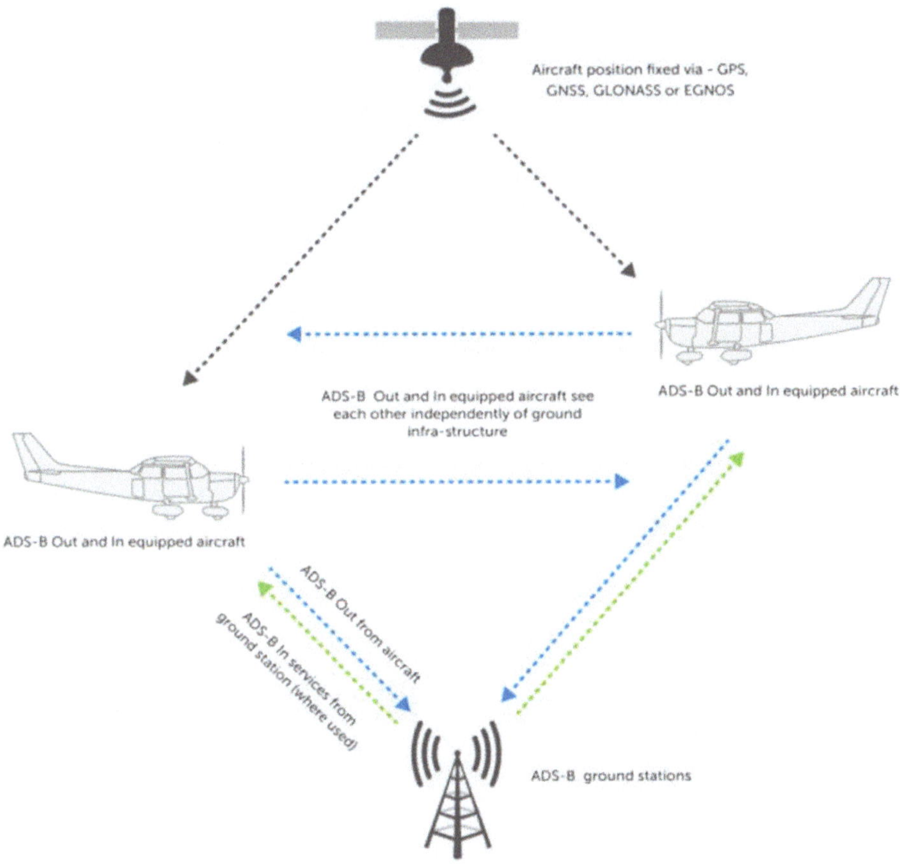

Fig. 3.5 ADS-B Out and In operational mode

ADS-B and associated FAA "Next Gen" technologies. This was due to the significant commercial aviation accident rate suffered in this harsh operating environment. Aircraft were fitted with ADS-B, GPS moving maps, and improved communications to enhance safety. In Southwest Alaska ADS-B (combined with these other initiatives) helped to reduce fatal accidents by 47%.

As far as ADS-B 1090ES OUT is concerned, the ICAO international standard for ADS-B is known as 1090 MHz or more usually 1090ES (Extended Squitter). This is the frequency used to transmit ADS-B information. Currently mandates for ADS-B airspace demand the installation of ADS-B Out equipment, but the use of ADS-B In equipment is at the moment optional driven by the International Civil Aviation Organization (ICAO).

ADS-B has been established in commercial aviation operating above 18,000 ft. for many years. Now, in common with airliners, GA aircraft fitted with an ADS-B-capable Mode S transponder will use something known as "Extended Squitter" to transmit ADS-B Out data. The Extended Squitter is in effect an extended portion of a

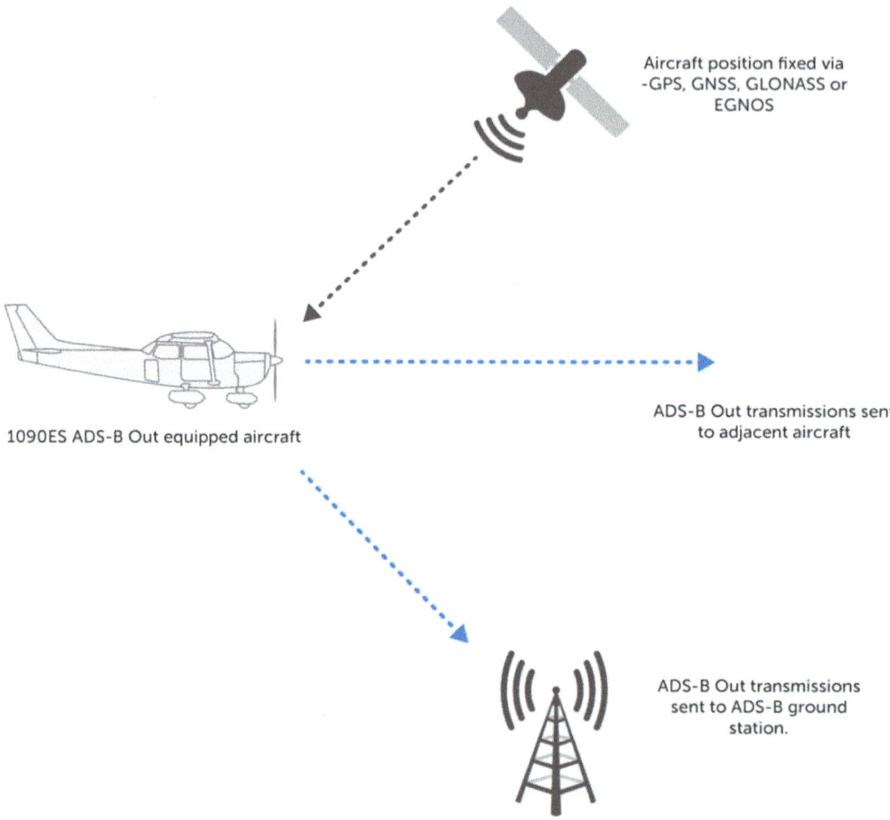

Fig. 3.6 ADS-B Out operational mode

transponder's transmission bandwidth. This contains a data packet of ADS-B information. This data packet holds unique identifying information about an aircraft and its flight position, speed, and profile. The ground station interprets this Extended Squitter transmission and can verify the aircraft against its database. Of course, it is necessary to always check the specific regulations on transponder use in your country. As a general principle, if you fly in airspace that currently requires a Mode A/C transponder, there's a good chance that you will need to operate with an ADS-B 1090ES Out solution if and when a mandate is required.

Uniquely, and only in the United States, another ADS-B system can also be used called User Acceptance Testing (UAT) technology. This system has limitations; it can only be used by pilots operating below 18,000 ft. and only within US airspace. A UAT transmitter uses 978 MHz, in contrast to 1090 MHz—the "international standard."

All Trig transponders are 1090 MHz compliant. As the US ADS-B network is a dual system which supports both 1090 MHz and 978 MHz, it is possible to install a variety of configurations of ADS-B equipment. This dual system does mean that

ADS-B ground stations have to re-broadcast ADS-B traffic information on both 1090 MHz and 978 MHz (UAT). This allows aircraft equipped with ADS-B In Traffic Receivers to *see* all aircraft irrespective of their own ADS-B Out/In equipment. Ground stations broadcast ADS-B information known as TIS-B (traffic information service-broadcast) and FIS-B (flight information service-broadcast)—this includes weather information, and FIS-B is only broadcast on 978 MHz (UAT).

To receive a traffic information service-broadcast (on either 1090 MHz or 978 MHz), the FAA requires that you must first have a certified ADS-B Out transmission. Fitting a Trig transponder provides the easiest upgrade path to secure a certified ADS-B Out signal.

Now the question is why the change when come to new technology from existing ground radar one to ADS-B innovative technology.

The answer then is countries such as the Americas, Australia, and Fiji are creating an ADS-B infrastructure. In the United States, oil and gas platforms in the Gulf of Mexico rely upon high levels of helicopter traffic, here ADS-B is being used to improve visibility, and traffic information is now used where no radar service exists. Congestion of airspace around the Eastern seaboard of America is another area where ADS-B is expected to bring genuine benefits, as it is implemented. In the Australian Outback, ADS-B will, for the first time, enable aircraft to retain a surveillance capability via direct 1090ES air-to-air communication.

The Federal Aviation Administration (FAA) estimates that without changes to the US air traffic infrastructure (of which ADS-B technology is one part), the cost to the US economy will be $22 billion in lost economic activity by 2022. The US initiative to overhaul the air surveillance infrastructure which includes ADS-B is known as "Next Gen." This initiative is predicted to make a positive environmental impact by 2018, with estimates of reduced fuel consumption of 1.4 billion gallons, reduced emissions of 14 million tons, and estimated savings of $23 billion in costs.

While these estimates are based upon commercial flight operations, it is undeniable that ADS-B will positively impact Global Acceptance (GA) too. Already ADS-B Out-equipped pilots may receive more direct routing with shorter flight times making time and fuel savings a reality for all. Thus, according to an article written by C4ISRNET [3] on September 29, 2019, this type of radar tracking system as a passive system will be a weak point for F-22, F-35, or any other stealthy airplane, when it comes to the feature of their stealth technology integrated to their manufacturing of them.

The entire story comes from the fact that camped out amid equines; engineers got word from the Schönefeld tower about when the F-35s were slated to take off. Once the planes were airborne, the company says it started tracking them and collecting data, using signals from the planes' ADS-B transponders to correlate the passive sensor readings. (highlight mine).

Indeed, passive radars are often mentioned as a preeminent anti-stealth technology. As opposed to traditional radars, which use a single transmitter and receiver and study the waves that reflect off flying objects, passive radars use reflections from non-cooperative sources of illumination (such as commercial broadcast and communications signals). A passive radar system is called bistatic because it relies on

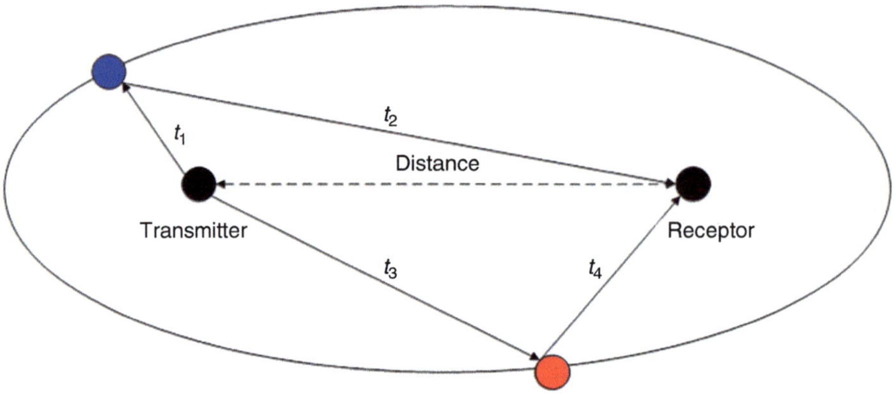

Fig. 3.7 The concept of passive radar system. (Source: Nutaq) [4]

signals transmitted from a different location: by calculating the delay between the signal received directly from the transmitter and the one received after being reflected off a flying object, a passive radar system is able to determine the distance of the target. However, since only the time delay can be calculated from this technique with one transmitter and one receiver, the single conclusion that can be drawn is that the detected object is located somewhere on an ellipse whose foci are the transmitter and the receiver as illustrated in Fig. 3.7.

By measuring bistatic Doppler shift of the echo and its direction of arrival, a passive radar system can calculate speed and heading of the target. Accuracy can be improved by using multiple transmitters and receivers and their geometry.

In order to go forward on matter of active versus passive radar, we provide a general technical background here generated by Nutaq [4].

As we described in Chap. 1, radar (RAdio Detecting And Ranging) systems, as their name implies, are systems used to detect objects and to evaluate the distance between them and a single antenna or a group of antennas. In this blog post, we'll explore two radar mechanisms: active and passive.

Given the radar definition here, then we look at brief definition of active and passive radar.

1. Active Radar

   Active radar is the type of radar most of us are familiar with. Its principle of operation is simple: a radio wave is emitted from an antenna and reflects off objects the wave encounters. The signal is reflected back to the emitter location, where a receiving antenna picks up the echoed signal. When the transmitter and the receiver of a radar system are collocated, the radar is said to be monostatic.

   Once the echo is received, the distance between the radar system and the object can be determined with a simple time-of-flight calculation. Since the speed of a radio-frequency (RF) wave in the air is the speed of light (3×10^8 m/s), and since the time between the emission of the wave and its reception takes into account a

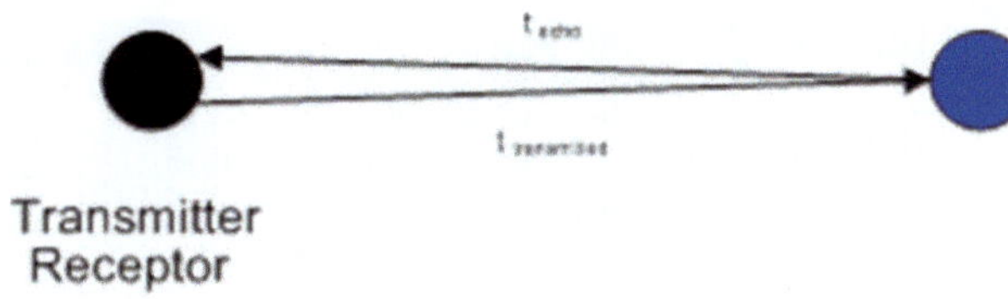

Fig. 3.8 Radio frequency transmitting and echoing back. (Source: Nutaq) [4]

round trip to the target and back, the distance to the object can be calculated by the simple formula as Equation (3.1):

$$D = t \times \left(\frac{c}{2}\right) \tag{3.1}$$

where: $D =$ the Distance in meter; $t =$ the time delay between the emission of the signal and its reception; $c =$ the speed of light ~3×10^8 m/s.

Figure 3.8 explains the basic mechanism of an active radar system. In this figure, the variable t mainly time delay equals the total time for the signal to be transmitted to the object and reflected back: $t_{\text{transmitted}} + t_{\text{echo}}$.

More details were given in Chap. 1 as well.

2. *Passive Radar*

Instead of using collocated transmission and reception antennas, a passive radar system relies on a signal transmitted from a different location. This type of radar system is called bistatic.

The ranging of this type of radar is done by calculating the delay between the signal received directly from the transmitter and the signal received after being reflected off a target.

Since only the time delay can be calculated from this technique with one transmitter and one receiver, the single conclusion that can be drawn is that the detected object is located somewhere on an ellipse whose foci are the transmitter and the receiver.

Figure 3.7 illustrates this concept. In this figure $t_1 + t_2 = t_3 + t_4$. This holds true for every object that is located on the ellipse of the figure. For both of the objects in the diagram, the time delay between the original signal and the reflected signal as seen by the receptor is exactly the same. Only by using multiple transmitters and receivers can this type of radar system precisely locate an object. The performance of the system is highly dependent on the number of transmitters and receivers and their geometry.

Continuing our discussion with ADS-B transponder, we can state that there are limitations to the (passive radar) technology. For one, it depends on the existence of radio signals, which may not be a given in remote areas of the globe. In addition, the technology is not yet accurate enough to guide missiles, though it could be used to send infrared-homing weapons close to a target. However, the published article of September 29, 2019, admits some limitation, but it continues to say:

Fig. 3.9 Stealth jet fighter F-35 with the two radar reflectors installed

> Hensoldt previously said its passive-radar detection works regardless of whether the targeted aircraft has radar reflectors (so-called Luneburg lenses) installed. Those features — little knobs on the roots of the F-35 wings — can be seen in photos released by the U.S. Defense Department on the occasion of the journey to Berlin.

Illustrated in Fig. 3.9 are the two radar reflectors installed on the right side of the F-35. The other two are on the other side.

These fighters (i.e., F-35) almost always fly with the radar reflector; photographs of the aircraft without the four notches (two on the upper side and two on the lower side of the fuselage) are particularly interesting: for instance, some shots taken on Jan. 24, 2018, and just released by the US Air Force show F-35As deployed to Kadena AB, Japan, in October as a part of the US Pacific Command's Theater Security Package program, preparing to launch without their Luneburg reflectors. Luneburg reflectors images are pointed out with red color arrow in Fig. 3.9 as well as image in Fig. 3.10, which clearly shows these reflectors on either side of airplane that is causing degradation of its stealthy features of this fighter.

Luneburg Reflector

The Luneburg reflector significantly increases the radar cross section (RCS) of any system which has little or none at all.

Its radar cross section is several hundred times the RCS of a metallic sphere of the same diameter.

The Luneburg Reflector Advantages

(continued)

The Luneburg reflector gives a homogeneous response inside a wide angle. It is an ideal passive responder, perfect for highlighting and eventually monitoring the radar target to which it is attached, with a high level of security.

The Luneburg lens is the most efficient passive radar reflector available.

The Luneburg reflector requires no power supply nor maintenance.

For more information refer to Appendix A of this book.

When the F-35 flies over friendly countries for overseas deployments, you may notice some strange tags on the body of the otherwise sleek jet.

Fig. 3.10 Image of F-35 with strange tags

Every angle and surface of the F-35 has been precisely machined to baffle radar waves, so little notches like the ones on the picture above would defeat the purpose of the weapons system that has cost about $400 billion so far.

However, US stealth jet fighters such as F-22 and F-35 are modified with augmentation of their Luneburg reflectors as illustrated in Figs. 3.9 and 3.10, in order to evade detection with these devices that supposedly makes them visible to Russian radar.

The US Air Force has been countering Russian efforts to detect its Lockheed Martin F-22 Raptor stealth jets flying over Syria by making these almost invisible fighters more visible to Russian radar systems.

This counterintuitive solution to foiling Russian radar spying consists of installing a device called a "Luneburg lens radar reflector" or a Luneburg reflector on American stealth fighters.

This device increases the radar cross section (RCS) (i.e., see next section for description of RCS) of the F-22—which appears like a steel marble on Russian radars—so the F-22 looks as large as an ordinary fourth-generation jet fighter to a radar.

Satisfied the aircraft on their radar screens isn't an F-22, Russian radar operators won't spend an inordinate amount of time tracking this aircraft and deducing its combat capabilities.

Some experts have described the Luneburg lens as the most efficient passive radar reflector available and one that doesn't require a power supply or maintenance. Figure 3.11 indicates the strange tags on F-35 Lightning stealth fighter.

Fig. 3.11 This strange F-35 modification kills its stealth

Fig. 3.12 A US Air Force F-35A Lightning II assigned to the 34th Fighter Squadron takes off at Yokota Air Base, Japan

The Air Force has also installed Luneburg reflectors (also called RCS enhancers) on a number of its Lockheed Martin F-35 Lightning II stealth fighters to pull off the same trick against the Russians in Europe and possibly against the Chinese.

American media has revealed that a number of Air Force F-35As operating in Eastern Europe near Estonia (one of the Baltic states) are equipped with Luneburg reflectors and are now conducting aerial patrols within range of Russian radars.

Since Russian radar systems in Eastern Europe are similar to the ones it operates in Syria, the reflectors will also hoodwink Russian radar operators in Europe into believing the planes they've detected aren't stealth fighters and aren't that much of a concern.

Since the reflectors exaggerate the RCS of the F-35, the device is preventing Russia from testing their sophisticated radar defenses against this supersonic stealth jet. It appears the Air Force has been testing Luneburg reflectors on F-35s and F-22s since 2010.

Furthermore, a Pentagon April 19, 2019 Published article and reported by James Rogers of Fox News [5] indicates that missing Japanese F-35 stealth fighter in the Pacific poses a major security headache for the United States if it falls into Russian or Chinese hands, if they locate the state-of-the-art fighter jet first, experts warn. See Fig. 3.12 showing a photo of a F-35A Lightning II assigned to the 34th Fighter Squadron which takes off at Yokota Air Base, Japan, on February 9, 2018, after supporting of the vice president's visit to Japan (US Air Force photo by Yasuo Osakabe).

Japanese defense officials say a search is underway for the fighter jet after it disappeared from radar during a flight exercise in Northern Japan. The plane's pilot is also missing.

There is no price too high in this world for China and Russia to pay to get Japan's missing F-35 if they can. There is no price too high in this world for China and Russia to pay to get Japan's missing F-35 if they can.

Both Russia and China maintain a significant naval presence in the region, sparking concerns that they could find the missing F-35, Business Insider reports.

"If one of Japan's F-35s is sitting at the bottom of the Pacific, we are probably about to see one of the biggest underwater espionage and counter-espionage ops since the Cold War. If it was operating without its radar reflectors pinpointing where it went in may be an issue," tweeted Tyler Rogoway, editor of The War Zone.

3.4 Radar Cross Section

The radar cross section σ is a specific parameter of a reflective object that depends on many factors and which has units of m^2. The calculation of the radar cross section is only possible for simple objects. The surface area of simple geometric bodies depends on the shape of the body and the wavelength or rather on the ratio of the structural dimensions of the object to the wavelength. If absolutely all of the incident radar energy on the target were reflected equally in all directions, then the radar cross section would be equal to the target's cross-sectional area as seen by the transmitter. In practice, some energy is absorbed, and the reflected energy is not distributed equally in all directions. Therefore, the radar cross section is quite difficult to estimate and is normally determined by measurement. For example, Fig. 3.13 is a presentation of the experimental radar cross section (RCS) of the B-26 aircraft at 3 GHz frequency as a function of azimuth angle.

The target radar cross-sectional area depends on:

- The airplane's physical geometry and exterior features
- The direction of the illuminating radar
- The radar transmitters frequency
- The used material types

The use of stealth technology to reduce radar cross section increases the survivability and decreases the target detection of military aircraft. But the stealth technology depends on the used radar transmitter frequency and has no effect against very-high-frequency (VHF) radars like P-12 or P-18, both used by Serbian air defense units during the Kosovo war.

Note that P-12 NP "Spoon Rest B" as depicted in Fig. 3.14 is the former Soviet Union, where they had considerable successes in the development of VHF radar units. The assortment of VHF radar units in the Army and the Air Force sufficed of the P-8 over the P-12 up to the P-18 radar sets. One of them is the mobile variant NP, a version of the P-12 installed into two caravans [7].

By the same talking, P-18 "Spoon Rest D", a Russian designator for 1RL131 "Terek"; Cyrillic as illustrated in Fig. 3.15.

1РЛ131 ("Терек") is a very fast moveable radar unit in the very-high-frequency (VHF) range, constructed into two all-terrain trucks (Ural) with two supporters. This radar is used predominantly in the East European space and in the third world to the target assignment for anti-air missiles ("Strela" and "Igla"). But it is also used for

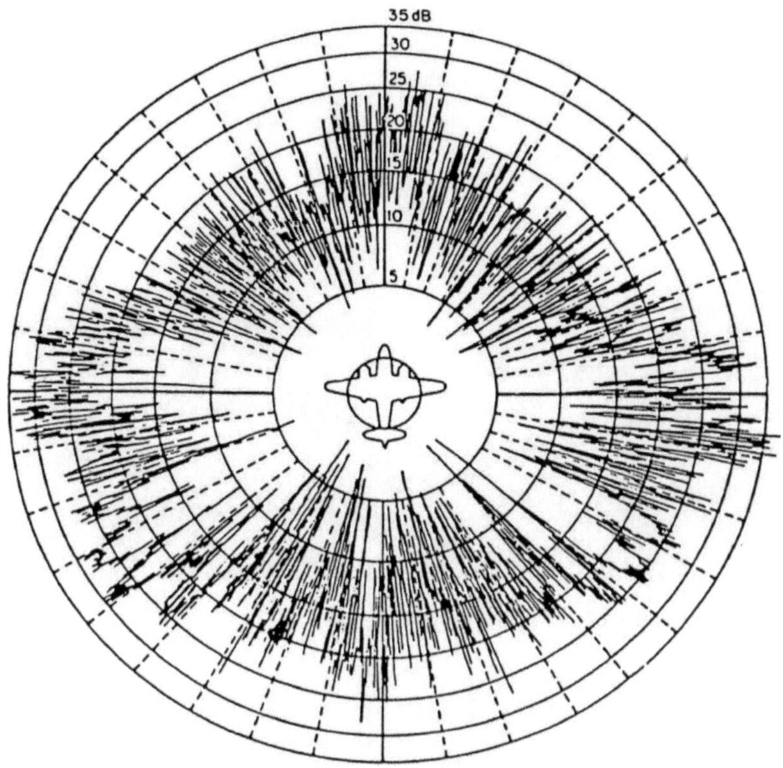

Fig. 3.13 Radar cross section of the B-26 aircraft at 3 GHz frequency [6]

larger rocket systems coupled with a height finder as a direct target assignment, e.g., for SA-2 "Guideline." [8]

3.4.1 Calculation of the Radar Cross Section

Radar cross section (RCS) is the measure of a target's ability to reflect radar signals in the direction of the radar receiver, i.e., it is a measure of the ratio of backscatter density in the direction of the radar from the target to the power density that is intercepted by the target. Since the power is distributed on the shape of a sphere, a small part of this ($4\pi r^2$) can be received by the radar.

Radar cross section σ is as defined as:

$$\sigma = \frac{4\pi r^2 \cdot S_r}{S_t} \tag{3.2}$$

where:

Fig. 3.14 Antenna truck of the "Spoon Rest A"

Fig. 3.15 P-18 image in grading

Table 3.1 RCS for geometrically bodies

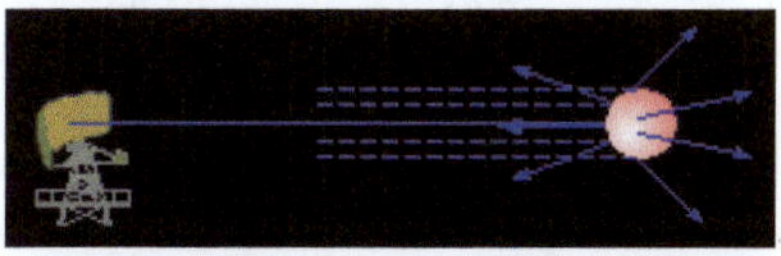 Reflected signal from a spherical shape	$\sigma_{\max} = \pi \cdot R^2$
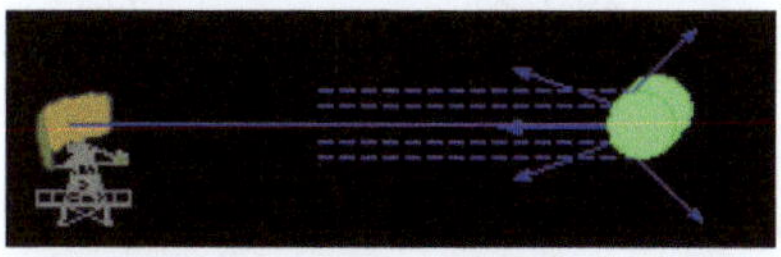 Reflected signal from a cylinder	$\sigma_{\max} = \frac{2\pi r h^2}{\lambda}$
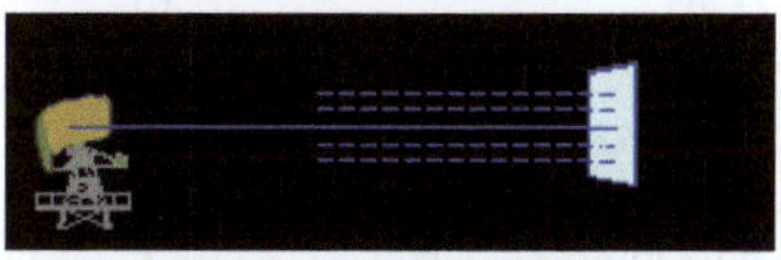 Reflected signal from a flat plate	$\sigma_{\max} = \frac{4\pi b^2 h^2}{\lambda^2}$
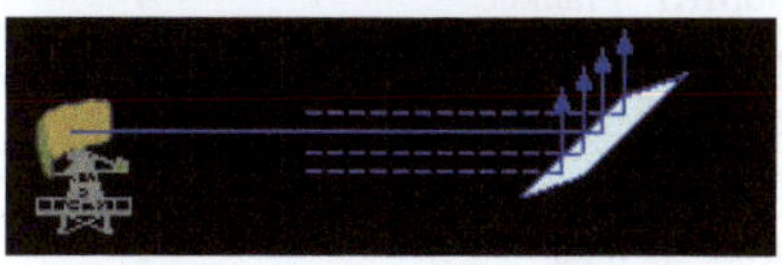 Reflected signal from a tilted plate	Real as the previous example. Unusual feature: The reflected energy is reflected in another direction. Well, the transmitting radar cannot receive this energy. Therefore, there are bistatic radars at which the transmitter and the receivers are separated from each other spatially

σ = measure of the target's ability to reflect radar signals in direction of the radar receiver, in m^2
S_t = power density that is intercepted by the target, in [W/m^2]
S_r = scattered power density in the range r, in [W/m^2].

The RCS of a target can be viewed as a comparison of the strength of the reflected signal from a target to the reflected signal from a perfectly smooth sphere of cross-sectional area of 1 m^2.

The following backscattering formulas in Table 3.1 from shapes occur in an optical independent of frequency region.

3.4.2 *RCS for Point-Like Targets*

Some targets have large values of RCS owing to their size and orientation and, consequently, reflect a large portion of the incident power. Table 3.2 shows the values of RCS for some targets at X-band.

A few educated estimates of radar cross section values of modern aircraft, available in the public domain, are listed in Table 3.3.

Values for RCS in Table 3.3 show that the RCS of modern stealthy aircraft has been reduced to a great degree but not as yet to zero. Figure 3.16 is also a

Table 3.2 RCS for point-like targets [6]

Targets	RCS [m^2]	RCS [dB]
Bird	0.01	−20
Man	1	0
Cabin cruiser	10	10
Automobile	100	20
Truck	200	23
Corner reflector	20,379	43.1

Table 3.3 RCS values for modern aircraft

Number	Typical target type aircraft	RCS in m^2
1	A typical car	100
2	B-52	100
3	B-1 (A/B)	10
4	F-15 Eagle	25
5	Su-27 "Flanker"	15
6	Cabin cruiser	10
7	Mig-29 A/B "Fulcrum"	5
8	Su-30MKI "Flanker"	4
9	Mig-21Bis "Fished"	3
10	F-16 A/B "Falcon"	5
11	F-16 C/D "Falcon"	1.2
12	An average man	1
13	F-18 E/F "Super Hornet"	1
14	Rafale "Israelian Plane"	1
15	B-2 "Spirit"	0.75
16	Eurofighter "Typhoon"	0.5
17	Tomahawk cruise missile	0.5
18	A-12/SR-71 "Blackbird"	0.01
19	A representative Bird	0.01
20	F-35 "Lightning" JSF	0.005
21	F-117 "Nighthawk"	0.003
22	Insect	0.0001
23	F-22 "Raptor"	0.0001

Source: "Radar Cross Section (RCS)," http://www.globalsecurity.org/military/world/stealthaircraft-rcs.htm (Accessed June 03, 2013)

presentation of some of the stealth aircraft RCS with best possible approximation in graphic format.

Hence, radar detection of these stealthy aircraft is delayed by the low RCS but not eliminated altogether. Therefore, it is clear that these stealthy aircraft will be picked up by radars that are powerful enough but at much lower ranges than those at which non-stealthy aircraft would have been detected.

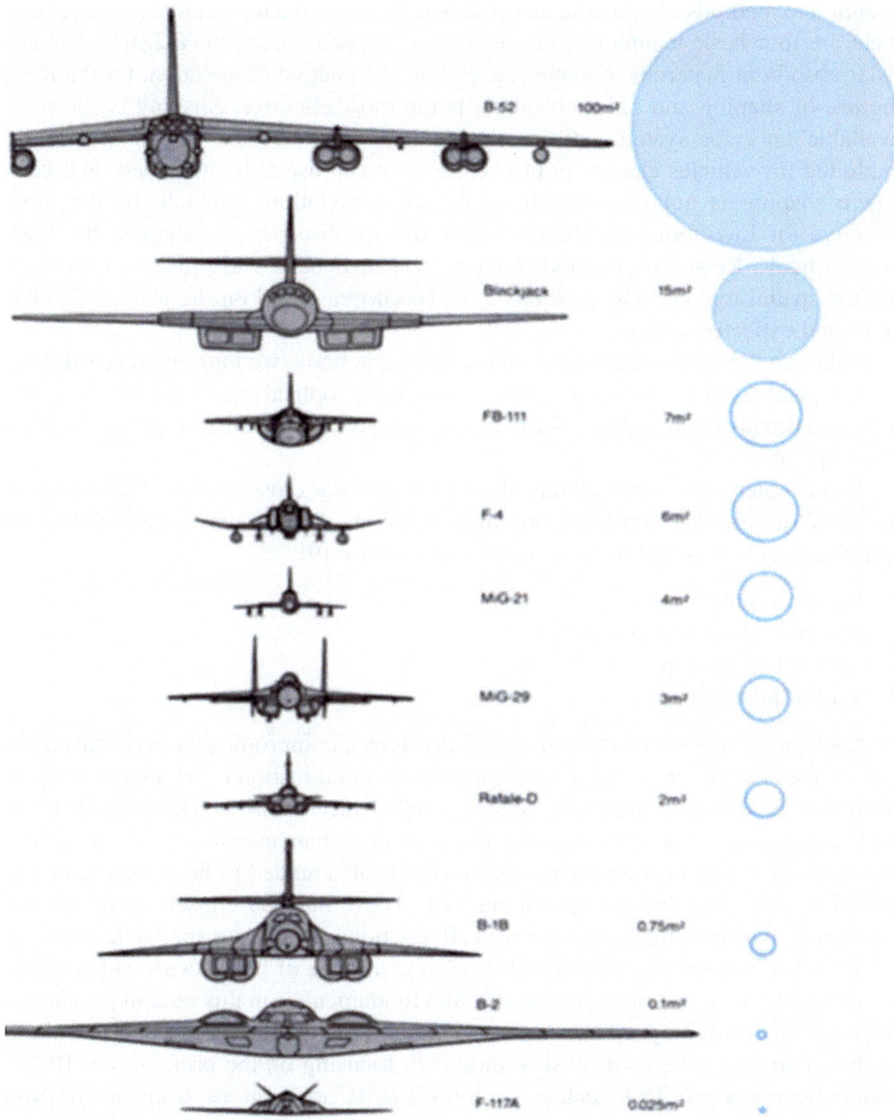

Fig. 3.16 The approximate RCS of aircraft. (Source: Doug Richardson, "Stealth Warplanes: Deception, Evasion, and Concealment in the Air," MBI Publishing Company, New York, 2001)

3.5 Radar Cross Section Reduction

The development of sophisticated detection systems threatens to reduce the mission effectiveness of weapon platforms. Therefore, increasing survivability by reducing detectability has become a very important subject for the designers giving deep attention to methods of reducing detectability. As far as radar signature is concerned, there are four basic techniques for radar cross section reduction (RCSR): shaping, radar-absorbent materials, passive cancelation, and active cancelation. Of the four, the use of shaping and radar absorbers is the most effective. Shaping is typically available only for systems still in the design stage, because it can seldom be exploited for vehicles already in production. We can use radar-absorbent materials where shaping is not efficient alone. Active cancelation seems to be the most effective for low-frequency RCSR, where use of absorber and shaping becomes very difficult. Reduction methods tend to be narrowband and effective only over limited spatial regions. The methods must be chosen based on the platform's missions and expected threats.

Although we focus on the radar signature in this book, we must also consider the other signatures (e.g., infrared, acoustic, magnetic, optical) and balance all signatures and threats for signature control as well; however these other issues are beyond the scope of this book.

This chapter examines methods of controlling radar cross section (RCS) and the trade-offs involved in implementing these methods. Radar cross section reduction techniques generally fall into one of four categories: [9]

1. Target shaping
2. Material selection and coatings
3. Passive cancelation
4. Active cancelation

Application of each of these methods involves a compromise in performance in other areas. For instance, there are limitations to modification of an aircraft's shape from the aerodynamic optimum. Sharply angled facets may be desirable from an RCS perspective, but they degrade the aircraft's maneuverability and handling characteristics. Until recently reduction methods also tended to be narrowband and effective only over limited spatial regions. They must be chosen based on the platform's missions and expected threats. Reduction methods are applied to maintain the RCS below a specified threshold level over a range of frequencies and angles.

The radar cross section reduction (RCSR) fundamentals in this section provide an overview of electromagnetic scattering and radar echo characteristics and how these characteristics may be controlled or modified, focusing on the principles of RCSR and basic concepts. This section explores RCSR applications from an aviation perspective and also touches on other vehicle types. The student will gain an understanding of the tools used in devising RCSR treatments, to include descriptions of computer modeling and laboratory measurements. Further detailed information can be found in book by David C. Jenn [9].

The concept of stealth or radar cross section (RCS) reduction and control has been a topic of interest since World War II. Attempts were initially made to reduce the detectability of the aircraft by employing wood and other composites as aircraft materials since they were less reflective to the radar waves than a metal. Following the initial systematization, one realized that shaping and coating [by radar-absorbent

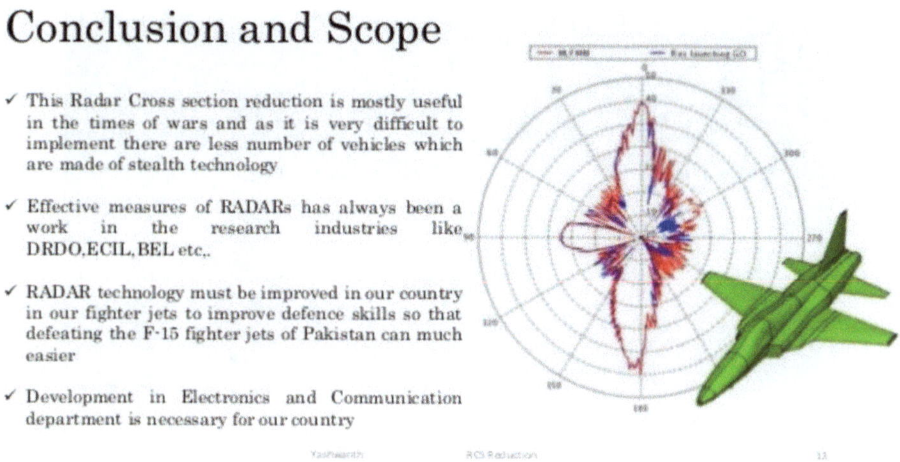

Fig. 3.17 RCSR conclusion and scope

materials (RAMs)] emerged as the primary techniques for the RCS reduction (RCSR).

RCSR through shaping is readily apparent in the case of stealth fighter aircraft such as F-22 Raptor, F-35 Lightning, and F-117 Nighthawk, although the original idea of stealth plane falls in arena of Recommence Airplane SR-71 Blackbird. The edges at principal and drooping ends of wings and rear end of the aircraft have similar angular sweep.

Further, the fuselage and canopy are smooth-surfaced with slopes at sides. The shapes of the surface interfaces, such as the doors at bay and the seam of the canopy, are saw-wave-type. The vertical airfoil of aircraft tails is slant. The front side of its engine is obliterated and includes a serpentine-shaped engine duct. Finally, all the weapons are stored within the aircraft itself.

These alterations in the conventional shape of the aircraft resulted in considerable RCSR of the aircraft. Figure 3.17 shows conclusion and scope of RCSR here.

In contrast to radar cross section reduction technique and four different approaches as mentioned above, we use coating and painting materials that have been used since the 1950s and were integrated as part of fourth-generation airplane such as Lockheed SR-71 Blackbird (Fig. 3.18), and we know them as radar-absorbent material (RAM) to achieve low-RCS aircraft design.

RAM was also useful in mitigating the coupling effect and cross talk between the antennas mounted on the surface of the aircraft. The reconnaissance airplane Lockheed U-2 (Fig. 3.19) and the fighter aircraft Lockheed F-117 Nighthawk stealth fighter (Fig. 3.20) are few examples where RAM has been used and integrated into frame structure of these planes for radar cross section reduction (RCSR).

Sufficient knowledge base was created over time regarding the scattering behavior of aircraft structures. The parameters that played a significant role in overall scattering characteristics of these structures were identified. For example, flat plates

Fig. 3.18 Super-Secret SR-71 reconences plane

Fig. 3.19 Lockheed U-12 spying plane

and cavities were observed to result in large radar return at normal incidence. Similarly, the inlet and exhaust systems of the fighter aircraft were identified as significant contributors toward the aircraft RCS in front-on and rear-on angles, while its vertical tail dominated the radar signature from other angles at the sides.

Numerical techniques were developed over the years for the quantitative estimation of scattering from different parts of the aircraft structure. This facilitated the balanced design of aircraft with optimum RCS. Such aircraft include the Lockheed F-117A Nighthawk (Fig. 3.21), Rockwell B-1 Lancer (Fig. 3.20), and Northrop Grumman B-2 Spirit (Fig. 3.22) type stealth aircraft.

The frontal RCS can be reduced by avoiding shapes and angles of high radar return. Multiple reflections are one of the important factors apart from orientation of the shape and polarization of the impinging wave. If the wave enters into a long,

Fig. 3.20 Lockheed F-117 stealth fighter

Fig. 3.21 Rockwell B-1 Lancer Bomber

closed perfect electric conductor enclosure, it undergoes multiple bounces and may result in large scattered field toward the radar source.

The field associated with radar return can be reduced by coating the inner surface of the enclosure with RAM or redesigning the shape of the enclosure. For example, a curved duct can be useful in increasing the reflections significantly, thereby attenuating the incident energy without any adverse effect on its aerodynamic

Fig. 3.22 Northrop Grumman B-2 Spirit Bomber

performance. Such a cavity, in particular, should have large cross-sectional aspect ratio. The SR-71 engine duct inlet as illustrated in Fig. 3.18 is an example of such multiple bounce low-RCS design. Reader should refer to a book by H. Singh and R. M. Jha [10] for more details and further information.

Furthermore, it is noteworthy to state that one purpose behind shaping of stealthy structure is as it is mentioned here. With purpose shaping, the shape of the target's reflecting surfaces is designed such that they reflect energy away from the source. The aim is usually to create a "cone-of-silence" about the target's direction of motion. Due to the energy reflection, this method is defeated by using passive (multistatic) radars, and these types of radar are described in Chap. 1 of this book, and briefly we can state it here as well.

Passive radar (PR) systems also referred to as passive coherent location (PCL) and passive covert radar (PCR) encompass a class of radar systems that detect and track objects by processing reflections from non-cooperative sources of illumination in the environment, such as commercial broadcast and communications signals. It is a specific case of "bistatic radar," the latter also including the exploitation of cooperative and non-cooperative radar transmitters.

Bistatic radar (Fig. 3.23) was described in Chap. 1 of this book and is the name given to a radar system comprising a transmitter and receiver that are separated by a distance comparable to the expected target distance. Conversely, a radar in which the transmitter and receiver are collocated is called a monostatic radar. A system containing multiple spatially diverse monostatic radar or bistatic radar components with a shared area of coverage is called multistatic radar. Many long-range air-to-air and surface-to-air missile systems use semi-active radar homing, which is a form of bistatic radar.

Note: Silent Sentry's passive coherent location (PCL) technology provides precise, real-time, all-weather detection and tracking ideal for air surveillance, missile tracking, and homeland security applications. Silent Sentry's innovative approach is totally passive, allowing targets to be tracked without generating any RF energy by

Fig. 3.23 Bistatic radar system

Fig. 3.24 Passive coherent location system

using existing broadcast signals from FM radio and TV (analogue and digital) transmitters across the globe. This virtually undetectable surveillance system has no safety or environmental impact. With no moving parts and a commercial off-the-shelf (COTS) approach, Silent Sentry is less expensive to acquire, operate, and maintain than traditional radar systems (Fig. 3.24).

Silent Sentry systems provide covert, robust performance featuring three-dimensional tracking with highly accurate horizontal position and velocity measurements. A modular, flexible, network-ready COTS design facilitates integration with legacy and emerging systems. Silent Sentry systems are compact, easily deployed, and configurable for a variety of surveillance applications.

Silent Sentry offers the following advantages:

- Fill gaps in active surveillance radar coverage (e.g., at low altitudes, in difficult terrain).
- Simultaneously provide coverage throughout large volumes of airspace at higher altitudes.
- Can be deployed covertly and non-provocatively.
- Update target position and velocity for every detected aircraft in real time.
- Can be used to cue active tracking radars.
- Cost less than traditional radar systems to buy and operate because it is COTS-based.
- Easily benefit from improvements in performance and reductions in cost of COTS components.

A passive sensor system is an attractive adjunct to a predominantly active air surveillance sensor suite.

As part of recent interest on "target image (TI)" and "automated target recognition (ATR)," researchers at the University of Illinois at Urbana-Champaign and Georgia Institute of Technology, with the support of DARPA and NATO C3 Agency, have shown that it is possible to build a synthetic-aperture image of an aircraft target using passive multistatic radar. Using multiple transmitters at different frequencies and locations, a dense data set in Fourier space can be built for a given target. Reconstructing the image of the target can be accomplished through an inverse fast Fourier transform (IFFT). Herman, Moulin, Ehrman, and Lanterman [11] have published reports based on simulated data, which suggest that low-frequency passive radars (using FM radio transmissions) could provide target classification in addition to tracking information.

These automatic target recognition (ATR) systems use the power received to estimate the RCS of the target. The RCS estimate at various aspect angles as the target traverses the multistatic system is compared to a library of RCS models of likely targets in order to determine target classification. In the latest work, Ehrman and Lanterman implemented a coordinated flight model to further refine the RCS estimate [11].

3.6 Ways to Track Low Observable (LO) Aircraft

As we have described so far, fifth generation of warplane are built around technology of stealth to make them as stealthy as possible, by reducing their radar cross section via technique of radar cross section reduction. However, given geopolitical that exists between, for example, the United States and Russia or China, they all challenge each other so openly; thus, they are looking at each other by threatening their military capabilities by showing them off. Since conflict between China and the United States in China Sea escalating between these two nations thus, the showoff their military arsenals is on rampage as well. With the introduction of Lockheed stealth fighter such as F-22 Raptor, following Lockheed built stealth fighter as F-35 Lightning, you expect countries with financial resources and technologies in hand to come with their own version of countermeasures against measure of stealth built into these planes.

Because of the military challenges that are existing between the abovementioned countries, you expect their state-run media in particular in Russia and China to blow their own horn by claiming that they can and are able to face up to our new and modern military hardware when it comes to defeating them in the real battlefield. Although the opportunity of such event is very rear and might not be even taking place because we all own nuclear weapon capabilities and can destroy each other so many times over, from this author's perspective, it is a guarantee that we never go into such extreme of fighting among each other under any circumstances, no matter what our differences are.

For example, around 2014 timeframe, in an article that was published by Dave Majumdar [12], in February 19, 2016, under the title "War Is Boring," he stated that the state-run Chinese media is claiming that the People's Liberation Army has been able to track the US Air Force's Lockheed Martin F-22 Raptor (Fig. 3.25) stealth fighters over the East China Sea. While the Chinese report might be easily dismissed as propaganda, it is not beyond the realm of possibility, given even recent claim by a German radar making company and their passive radar capability they have managed to track a F-35 Lighting for almost 100 mi during their departure from an air show that took place in Barling during 2018 by utilizing their newly built Passive Tracking Radar (PTR). See Sects. 3.2 and 4.1 of this book.

Fig. 3.25 A Lockheed F-22 Raptor stealth in flight mode

In fact it's very possible that China can track the Raptor. Stealth is not a cloak of invisibility, after all. Stealth technology simply delays detection and tracking.

First off, if a Raptor is carrying external fuel tanks—as it often does during "ferry missions"—it is not in a stealth configuration. Moreover, the aircraft is often fitted with a Luneburg lens device on its ventral side during peacetime operations that enhances its cross section on radar.

That being said, even combat-configured F-22s are not invisible to enemy radar, contrary to popular belief. Neither is any other tactical fighter-sized stealth aircraft with empennage surfaces such as tailfins—the F-35, PAK-FA, J-20, or J-31. That's just basic physics.

As you can see in Sect. 3.11 of this book, there are few technical steps that are involved in stealth technology, and mainly they are:

1. Target shaping
2. Material selection and coatings, such as radiation-absorbing material (RAM)
3. Passive cancelation
4. Active cancelation
5. Speed

Thus, having anything external makes the stealth plane visible to any radar beam targeting that plane like F-22 Raptor with its external fuel tanks vulnerable to radar transmission and reflection beam.

Furthermore, the laws of physics essentially dictate that a tactical fighter-sized stealth aircraft must be optimized to defeat higher-frequency bands such the C-, X-, and Ku-bands (see Chap. 1 of this book for the description of these bands) and the top part of the S-bands. There is a "step change" in a low observable aircraft's signature once the frequency wavelength exceeds a certain threshold and causes a resonant effect. Typically, that resonance occurs when a feature on an aircraft—such as a tail fin—is less than eight times the size of a particular frequency wavelength.

Effectively, small stealth aircraft that do not have the size or weight allowances for 2 ft. or more of radar-absorbent material coatings on every surface are forced to make trades as to which frequency bands they are optimized for.

Therefore, a radar operating at a lower-frequency band such as parts of the S- or L-band—like civilian air traffic control (ATC) radars—are almost certainly able to detect and track tactical fighter-sized stealth aircraft. However, a larger stealth aircraft like the Northrop Grumman B-2 Spirit, which lacks many of the features that cause a resonance effect, is much more effective against low-frequency radars than, for example, an F-35 or F-22. Typically, however, those lower-frequency radars do not provide what Pentagon officials call a "weapons quality" track needed to guide a missile onto a target.

However, according to the US Air Force official, "Even if you can see a Low Observable (LO) strike aircraft with ATC radar, you can't kill it without a fire control system."

So far, what is stated at the beginning of this section, Russia, China, and others are developing advanced UHF and VHF band early warning radars that use even longer wavelengths in an effort to cue their other sensors and give their fighters some

Fig. 3.26 Northrop Grumman E-2 Hawkeye. (Source: www.wikipedia.com)

idea of where an adversary stealth aircraft might be coming from. But the problem with VHF and UHF band radars is that with long wavelengths come large radar resolution cells. That means that contacts are not tracked with the required level of fidelity to guide a weapon onto a target.

Traditionally, guiding weapons with low-frequency radars has been limited by two factors. One factor is the width of the radar beam, while the second is the width of the radar pulse—but both limitations can be overcome with signal processing. Phased array radars—particularly active electronically scanned arrays (AESA)—solve the problem of directional or azimuth resolution because they can steer their radar beams electronically. Moreover, AESA radars can generate multiple beams and can shape those beams for width, sweep rate, and other characteristics.

Indeed, some industry experts suggested that a combination of high-speed data links and low-frequency phased array radars could generate a weapons quality track.

The US Navy and Lockheed may have already solved the problem. The service openly talks about the Northrop Grumman E-2D's Hawkeye (Fig. 3.26) role as the central node of its NIFC-CA battle network to defeat enemy air and missile threats. Rear Adm. Mike Manazir, the Navy's director of air warfare, described the concept in detail at the US Naval Institute just before Christmas in 2013.

Under the NIFC-CA "From the Air" construct, the APY-9 radar would act as a sensor to cue Raytheon AIM-120 AMRAAM air-to-air missiles for Boeing F/A-18E/F Super Hornets fighters via the Link-16 datalink. Moreover, the APY-9 would also act as a sensor to guide Raytheon Standard SM-6 missiles (Fig. 3.27) launched from Aegis cruisers and destroyers against targets located beyond the ships' SPY-1 radars' horizon via the Cooperative Engagement Capability data link under the

Fig. 3.27 Raytheon SM-6 intercepts ballistic missile. (Source: www.wikipedia.com)

NIFC-CA "From the Sea" construct. In fact, the Navy has demonstrated live-fire NIFC-CA missile shots using the E-2D's radar to guide SM-6 missiles against over-the-horizon shots—which by definition means the APY-9 is generating a weapons quality track.

That effectively means that stealthy tactical aircraft must operate alongside electronic attack platforms the like Boeing EA-18G Growler.

It is also why the Pentagon has been shoring up American investments in electronic and cyber warfare. As one Air Force official explained, stealth and electronic attack always have a synergistic relationship because detection is about the signal-to-noise ratio. Low observables reduce the signal, while electronic attack increases the noise. "Any big picture plan, looking forward, to deal with emerging A2/AD threats will address both sides of that equation," he said.

As we have seen through public news, the Pentagon has poured billions of dollars into the Lockheed Martin F-22 Raptor and F-35 lighting stealth fighters, however with all the news we hear and some was described here in this book, and details are emerging from the news, how low-frequency (LF) passive radars. Pentagon and industry officials concede that radar operating in the very-high-frequency (VHF) band can detect and even track most low observable (LO) aircraft, but conventional wisdom has always held that such systems cannot generate a "weapons quality" track—they are unable to guide a missile onto a target. Even one Navy official goes on to ask that "is it OK if the threat sees it but can' do anything about it?" rhetorically.

Today, technology may have alleviated VHF's weaknesses of the past and with more thrive in technology of signal processing and enhancement combined with a missile with a large warhead and its own terminal guidance system integrated on board system could allow VHF radars to engage a tactical fighter-sized stealth aircraft. Although low-frequency radars do not kill human being, but augmented with a missile-guided system, then they kill people and designated targets. Factors

Fig. 3.28 Russian's P-14 air defense system deployed around September 2017 in Syrian battlefield front. (Source: www.wikipedia.com)

limiting accuracy including width of both the radar beam and the radar pulse can be overcome with digital signal processing (DSP) techniques (see Appendix C on the subject of digital signal processing) that are in existence today, and they will do get enhanced if we take advantages of artificial intelligence (AI) and its sub-system machine learning (ML) into account [13].

Note: The width of the beam is directly related to the size of the antenna. Early low-frequency radars, such as the Soviet-built P-14 Tall King as illustrated in Fig. 3.28, were used with enormous semi-parabolic antennas to generate a narrow beam in battlefield front of Syria.

The later P-18 Spoon Rest used a compact, folding Yagi-Uda array, a bedstead-like framework carrying multiple antennas (Fig. 3.29).

Still, early low-frequency (LF) radars had limitations in determining the altitude, range, and precise direction of a contact. Moreover, the beams produced by these radars are several degrees wide in azimuth and tens of degrees wide in elevation.

Another limitation of very-high-frequency (VHF) radars is that their pulse width is long and they have a low pulse width is long and they have a low pulse repetition frequency, resulting in poor range resolution. A pulse width of 20 μs yields a pulse that is roughly 19,600 ft. long. Range resolution is half the length of the pulse, and in this case, the range cannot be determined accurately within 10,000 ft., and two

Fig. 3.29 P-18 Spoon Rest system. (Source: www.wikipedia.com)

targets with that separation or less cannot be distinguished as separate contacts according to an official Mr. Mike Pietrucha and electronic warfare (EW) and Air Force officer who flew a McDonnell Douglas F-4G Wild Weasel and Boeing F-15E Strike Eagle.

As early as in the 1970s, signal processing helped with range resolution, and the key is a process called frequency modulation on pulse; this is what also Mr. Pietrucha states that "It takes a pulse and modulates the frequency as it goes out," which is basically called a chirp, because that is what it sounds like acoustically. When the pulse is received back, it is run through a special chip, which decompresses it. More details can be found in Ref. [14] written by Dave Majumdar.

3.7 Stealthy No More

As we stated in prior sections of this chapter, with the published article by C4ISRNET [3] on September 29, 2019, under the title of "Stealthy no more, where *a German radar vendor says it tracked the F-35 jet in 2018 — from a pony farm*" and written by Sebastian Sprenger, the article explains how a German firm used passive radar technology to track two fifth-generation stealth aircraft. All these issues were to some degree explained in the previous section; then we can further explain that:

> Stealth aircraft, such as the F-22 Raptor or the F-35 Lightning II 5th generation jets are equipped with Luneburg (or Luneburg) lenses: radar reflectors used to make the LO (Low Observable) aircraft (consciously) visible to radars. These devices are installed on the aircraft on the ground are used whenever the aircraft don't need to evade the radars: during ferry flights when the aircraft use also the transponder in a cooperative way with the ATC (Air Traffic Control) agencies; during training or operative missions that do not require stealthiness; or, more importantly, when the aircraft operate close to the enemy whose ground or flying radars, intelligence gathering sensors.

This is what we explained explaining how the Israeli the heavy presence of Russian radars and ELINT platforms in Syria cause some concern to the Israeli F-35 Adir recently declared Initial Operating Capability (IOC):

The initial operational capability (IOC) declaration comes after the Navy's first F-35C squadron, Strike Fighter Squadron (VFA) 147, conducted aircraft carrier qualifications aboard USS *Carl Vinson* (CVN-70) in early December, received its safe-for-flight operations certification on Dec. 12, and spent the intervening weeks working with the Navy's test community to prove it could operate and maintain the new stealthy jets.

Before the Navy would declare VFA-147 operationally capable, the squadron had to prove several things; Joint Service Fighter Wing commodore Capt. Max McCoy told USNI News in November: the squadron had to be fully manned, with all pilots qualified for shore-based operations and then carrier operations from *Vinson*; the pilots had to prove they could conduct a range of operations and maneuvers; the maintainers had to prove they could keep the new planes flying; and the Navy had to prove it could sustain the squadron through a mature logistics system (Fig. 3.30).

In an article written by Dave Majumdar in *The National Interest* November 8, 2018, issue [15] under the title "How Russia Could Someday Shootdown an F-22, F-35 or B-2 Stealth Bomber," he goes on to explain the theory behind it, and we reflect his article here vibratome.

He also goes on to say, "'*Just because something is technically possible doesn't make it tactically feasible,' one Air Force official with extensive stealth aircraft experience explained.*"

> As tensions between Washington and Moscow flare, the Russian military is warning the United States that it has the ability to target stealth aircraft such as the Lockheed Martin F-22 Raptor, F-35 Joint Service Fighter and Northrop Grumman B-2 Spirit that might be operating over Syria with the Almaz-Antey S-400 (NATO: SA-21 Growler) and the recently arrived S-300V4 (NATO: SA-23 Gladiator) air and missile defense systems (See the description in sections below). However, Western defense officials and analysts are skeptical and note that both the F-22 and the F-35 were specifically designed to counter those Russian-developed weapons.

"Russian S-300, S-400 air defense systems deployed in Syria's Hmeymim and Tartus have combat ranges that may surprise any unidentified airborne targets," Russian Defense Ministry spokesman Major General Igor Konashenkov told the Russian state media outlet Sputnik. "Operators of Russian air defense systems won't

Fig. 3.30 Three F-35C Lightning II—one each attached to the "Argonauts" of Strike Fighter Squadron (VFA) 147

have time to identify the origin of airstrikes, and the response will be immediate. Any illusions about 'invisible' jets will inevitably be crushed by disappointing reality."

However, while Moscow makes bold claims about the counter-stealth capabilities of their S-400 and S-300 V4 air defense systems, the fact remains that even if Russian low-frequency search and acquisitions radars can detect and track tactical fighter-sized stealth aircraft such as the F-22 or F-35, fire-control radars operating in C-, X-, and Ku-bands cannot paint low observable (LO) jets except at very close ranges. Stealth is not—and never has been—invisibility, but it does offer greatly delayed detection so that a fighter or bomber can engage a target and leave before the enemy has time to react.

Tactical fighter-sized stealth aircraft must be optimized to defeat higher-frequency bands such the C-, X-, and Ku-bands—that's just a simple matter of physics. There is a "step change" in an LO aircraft's signature once the frequency wavelength exceeds a certain threshold and causes a resonant effect. Typically, that resonance occurs when a feature on an aircraft—such as a tail fin or similar—is less than eight times the size of a particular frequency wavelength. Fighter-sized stealth aircraft that do not have the size or weight allowances for 2 ft. or more of radar-absorbent material coatings on every surface are forced to make trades as to which frequency bands they are optimized for.

That means that radars operating at a lower-frequency band such as parts of the S- or L-band are able to detect and track certain stealth aircraft. But ultimately, to counter lower-frequency radars, a larger flying-wing stealth aircraft design like the

Northrop Grumman B-2 Spirit or the B-21 Raider—which lacks many of the features that cause a resonance effect—is a necessity. But at the ultra-high-frequency (UHF) and very-high-frequency (VHF) band wavelengths, designers are not trying to make the aircraft invisible—rather engineers hope to create a radar cross section that will blend in with the background noise that is inherent to low-frequency radars.

Low-frequency radars can be used to cue fire-control radars, however. Additionally, some US adversaries have started to make an effort to develop targeting radars that operate at lower frequencies. However, those lower-frequency fire-control radars exist only in theory and are a long way off from being fielded.

"Stealth is 'delayed detection' and that delay is getting shorter. Surface-to-Air Missile (SAM) radars are shifting their frequencies into lower frequency bands where U.S. stealth is less effective," said Mark Gammon, Boeing's F/A-18E/F and EA-18G program manager for advanced capabilities, to Dave Majumdar, the author of this article some time ago [16].

> Early warning radars are in the VHF spectrum where stealth has limited if any capability. These radars are networked into the SAM radars giving the SAM radars cued search.

But low-frequency radars do not themselves provide a "weapons quality" track that is needed to guide a missile onto a target. There are various techniques that have been proposed to use low-frequency radars for such purposes, but none of those are likely to prove viable. US Air Force Col. Michael Pietrucha had described one possible approach to Mr. Dave Majumdar to accomplish such a feat in an article I wrote for Aviation Week & Space Technology [14] a few years ago. However, US Air Force officials were dismissive of the technique. "Just because something is technically possible doesn't make it tactically feasible," one Air Force official with extensive stealth aircraft experience explained.

Meanwhile, operational Raptor pilots tell me "it would be really classified to discuss specific SAM counter tactics," however, the F-22 is more than capable of defeating any of the current Russian surface-to-air missile systems that are currently or projected to be fielded. Hopefully, we will not have to find out the how effective the Raptor truly is during a shooting war over Syria—since conflicts can rapidly escalate out of control, as history loves to teach us over and over again.

Dave Majumdar is the former defense editor for The National Interest.

Another article written by Mr. Dave Majumdar in September 17, 2005, issue of *The National Interest* under the title of "China's Master Plan To Destroy the Stealthy Jet Fighter F-22 and F-35 in Battle" goes on to explain such situation and it is noteworthy to re-write his article here as well and the way he reported word-by-word.

China's Shengyang J-11 unlicensed derivative of the Russian-developed Su-27 Flanker has become the mainstay of the People's Liberation Army Air Force (PLAAF). While the Chinese-built jets are not able to match U.S.-built fighters one-for-one, China is building a lot of them. Down the road, advanced derivatives of the J-11 might become every bit as capable as the most advanced versions of American and allied fourth-generation fighters like the F-15 or F-16. Even fifth-

(A) F-22 (B) F-35

Fig. 3.31 Stealth jet fighters F-22 and F-35. (**a**) F-22. (**b**) F-35

generation Lockheed Martin F-22 Raptors and F-35 Lightning Joint Strike Fighters (i.e., Fig. 3.31a, b) might be overwhelmed by the sheer numbers of Chinese jets and the problems associated with the lack of bases in the Western Pacific.

There have been many iterations of the J-11. Those range from the original license-built models to the "indigenously" produced A-model to the upgraded B/BS-model, which uses a host of Chinese upgrades and avionics hardware. China continues to develop other versions of the J-11 including the advanced J-15, which is designed to operate off China's lone aircraft Liaoning (Fig. 3.32), which was purchased incomplete as a derelict from the Nikolayev shipyards in Crimea. Shenyang was aided in the development of the J-15 through the purchase of a Su-33 Flanker prototype from Ukraine. See Chap. 4 for description of airplane generations.

The J-15, however, was more than just a reverse engineered copy of the original Russian Flanker design. The carrier-based aircraft is expected to feature a host of advanced avionics, including a phased array radar and new infrared search and track system. But while the carrier variant has gotten a lot of attention, a parallel development that features many of the same advancements seems to be making headway.

The J-11D, which is currently in development, is arguably the most advanced land-based single-seat Chinese version of the Flanker. While it probably is not quite as potent as the Russian Su-35S, it is very comparable in a lot of respects. While almost all information concerning Chinese hardware is suspect, the new J-11D allegedly made its first flight sometime in April. The new variant is purportedly equipped with a new electronically scanned radar—possibly an active electronically scanned array (AESA). But China wouldn't need the Su-35 if it had developed a working, producible AESA. That could be why China and Russia have been taking so long to work out a deal to buy the Su-35—the People's Republic has reached a point where it doesn't need the Russians as much as they used to.

The J-11D is also purported to use radar-absorbent materials to help reduce the jet's signature, possibly a new infrared search and track (IRST) system, and revamped electronic warfare systems. It also allegedly features an improved version of China's WS-10 jet engine—but the Chinese have had a lot of difficulties with producing a reliable motor for their aircraft. One reason China is interested in the Su-35 is because of that plane's engines.

Fig. 3.32 Chinese first aircraft carrier Liaoning at sea

However, with these advanced technical approaches to stealthy warplanes, would the jets ever meet in the skies over Asia?

While it is certainly important to consider all of the various possible US-China fighter matchups, we must consider another possibility: there are important data points that suggest these planes may never meet in the skies above Asia.

Given the vast distances of the Pacific, land-based Chinese fighters have limited ability to strike at their more distant neighbors, but there is likely to be an "access" problem for US forces in the event of a conflict, especially if used in conjunction with an integrated air defense system.

If there were to be a war in the Western Pacific, the massive air battles that many might envision are not likely to take place because the United States and our allies have few bases in the region to host tactical fighters like the F-35. More problematic is that even if jets were to take off from bases in Japan like Kadena or Andersen Air Force Base on Guam, the distances are vast. Tankers would come at a premium and would likely to be among the first to be targeted. Moreover, the Chinese are almost certain to attack those air bases with massive barrages of cruise and ballistic missiles—potentially rendering them useless even if structures on the facilities are hardened.

Even if US fighters like the F-22 and F-35 are superior to their Chinese counterparts (and they are), it is meaningless if they don't have bases to operate from or tankers to refuel from. Further, without intelligence, surveillance, and reconnaissance assets, those jets couldn't be properly supported—and it becomes even more difficult when the Chinese attack the space assets and data networks that hold America's fighting forces together (Fig. 3.33).

The question shouldn't be if the F-35 would be able to hold its own in a dogfight; the real question should be: Are short-range tactical fighters relevant in the Pacific theater?

Dave Majumdar is the defense editor for The National Interest. You can follow him on Twitter: @davemajumdar.

Moreover, the abovementioned published article by C4ISRNET [3] on September 29, 2019, under the title of "Stealthy no more" is not the only one that is reported. On November 30, 2001, in Volume 121, Issue 63, The Tech Online Edition written by Tao Yue 14 a Staff Write wrote '*Detection of the B-2 Stealth Bomber And a Brief History on "Stealth"'* and *Scouting For Surveillance.* He claimed that "Cell phones uncover stealth bombers." See picture of Fig. 3.28 of all the US Air Force stealth bombers and fighters.

In early June, the news was filled with headlines such as this one. Newspapers put them at the top of the front page, magazines printed colorful diagrams, and television networks ran the story as the lead on their evening news broadcasts.

And why not? The story was irresistible. Stealth technology is the most potent symbol of America's military supremacy in the post-Cold War world. Though other nations have worked on similar technology, so far none have been as successful as the United States. For something as commonplace as cellular telephones to bring down this symbol of America's military-industrial complex was simply too ironic for the media to resist. In almost all accounts, the technology was described as new and revolutionary, and numerous analogies to David and Goliath were drawn.

Within a week, though, the story had practically disappeared from the media. The US military did not launch any crash program to counter this threat.

No systems were sold. We are left wondering: "What happened?"

In Chap. 4 of this book, we are discussing the stealth technology with more details; however, to go forward with this section, we talk about an overview of stealth technology for the time being.

Stealth technology was developed at Lockheed Martin's legendary Skunk Works research facility. This facility had produced aircraft such as the P-80, America's first jet fighter; the U-2, the high-altitude reconnaissance aircraft made famous by photographing Soviet nuclear missiles being installed in Cuba in 1962; the SR-71, still the fastest operational jet aircraft ever built; and the F-117 Nighthawk, the stealth fighter that captured the world's attention.

Even before the stealth fighter's existence had been publicly announced, rumors circulated in the aerospace and defense community. Tom Clancy featured the stealth fighter in his novel *Red Storm Rising*, a political-military thriller describing a conventional war between the Warsaw Pact and NATO. Testors, maker of accurate scale models of cars, ships, and aircraft, even went so far as to sell a model, based upon alleged sightings of the F-19, the logical designation for this new aircraft.

When the F-117 was publicly announced, more than just its designation was surprising. The plane itself simply didn't look like a modern jet fighter. Instead of a

Fig. 3.33 Stealth airplane in US Air Force inventory and arsenal

sleek, aerodynamic profile optimized for supersonic performance, the F-117 was blocky and featured many flat surfaces. Its wing was swept so sharply back that the plane had difficulty developing enough lift to take off.

There was a reason for this. Stealth technology had begun with coatings that reflect less radar than the aluminum commonly used on airplanes. In fact, the now 30-year-old SR-71 reconnaissance aircraft made use of radar-absorbent coatings to help lower the risk of detection. But there is no perfect absorber of radar. Skunk Works went a step further by shaping the F-117 so that a radar beam would be bounced in direction different from the one in which it originated.

Due to the limited computing power available in the 1970s, the plane was designed using flat surfaces to reduce the number of calculations needed. Each flat surface would add an extra direction in which radar could be reflected, so the number of surfaces used was kept to a minimum. This made the plane aerodynamically unstable about all three axes, so fly-by-wire capability was required to allow the pilot to control the airplane. Enclosed bomb bays, special pilot canopies, special seals at all joints, and special cooling vents for the engines also helped make the plane stealthier.

The F-117 had a radar signature about a hundredth as large as that of conventional airplanes, making it appear a little larger than a bird on radar scopes. The B-2 Stealth Bomber, which followed the F-117, benefited from greater computing power with a contoured shape that further reduced its radar signature. The newest fighter to enter the US armada, the F-22, uses a still more advanced shape.

However, countering stealth is a difficult task and, technologically, requires augmentation of more sophisticated and more advanced passive radar system such as monostatic or bistatic or even conventional passive system, as they were described in previous sections.

Stealth required years of research and massive computing power to develop. Defeating it was a similarly daunting task. F-117 stealth fighters flew over 1300 sorties in the Gulf War without a single one being shot down. A stealth airplane was not lost in combat until 1999, when Yugoslav forces in Kosovo shot one down. This feat was, however, not repeated.

Since the beginning, though, it has been recognized that stealth is not invulnerable. Stealth relies not only on its ability not to be detected by radar but also on its ability not to be detected by other means. This is why stealth aircraft typically do not use radar or send any radio communications while in combat. However, the engines, while cooled to minimize their infrared signatures, still emit more heat than ambient air, a vulnerability that permitted Russian-made SA-3 infrared air-to-air missiles to lock onto the aircraft shot down over Yugoslavia. In addition, stealth aircraft show up visually over a bright sky, making them usable only at night.

Those problems can be solved operationally, though, by limiting the use of stealth warplanes to favorable military situations. A more serious problem is the inherent imperfection of the surfaces of the airplane. No matter how precisely they are manufactured, they will degrade naturally during flight as a consequence of atmospheric friction. Dust in the air and rain affect it even more. Despite special techniques for repairing nicks and scratches and sealing joints where one manufactured part is attached to another, these are done by maintenance crews working under time pressure to get each plane out for another attack run. All of these contribute to the fact that a stealth plane will always reflect some amount of radar.

Furthermore, looking on issue of *stealthy no more* issues from the "Roke Manor system" (i.e., Roke Manor Research Limited is a UK company based at Roke Manor in Romsey, Hampshire. It is a contract research and development business for communications, networks, and electronic sensors) point of view, this UK company expresses the following.

The stealth-detecting system announced over the summer was developed at Roke Manor Research, a British defense firm based in Romsey, Hampshire. It does not try to detect emissions from careless stealth aircraft, a half-hearted and easily countered move.

Instead, it attacks the stealth system itself by detecting the radar waves that do reflect off it.

John Hansman, a professor of Aeronautics and Astronautics at MIT, explains, "Some stealth aircraft, like the F-117, are specifically designed to have a low radar cross section to monostatic, or conventional, radars. They are not stealthy to some bi-static configurations."

Conventional monostatic radar places the transmitter and receiver in the same location, making it simple to locate a plane when spotted. Bistatic, or multistatic, radar would position the receiver at a different position from the transmitter. This makes it more difficult to compute the location of the aircraft.

However, since stealth aircraft do reflect some radar, but away from the transmitter, bistatic radar could conceivably receive the reflection and detect the stealth aircraft.

The problem then becomes one of scale and coordination. The stealth aircraft will be visible only if ideal alignment exists so that the transmitter bounces a signal off the stealth aircraft to the receiver. Stealth aircraft, however, are vulnerable from a very small subset of possible combinations of angles that they bank during their flight and operational action in the air.

The Roke Manor system solves that problem with computing power and some creative thinking. Building a radar every few miles to solve the first problem is prohibitively expensive. However, radar is simply an application of radio, and in today's wireless age, radio waves surround us. In particular, in industrialized nations, cell phone towers can be found every few miles, sometimes every 100 ft. Telephone companies also know exactly where the towers are located and have telephone lines hooked up to them, facilitating communication.

In effect, the Roke Manor researchers have envisioned the use of cell phone towers as an extremely dense network of radar transmitters and receivers, interconnected via communications links. The sheer number of cell phone towers makes detection much easier than with solitary radar sites.

"A lot of stealth technology deals with redirecting radar waves," said Greg Duckworth, a Principal Scientist at BBN working on underwater acoustics in an area very much analogous to radar. "It's very effective against monostatic radars. However, if you have bistatic radars, in particular a very large number of sources, so that you excite the target from a wide range of angles, and you have a multiplicity of receivers in many locations, you essentially will get around the stealth target's redirection capabilities. It is highly likely that an incident wave from a cell tower will be redirected towards one or more receivers" [17].

Having gotten around the stealth aircraft's redirection capabilities, the system then puts together all the data from the cell phone towers. Until recently, this was not possible. However, increased computational power and advanced signal processing techniques have made it possible to sort through all the signals and form a coherent radar picture. Ironically, the further development of the same computing technology that originally made stealth possible has now made it possible to detect stealth aircraft.

The implications of Roke Manor go on to say, given a cell phone network, massively parallel computers, and the Roke Manor software, how much can one determine about a plane? Quite a bit, as it turns out.

"If you can get a radar return, you can get all kinds of information from the return signal if you can process it sufficiently," Hansman said. "For example, if you take a look at the Doppler shift of the returned signal, you can get aircraft velocity. If you are sensitive enough, you can see frequency effects, such as engine rotation or structural vibration. If you have several receivers or different imaging angles, you can begin to reconstruct an image of the target" [17].

These data further reduce the effectiveness of stealth technology. While stealth has always returned a small signal, even to monostatic radars, that signal is so small that it is usually filtered out either by the radar scope or by the operator. However, with velocity and shape information, as well as software specifically designed to detect the inconsistencies that give away a stealth airplane, it becomes considerably easier to separate planes from birds in the sky.

Ernie Rockwood, a researcher for Sensis Corporation, a company that specializes in air traffic and air defense, said that he was "not surprised" by this development. "Some of my co-workers and I worked on novel bistatic battlefield radar techniques to improve survivability. We also submitted a proposal to Rome Labs for an operational concept using multistatic techniques" [17].

Defense researchers and experts in the defense industry also seem to agree that the technology is sound. Some believe this to be a natural development in radar technology.

"Underwater, they have already gone to multistatic systems because the reflectively of targets is such that they don't naturally bounce stuff back," said Greg Duckworth. "Not because they tried to, as was the case with stealth technology, but because the physics makes them do that naturally."

Duckworth also drew an analogy between cell phone towers and television transmissions.

"Televisions have improved quite a bit, and comb filters have gotten better," said Duckworth. "On older TV sets, though, when an airplane goes over your house, a reflective wave from the aircraft ends up interfering at your antenna, and you see lines and artifacts on your screen. To the extent that a stealth aircraft does not absorb the wave, the remnants of it still interact with the airplane and result in detectable interference patterns."

The television analogy is particularly apt, since Lockheed has been working on a project that operates on the same principles as Roke Manor's anti-stealth system. In this project, called Silent Sentry, FM radio stations and VHF television broadcasts are used to provide the dense network of radio waves that interacts with stealth aircraft. While there are fewer FM and VHF transmission towers than cell phone towers, each individual station transmits much more powerfully. The smaller number of stations would also reduce the computational requirements of the system [17].

So now the question is that: What are the *consequences of anti-stealth*? The answer is as follows.

How far-reaching are the implication of this anti-stealth technology? As with all military technologies, it depends on the particular application.

Owen Cote, Associate Director and Principal Research Scientist of MIT's Security Studies Program, explained, "Even if this system works, it wouldn't be useful if you couldn't shoot the aircraft down. You'd have to find some way of guiding a missile very close to the target before an infrared or illuminating radar could achieve a lock on the aircraft" [17].

"This is not very mobile technology," he continued. "Your cell phone towers are in fixed locations. While it would be close to impossible to destroy them all, they are susceptible to jamming just like conventional radar. Stealth might very well be a technology with a very short half-life. However, against foes such as Serbia or Iraq whose technology is not yet competitive with ours, I see stealth as having a much longer life. As a proof of concept, this bistatic technology sounds right. The actual implementation, though, is another matter."

Still, Dr. Cote saw some long-term effects of a successful system. "No offensive advantage lasts," he said. "Often there is a relatively cheap defense counter to match new offensive technology. We may find ourselves moving further away from manned delivery platforms and focusing more on cruise missiles, tactical ballistic missiles, and short range missiles with incredible accuracy" [17].

The technology is widely acknowledged to be feasible, and Roke Manor claims to have been working prototypes. However, bistatic radar is neither a miracle nor a disaster that renders worthless decades of stealth research. It is yet another battle in the war between armaments and armor [17].

3.8 S-300V4 (NATO: SA-23 Gladiator)

The S-300 (NATO reporting name SA-10 Grumble) is a series of initially Soviet and later Russian long-range surface-to-air missile (SAM) systems produced by NPO Almaz, based on the initial S-300P version as illustrated in Fig. 3.34.

Figure 3.28 is a picture of S-300 anti-aircraft missile system at the Victory Parade, Red Square, May 9, 2009.

The S-300 system was developed to defend against aircraft and cruise missiles for the Soviet Air Defense Forces. Subsequent variations were developed to intercept ballistic missiles. The S-300 system was first deployed by the Soviet Union in 1979, designed for the air defense of large industrial and administrative facilities and military bases and control of airspace against enemy strike aircraft.

The system is fully automated, though manual observation and operation are also possible systems produced by NPO Almaz, based on the initial S-300P version. The S-300 system was developed to defend against aircraft and cruise missiles for the Soviet Air Defense Forces. Subsequent variations were developed to intercept ballistic missiles (IBM) as illustrated in Fig. 3.35.

The S-300 system was first deployed by the Soviet Union in 1979, designed for the air defense of large industrial and administrative facilities and military bases and control of airspace against enemy strike aircraft. The system is fully automated, though manual observation and operation are also possible [18].

Components may be near the central command post or as distant as 40 km. Each radar provides target designation for the central command post. The command post compares the data received from the targeting radars up to 80 km apart, filtering false targets, a difficult task at such great distances. The central command post features both active and passive target detection modes.

The S-300 is regarded as one of the most potent anti-aircraft missile systems currently fielded. An evolved version of the S-300 system is the S-400 (NATO reporting name *SA-21 Growler*), which entered limited service in 2004 [18].

Fig. 3.34 S-300 anti-aircraft missile system. (Source: www.wikipedia.com)

Fig. 3.35 Photo of intercept ballistic missile. (Source: www.wikipedia.com)

3.9 S-400 Almaz-Antey (NATO: SA-21 Growler)

The S-400 *Triumph* (Russian, С-400 Триумф, *Triumph*; North Atlantic Treaty Organization (NATO) reporting name, SA-21 Growler), previously known as the S-300PMU-3, is an anti-aircraft weapon system developed in the 1990s by Russia's Almaz Central Marine Design Bureau as an upgrade of the S-300 family. It has been in service with the Russian Armed Forces since 2007. In 2017 the S-400 was described by *The Economist* as "one of the best air defense systems currently made." According to Siemon Wezeman, Senior Researcher of SIPRI, the S-400 "is among the most advanced air defense systems available" [19].

The 30K6E is an administration system which manages eight divisions (battalions). The 55K6E is a command and control center based on the Ural-532,301. The 91N6E is a panoramic radar detection system (range 600 km) with protection against jamming which is mounted on an MZKT-7930. The S-band system can track

Fig. 3.36 Photo of S-400 along with integrated 92N6A radar. (Source: www.wikipedia.com)

300 targets. Six battalions of 98ZH6E surface-to-air missile systems (an independent combat system) can track no more than six targets on their own, with an additional two battalions if they are within a 40 km (25 mi) range. The 92N6E (or 92N2E) is a multifunctional radar with a 400 km (250 mi) range which can track 100 targets (Fig. 3.36).

The 5P85TE2 launcher and the 5P85SE2 on a trailer (up to 12 launchers) are used for launch. The 48N6E, 48N6E2, 48N6E3, 48N6DM, 9M96E, 9M96E2, and the ultra-long-range 40N6E are authorized by a Russian presidential decree. According to the Russian government, the S-400 utilizes an active electronically scanned array [19].

Comparing Russian-built S-400 and America Terminal High Altitude Area Defense (THAAD) system as illustrated in Fig. 3.37, one can state that the US-built THAAD is an effective missile defense system whose capabilities of downing ballistic missiles in terms of intercept altitudes and ranges surpass its rivals.

However, it is strictly an antimissile system, which can hit targets only at very high altitudes (minimum 40–50 km) which makes it useless against fighter jets or long-range strategic aircraft. It is not an air defense missile like S-400 or Patriot.

"Countries seeking effective defense against aircraft and missiles will have to buy two costly American systems – Patriot and THAAD, while the Russian S-400 can unite their functions.

S-400 can also hit difficult ballistic targets at distances up to 60 km, The ability to shoot down the high-speed targets of S-400 almost equals THAAD (around 17 km/h)," a defense industry source said [20].

3.10 S-500 Missile System Triumfator-M

The S-500 Prometey (Russian: С-500 Прометей, lit. "Prometheus") as illustrated in Fig. 3.38, also known as 55R6M "Triumfator-M" [21], is a Russian surface-to-air missile/anti-ballistic missile system intended to replace the A-135 [22] missile system (i.e., Russian anti-ballistic missile (ABM)) currently in use and supplement

Fig. 3.37 Terminal High Altitude Area Defense (THAAD) system

Fig. 3.38 Russian S-500 attacks cruise missiles. (Source: Global Affairs Press)

the S-400. The S-500 is under development by the Almaz-Antey Air Defense Concern. Initially planned to be in production by 2014, it is currently targeting 2020 for deployment. With its characteristics, it will be very similar to the US Terminal High Altitude Area Defense (THAAD) system.

The S-500 is a new-generation surface-to-air missile system. It is designed for intercepting and destroying intercontinental ballistic missiles (ICBM) (Fig. 3.39), as well as hypersonic cruise missiles (Fig. 3.34) and aircraft, for air defense against airborne early warning and control, and for jamming aircraft. With a planned range of 600 km (370 mi) for anti-ballistic missile (ABM) and 400 km (250 mi) for the air defense, the S-500 would be able to detect and simultaneously engage up to ten ballistic hypersonic targets flying at a speed of 5 km/s (3.1 mi/s; 18,000 km/h; 11,000 mph) to a limit of 7 km/s (4.3 mi/s; 25,000 km/h; 16,000 mph).

Mach 5, as well as spacecraft. The altitude of a target engaged can be as high as 180–200 km (110–120 mi). It is effective against ballistic missiles with a launch

Fig. 3.39 Titan II ICBM in flight mode

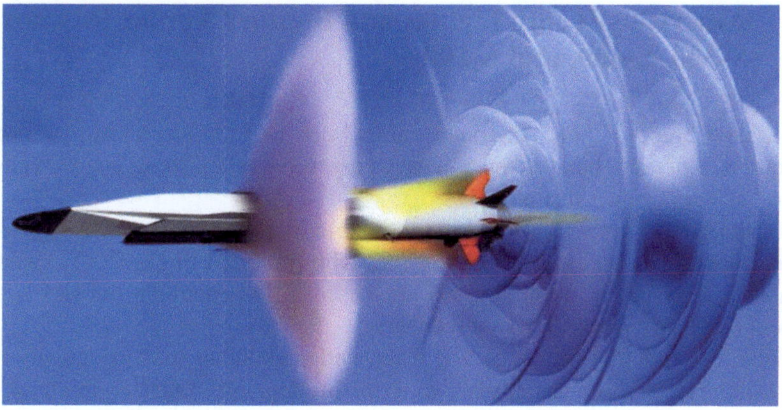

Fig. 3.40 Artistic illustration of hypersonic velocity object

range of 3500 km (2200 mi); the radar reaches a radius of 3000 km (1300 km for the EPR 0.1 m^2). Other targets it has been announced to defend against include unmanned aerial vehicles, low Earth orbit satellites, and space weapons launched from hypersonic aircraft, drones, and hypersonic orbital platforms (Fig. 3.40).

It also aims at destroying hypersonic cruise missiles and other aerial targets at speeds of higher than.

The system will be highly mobile and will have rapid deployability. Experts believe that the system's capabilities can affect enemy intercontinental ballistic missiles at the middle and end portions of flight, but reports by Almaz-Antey say that the external target designation system (RLS Voronezh-DM and missile defense system A-135 radar Don-2 N) will be capable of mid-early flight portion interceptions of enemy ballistic missiles, which is one of the final stages of the S-500 project. It is to have a response time of less than 4 s (compared to the S-400's less than 10) [21].

Note that the hypersonic velocity object is a very new and advanced generation of weapon system that both Russia and China are claiming they are in possession of and presently such weapon is a tremendous threat for US defense systems to the point neither Patriot nor THAAD missile defense systems can track and shoot down this incoming object.

Due to the nature of supersonic speed (5 Mach–15 Mach) that these objects are traveling and maneuvering with, tracking them with any existing radar system is an impossible task. Presently within US arsenal system, we do not have a reliable countermeasure against such measure.

How this author (Zohuri) has published a paper under the title of "New Weapon of Tomorrow's Battlefield Driven by Hypersonic Velocity" indicates that speed is the new stealth technology and along with his co-authors of this paper suggests a new defensive mechanism [23].

3.11 American Stealth Fighter and Bomber Versus Russian S-300, S-400, and S-500

With a new breed of new American fighter such as F-35 stealth airplane of fifth generation, Russians have taken a new measure by establishing a new viable air defense system against this aircraft.

Russian air defenses may appear formidable as part of Moscow's increasingly sophisticated Anti-Access/Area Denial (A2/AD) (pictured in Fig. 3.41) capability, but areas protected by these systems are far from impenetrable bubbles or "Iron Domes" as some analysts have called them.

While it is true that a layered and integrated air defense may effectively render large swaths of airspace too costly—in terms of men and materiel—to attack using conventional fourth-generation warplanes such as the Boeing F/A-18E/F Super Hornet or Lockheed Martin F-16 Fighting Falcon, these systems have an Achilles' heel. Russian air defenses will still struggle to effectively engage fifth-generation stealth aircraft such as the Lockheed Martin F-22 Raptor or F-35 Lightning Joint Service Fighter (JSF). See Fig. 3.42.

"In terms of establishing viable air defenses against opponents with fifth generation aircraft, it's quite clear how Russia is trying to tackle the problem of stealth," said Mike Kofman, a Research Scientist specializing in Russian military affairs at CNA Corporation during an interview with *The National Interest*. "Russia's advanced radar, variety of capable missiles and systems that try to integrate large amounts of data for a more potent air defense will increasingly segregate Western air forces into two benches. In a future where these systems have proliferated to China, Iran and other regional powers there will be those that can penetrate and survive

Fig. 3.41 Russian A2/AD system image

Fig. 3.42 An F-35Bs (JSFs) taking off from the USS America. (Source: Lockheed Martin)

against advanced air defenses in a high end fight, and those whose job it is to bomb ISIL or its successor."

Kofman notes that advanced Russian-built air defenses like the S-300, S-400, and forthcoming S-500 family come with systems designed to detect and track the presence of low observable (LO) aircraft such as the F-22 and F-35. That's just a function of physics (see Sect. 4.5 of Chap. 4 in this book). The problem for Moscow is that while Russian early warning and acquisition radars operating in the VHF, UHF, and L- and S-bands can detect and even track a tactical fighter-sized stealth aircraft, those systems don't deliver a weapons quality track. "Russia has invested in low-band early warning radars, with some great variants out there, but can it use these to put a good picture together, and process it to develop a track against low-observation aircraft?" Kofman asked rhetorically.

Physics dictate that a tactical fighter-sized stealth aircraft must be optimized to defeat higher-frequency bands such the C-, X-, and Ku-bands, which are used by fire-control radars to produce a high-resolution track. Industry, Air Force, and Navy officials all agree that there is a "step change" in a low observable (LO) aircraft's signature once the frequency wavelength exceeds a certain threshold and causes a resonant effect—which generally occurs at the top part of the S-band.

Typically, that resonance effect occurs when a feature on an aircraft—such as a tail fin—is less than eight times the size of a particular frequency wavelength. Effectively, small stealth aircraft that do not have the size or weight allowances for 2 ft. or more of radar-absorbent material coatings on every surface are forced to make trades as to which frequency bands they are optimized for. That means that stealthy tactical fighters will show up on radars operating at lower-frequency bands—such as parts of the S- or L-band or even lower frequencies. Larger stealth aircraft such as the Northrop Grumman B-2 Spirit or forthcoming B-21 don't have many of the airframe features that cause a resonance effect and are, as such, much more effective against low-frequency radars.

For the Russians, solving the problem of targeting a low observable aircraft is something that they continue to work on—but it is doubtful the Moscow has resolved the issue. Russia's strong investment in layers of air defenses tells us that the Kremlin believes the primary threat to its ground forces comes from US airpower. As such, defeating stealth technology is one of Moscow's top priorities, Kofman notes, and the Kremlin has dedicated a lot of resources to that end.

However, per recent report from a German radar making company revealed that they managed to track and F-35 stealth fighter using their newly designed passive radar system for couple hundred miles makes this fifth generation of airplane obsolete when it comes to being stealthy, thus the question arises would American stealth fighter and bomber can survive against the Russian newly developed and being developed surface-to-air missile (SAM) of S-series system such as S-300, S-400, and S-500 are really series threat. Thus, in this case "Is the Stealth Really Stealthy," the subject that this author has imposed here.

As we mentioned in Sect. 3.4 of this chapter, in order to be stealth, we need to reduce the radar cross section (RCS) of flying object and examine the methods of controlling radar cross section (RCS) and the trade-offs involved in implementing these methods. Radar cross section reduction (RCSR) techniques generally fall into one of four categories: [9]

1. Target shaping
2. Materials selection and coatings, such as radiation-absorbing material (RAM)
3. Passive cancelation
4. Active cancelation

However, one other step that can be added to the above four steps as a fifth step is *speed* by utilizing advantages hypersonic velocity functional at least somewhere between Mach 5 and Mach 15 as it is augmented into the new weapon system [23].

However, integrating stealth technology into a generation airplane comes with the penalty of speed degradation. See Chap. 4 Sect. 4.4 for more detailed information.

Russia has tried a number of different techniques to defeat stealth technology. Among those is trying to develop a tight integrated air defense network with multiple radars trying to look at the same aircraft from different directions—but how effective those efforts have been is an open question. "It's great being able to see an aircraft, or parts of it, but getting accuracy such that you can confidently get a missile near the target is the primary challenge," Kofman said.

While the Russians—and the Chinese—have not yet cracked the problem, it is clear that stealth is becoming much less of an advantage over time, though perhaps no less expensive an acquisition. Eventually, Moscow will find a solution to the stealth problem as the cyclical struggle between offense and defense continues ad infinitum—it is just a matter of time.

References

1. http://www.popsci.com/technology/article/2010-07/stealth-paint-turns-any-aircraft-radar-evading-stealth-plane
2. Swapnil Vasant Ghuge and Prof M.BN. Fanisam, "COMPOSITE MATERIAL AND NANOMATERIALS ON STEALTH TECHNOLOGY", Scientific Journal Impact Factor (SJIF), ISSN (PRINT): 2393 – 8161 & ISSN (ONLINE): 2349 – 9745
3. https://www.c4isrnet.com/intel-geoint/sensors/2019/09/30/stealthy-no-more-a-german-radar-vendor-says-it-tracked-the-f-35-jet-in-2018-from-a-pony-farm/?utm_source=twitter.com&utm_medium=social&utm_campaign=Socialflow+DFN
4. https://www.nutaq.com/blog/active-vs-passive-radar
5. https://www.foxnews.com/tech/missing-japanese-f-35-poses-major-security-headache-for-us-if-it-falls-into-russian-or-chinese-hands
6. M. Skolnik, "Introduction to radar systems", 2nd Edition, McGraw-Hill, Inc 1980.
7. http://www.radartutorial.eu/19.kartei/11.ancient/karte046.en.html
8. http://www.radartutorial.eu/19.kartei/11.ancient/karte049.en.html
9. David C. Jenn, "Radar and Laser Cross Section Engineering", Second Edition, Published by American Institute of Aeronautics and Astronautics, Inc (AIAA), Education Series, Reston, Virginia 2019
10. Hema Singh and Rakesh Mohan Jha, "Active Radar Cross Section Reduction, Theory and Applications", Cambridge University Press, published in 2015.
11. http://www.ifp.illinois.edu/~smherman/darpa/
12. https://medium.com/war-is-boring/how-to-detect-a-stealth-fighter-d504f0cb8fbb
13. B. Zohuri and F. M. Rahmani, "Artificial Intelligence Driven Resiliency with Machine Learning and Deep Learning Components", *International Journal of Nanotechnology & Nanomedicine, 26 Aug 2019,* Volume 4, Issue 2 pp 1–8.
14. https://aviationweek.com/defense/ways-track-low-observable-aircraft
15. https://nationalinterest.org/blog/buzz/how-russia-could-someday-shootdown-f-22-f-35-or-b-2-stealth-bomber-35512
16. https://news.usni.org/2014/04/21/stealth-vs-electronic-attack
17. http://tech.mit.edu/V121/N63/Stealth.63f.html
18. https://en.wikipedia.org/wiki/S-300_missile_system
19. https://en.wikipedia.org/wiki/S-400_missile_system
20. https://www.defenseworld.net/feature/20/Battle_of_the_Air_Defense_Systems__S_400_Vs_Patriot_and_THAAD#.XZu7dWdYZhE
21. https://en.wikipedia.org/wiki/S-500_missile_system
22. https://en.wikipedia.org/wiki/A-135_anti-ballistic_missile_system
23. B. Zohuri, P. McDaniel, J. Lee, and C. J. Rodgers, "New Weapon of Tomorrow's Battlefield Drive by Hypersonic Velocity" Journal of Energy and Power Engineering 13 (2019) 177–196.

Chapter 4
Stealth Technology

The rules of battle have changed over the entirety of military history. Tools such as technology, strategy, tactics, and weapons have been the principal elements determining what kind of rules apply to the battlefield. What can constitute to a sixth-generation fighter jet?—that is the question we should be asking ourselves since the past week. Perhaps, it might be too early to think of these questions, when even planes like Joint Service Fighter (JSF), PAK-FA, F-22, or F-35 are not even fully operational. The contemporary military rivalry is driven mostly by the ongoing military technical revolution. In particular, the weapons used on the future battlefield will play an important role in military affairs. Which weapons can play a key role in the future?

4.1 Introduction

On September 30, 2019, Stacy Liberatore of Daily Mail reported [1] that "Experimental German radar 'tracked two U.S. F-35 stealth jet (i.e. Fig. 4.1) for 100 Miles' after lying in wait on a pony farm to catch them flying home from airshow."

She went on to say, "The F-35 stealth fighter is lauded by the United States Air Force as almost invisible to radar, which is why it has spent $100 million on each of the jets." She goes on to say, however, a German radar maker claims to have tracked these two jets from a pony farm for nearly 100 miles using an innovative and emerging generation of sensors and processors.

This new innovation has used a new "passive radar" technological system that analyzes how a civilian communication such as radio and TV broadcasts and mobile phone station radiation bounces off airborne objects.

This German firm claims, renders the jet's stealth technology, that is designed to absorb ground based radar in form of materials such as Radar Absorbing Material (RAM) or by its physical shape to reduce the Radar Cross Section (RCS) to stop it reflecting back the incoming radar electromagnetic beam is totally redundant.

B. Zohuri, *Radar Energy Warfare and the Challenges of Stealth Technology*,
https://doi.org/10.1007/978-3-030-40619-6_4

Fig. 4.1 A F-35 in the process of landing

Reporting by the radar maker of German firm and their claims that they managed to track these two fighter jets (i.e. F-35s) for almost 100 miles after being flown to Germany from Luke Air Force Base in Arizona for the Berlin airshow, in 2018. Apparently, these two jets never were taken to the sky during the show time, meaning that the passive radar, based in a corner of the airfield, could not be tested on them, so the secret of their RAM or RCS technologies will not be compromised.

However, the radar maker of this German firm kept an eye on the F-35s, and once they knew the jets were gearing up to head back home, they set up their newly built system of passive radar tracking (PRT) called "TwInvis" at a nearby pony horse farm.

Report of such tracking incident by a manufacture of a passive radar in Germany to be able to observe and track the most secretive and stealthy fighter jet of fifth generation that is known as F-35 is considered in analogy to the "cat-and-mouse game between aircraft – designed to be undetectable by radar – and sensor makers seeking to undo that advantages." It appears that in the case of this generation of aircraft such as F-35, the promise of invisibility to a radar by US Air Force that is lauded to justify the cost of $100 million per aircraft is no longer a valid claim, if such claim of tracking is really true.

It is but a given fact that each innovative technology only lasts for long until another technology comes that makes the other one obsolete. Considering that for any measure there will be an existing countermeasure, and given the time and modern technology that is progressing so rapidly, the Russians and Chinese are known to be working on technology aimed at preventing whatever leg up NATO countries have tried to build for themselves. As a matter of fact, the recent generation of surface-to-air missile (SAM) known as S-400 and S-500 missile systems is capable of tracking and shooting down any stealthy and fast moving aircraft such as F-22 Raptor and F-35 Lightning generations (see Fig. 4.2). This means the radar

Fig. 4.2 Fifth generation of fighter jets in US Air Force. (**a**) F-22 raptor. (**b**) F-35 lightning

tracking system on this new generation of surface-to-air missile system allegedly with a pinpoint accuracy is able to track these airplanes and shoot them down from the sky.

In fact the capability of this radar tracking the S-400 and S-500 (see the section on these systems later on in this chapter of this book) and what now the German radar maker Hensoldt claims to have tracked two F-35s for 100 miles or 150 km following the 2018 Berlin Air Show in Germany in late April of that year from a pony barn, when these jets were ready to go back to their home bases in United States, seem to be coinciding.

The company's passive radar system, named TwInvis, is but one of the emerging generations of sensors and processors so sensitive and powerful that it promises to find previously undetectable activities in a given airspace.

In fact based on this report, what happened in Berlin was a rare chance to subject the aircraft—with stealthy design features such as reduction of radar cross section (RCS), special coating such as radar-absorbent material (RAM) and all—to a real-life trial to see if the promise of low observability still holds true.

Stories about the F-35 vs. TwInvis matchup had been swirling in the media since Hensoldt set up a shop on the tarmac at Berlin's Schönefeld Airport, its sensor calibrated to track all flying demonstrations by the various aircraft on the flight line. Media reports had billed the system, which comes packed into a van or SUV and boasts a collapsible antenna, as a potential game changer in aerial defense. See Fig. 4.3 for an illustration of the air situation picture provided by Hensoldt's passive radar tracking system, which covers the airspace of Southern Germany.

During a system demonstration by Hensoldt at the exhibit, company engineers convened around a large TwInvis screen showing the track of a Eurofighter performing a thundering aerial show nearby. But the prized target of opportunity, the two F-35s, remained sitting on the tarmac (i.e., Fig. 4.3) [2].

As the event ended, Hensoldt kept a close eye on any movement of the heavily guarded F-35s on the airfield. As exhibitors began to clear out, it looked like the chance of catching the planes during their inevitable departure back home would be lost. But in Hensoldt's telling, someone had the idea of setting up TwInvis outside

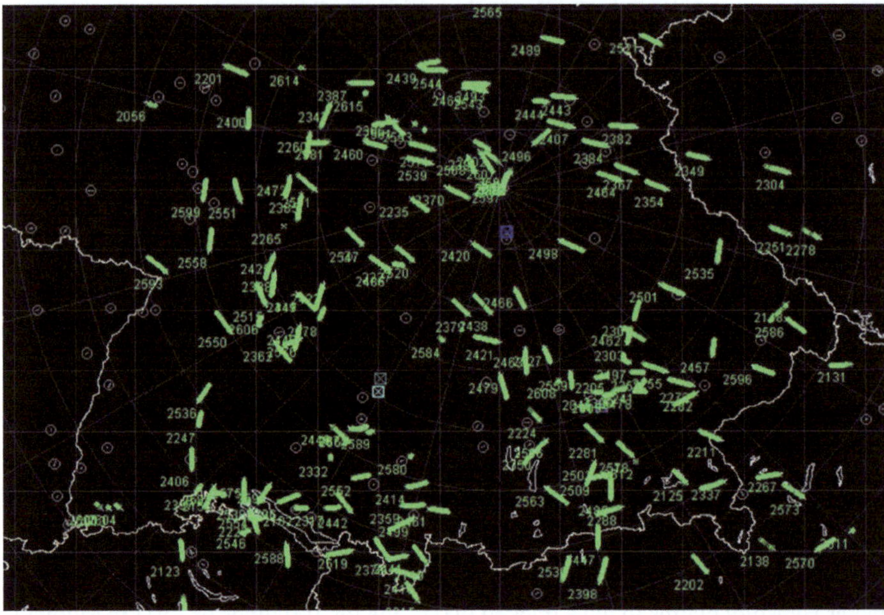

Fig. 4.3 Air situation picture provided by Hensoldt's passive radar tracking system

the airport, which ended up being at a nearby horse or pony farm. Thus, engineers got a word from the Schönefeld tower about when the F-35s were slated to take off. Once the planes were airborne, the company says it started tracking them and collecting data, using signals from the planes' ADS-B transponders to correlate the passive sensor readings.

Hensoldt's engineers are also claiming that various radio station broadcasts in the area, especially a bunch of strong Polish FM emitters broadcasting deep into Germany, improved TwInvis calibration during the Berlin show. The border is about 70 km away from the Schönefeld Airport.

That step came after the Defense Ministry sponsored a weeklong "measuring campaign" in Southern Germany last fall aimed at visualizing the entire region's air traffic through TwInvis. Also, noteworthy, in the year and a half that followed, the air show, emphasis on stealth features for the Franco-German-Spanish Future Combat Air System program, meant to be Europe's next-generation warplane, shifted.

4.2 Fifth Generation of Warplanes (1995–2025)

The fifth-generation fighter is the current standard naming convention for "next-generation" fighter aircraft. While there are no truly defined, widely accepted features of this breed, the aircraft considered for the title often utilize proven features

Fig. 4.4 Sukhoi T-50 PAK-FA

of fourth-generation fighter designs and introduce all new levels of performance, stealth profiles, and advanced avionics with integrated all-digital flight systems. Primary armament is held in an internal bay to further preserve the airframe's radar signature. Several fifth-generation fighters are in development including the American Lockheed F-35 Lightning II and the Russian T-50 PAK-FA (Fig. 4.4), while others are in discussion. The United States Air Force's Lockheed F-22 Raptor, introduced in 2005, remains the only fully operational fifth-generation fighter to date.

The FGFA is being developed by India and Russia that is based on the Russian PAK-FA as it is illustrated in Fig. 4.4. The PAK-FA is a fifth-generation fighter aircraft program of the Russian Air Force. Moreover, the PAK-FA lays the foundation for the FGFA which is being co-developed by Sukhoi and Hindustan Aeronautics Limited for the Indian Air Force. It will have stealth feature and advanced avionics built into it. It is a twin-engine fighter jet which is planned for attack missions.

The FGFA will be tailor-made for the Indian requirements and will have 40 improvements over the Russian version. India will invest $4 billion to develop the FGFA and wants to produce more than 100 such jets. President Putin also said that India and Russia will work on developing the FGFA and sharing critical technologies. India and Russia hope to conclude the negotiations by year-end and sign the contract for jointly manufacturing the FGFA.

Even the Chinese Chengdu J-20 (Fig. 4.5) also known as *Mighty Dragon* is a single-seat, twin-jet, all-weather, stealth fifth-generation fighter aircraft developed by China's Chengdu Aerospace Corporation for the People's Liberation Army Air

Fig. 4.5 J-20 photo flight at Airshow China 2016

Force (PLAAF). The J-20 is designed as an air superiority fighter with precision strike capability; it descends from the J-XX program of the 1990s.

The J-20 made its maiden flight on January 11, 2011, and was officially revealed at the 2016 China International Aviation and Aerospace Exhibition. The aircraft was introduced into service in March 2017 and began its combat training phase in September 2017. The first J-20 combat unit was formed in February 2018. The J-20 is the world's third operational fifth-generation stealth fighter aircraft after the F-22 and F-35.

There are a total of 14 fifth-generation fighter aircraft in the military factory. Entries are listed below in alphanumeric order (1 to Z). Flag images are indicative of country of origin and not necessarily the primary operator.

Note that fourth-generation fighter is the modern standard in combat warplanes. The term is used for those aircraft designs bridging the gap between the developments of the 1960s and 1970s and those appearing today under the fifth-generation fighter classification. Fourth-generation fighter types include the American F-16 Fighting Falcon (Fig. 4.6), the F/A-18 Hornet (Fig. 4.7), the Chengdu J-10 (Fig. 4.8), and the MiG-29 Fulcrum (Fig. 4.9). Fourth-generation types were the first combat warplanes to make regular use of fly-by-wire (FbW) control systems and had increased reliance on digital processing to achieve advanced flying characteristics and performance—meaning designers could make just about any form flyable (case in point, the F-117 Nighthawk stealth fighter (Fig. 4.10)).

Even country like Japan through Mitsubishi company is playing within the domain of fifth-generation fighter jet nations (Fig. 4.11).

The Mitsubishi X-2 Shinshin (formerly the ATD-X) is a Japanese experimental aircraft for testing advanced stealth fighter aircraft technologies. It is being

Fig. 4.6 General dynamics F-16 Fighting Falcon

Fig. 4.7 McDonnell Douglas F/A-18 Hornet

developed by the Japanese Ministry of Defense Technical Research and Development Institute (TRDI) for research purposes. The main contractor of the project is the Mitsubishi Heavy Industries. Many consider this aircraft to be Japan's first domestically made stealth fighter. ATD-X is an acronym for "Advanced Technology Demonstrator—X." The aircraft is widely known in Japan as Shinshin (心神,

Fig. 4.8 Chengdu J-10 Multirole Fighter

Fig. 4.9 Mig-29 Fulcrum

meaning "one's mind" or "Mount Fuji") [3] although the name itself is an early code name within the Japan Self-Defense Forces and is not officially in use. The aircraft's first flight was on April 22, 2016.

Figure 4.12 illustrates all the existing fifth-generation warplanes and fighter jets presently in the arsenal of different countries that are manufacturing them.

Among all these generations, two of them, mainly F-22 and F-35, with their aircraft specifications, are noteworthy to mention as illustrated in Figs. 4.13 and 4.14.

Fig. 4.10 Lockheed F-117 Nighthawk

Fig. 4.11 X-2 Shinshin during its maiden flight

In the case of F-35, its specifications, as it is available in open and public domain, are:

- The jet measures 51.2 ft in overall length, has a wingspan of 35 ft, and a height of 14.3 ft.
- It has a top speed of 1.6 Mach or 1200 mph, a Max G rating of 7G, and a combat radius of 518 miles.

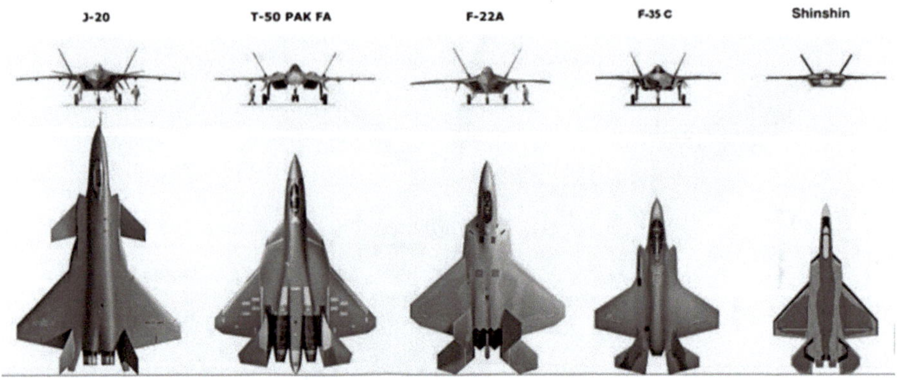

Fig. 4.12 Existing fifth-generation warplanes. (Source: RemeberSky.com)

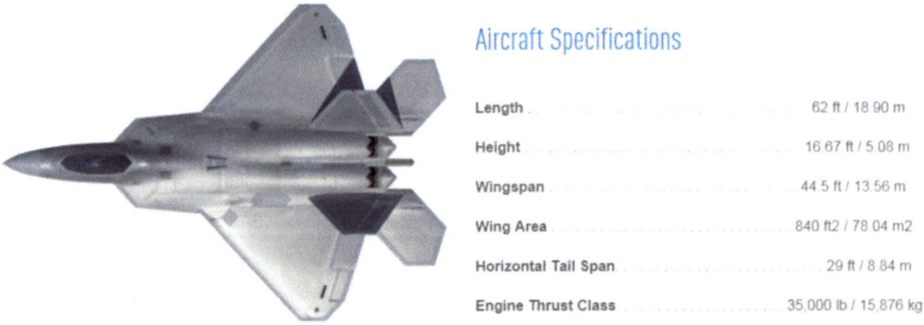

Fig. 4.13 F-22 Raptor specification available in open literature

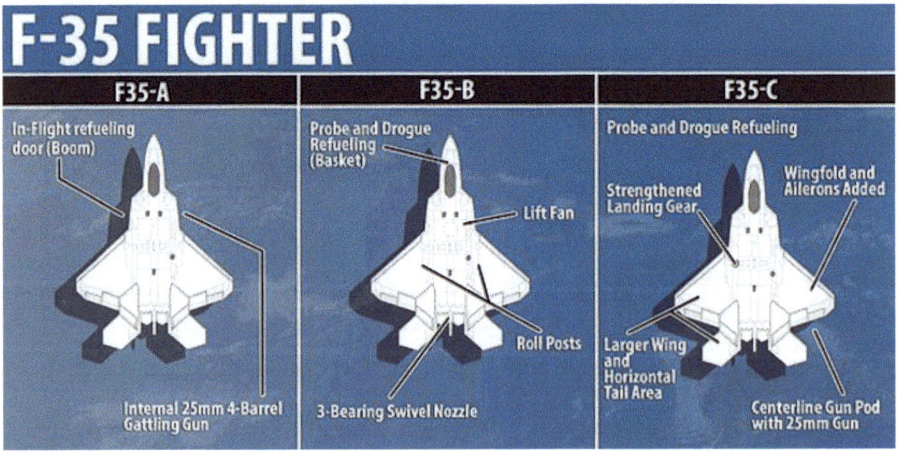

Fig. 4.14 F-35 fighter specification and configuration

- Lockheed Martin, who built the jet, describes its stealth capabilities as "unprecedented." Its airframe design, advanced materials, and other features make it "virtually undetectable to enemy radar."
- The F-35B jets are built from more than 300,000 individual parts.
- There are six distributed aperture system sensors around the jet—two underneath, two on top of the aircraft, and one on the either side of the nose. These infrared cameras feed real-time information and images into the pilot's helmet, allowing them to see through the airframe.
- All variants of the jets are mainly constructed on Lockheed Martin's mile-long production line in Fort Worth, Texas.
- It takes 58,000 man hours to build each F-35B.
- The F-35 can launch from land and will take off from HMS Queen Elizabeth via the ski-jump ramp, which has been designed to optimize the launch.
- Maximum thrust tops 40,000 lb and the jet has a range of 900 nautical miles.
- The jet is capable of two types of ship landing—vertically on to the deck and also through the shipborne rolling vertical landing, which, using forward airspeed, allows the aircraft to bring back several thousand pounds of extra weight to the ship.

However, bear in mind the other issue that the radar's ability to spot the jets relies on signals from civilian transmitters, and many war zones are wastelands with not a civilian in sight.

A quantum improvement in the fighter's lethality and survivability has been a qualifying requirement to achieve generational change, and the fifth-generation fighters personify these traits. The advances over earlier generation of fighters include nose-to-tail low observable or stealth technologies as part of the aircraft's design that make it almost impossible for even other fifth-generation fighters to detect them; improved situational awareness through having multispectral sensors located across all aspects of the airframe which allows the pilot to "look" through the airframe of the aircraft without having to maneuver the fighter to obtain a 360° picture enhances the aircraft's ability to use its suite of weapons to engage and neutralize an adversary without the adversary even being aware of the threat.

Modern electronic warfare (EW) technologies thriving rapidly, as we have seen in the reports above, make the fifth generation of fighter planes already obsolete, even with their existing capabilities such as radar-absorbent material (RAM) and radar cross section (RCS) augmented. So, it seems this fifth-generation of warplanes not even completely deployed into frontline needs to go to mothball storage or countries involved with building them to put their manufacturing into halt, considering that use of low observable "stealth" technology is the primary goal for designation as a fifth-generation fighter.

4.3 Proposed Sixth Generation of Warplanes

A sixth-generation jet fighter (Fig. 4.15) is a conceptualized class of fighter aircraft design more advanced than the fifth-generation jet fighters that are currently in service and development. Several countries have announced the development of a sixth-generation aircraft program, including the United States, China, the United Kingdom, Russia, Italy, Japan, Germany, Spain, Taiwan, and France.

The United States Air Force (USAF) and United States Navy (USN) are anticipated to field their first sixth-generation fighters in 2025–2030. The USAF is pursuing development and acquisition of a sixth-generation fighter through the Penetrating Counter Air to replace its existing air superiority aircraft such as the McDonnell Douglas F-15 Eagle and complement existing platforms in service such as the Lockheed Martin F-22 Raptor. The USN is pursuing a similar program called the Next Generation Air Dominance, likewise, intended to complement the smaller Lockheed F-35 and replace its existing aircraft such as the Boeing F/A-18E/F Super Hornet.

One of the companies within United States of America that does a lot of research on military weaponry is the **R**esearch **AN**d **D**evelopment (RAND) Corporation, and they have recommended that the US Military Services avoid joint program for the development of the design of a sixth-generation fighter (Fig. 4.16). The research studies by RAND Corporation have found that in previous joint programs, different

Fig. 4.15 Conceptual image of a sixth-generation fighter

Fig. 4.16 Sixth-generation conceptual configuration of a Joint Service Fighter. (Source: Rememberedsky.com)

service-specific requirements for complex programs have led to design compromises that raise costs far more than normal single-service programs.

It is noteworthy to state that with the birth of technologies such as artificial intelligence (AI) and super artificial intelligence (SAI) augmented with the function of machine learning (ML) and deep learning (DL), the rules of battle have changed over the entirety of military history [4, 5].

Tools such as technology, strategy, tactics, and weapons have been the principal elements determining what kind of rules apply to the battlefield. What can constitute to a sixth-generation fighter jet?—that is the question I am asking myself since the past week. Perhaps, it might be too early to think of these questions, when even planes like JSF, PAK-FA, or F-22 are not even fully operational. The contemporary military rivalry is driven mostly by the ongoing military technical revolution. In particular, the weapons used on the future battlefield will play an important role in military affairs. Which weapons can play a key role in the future? I will try not to be too technical, such that the article is applicable to general public as well; however, I have included the research papers and appropriate links for those intending to explore more about E-bombs or electromagnetic weapon systems.

Sixth-generation jet fighters are currently conceptual and expected to enter service in the United States Air Force and United States Navy in 2025–2030 timeframe, and the new breed of this generation is looking into the type of aircraft that will fly in an autonomous mode, ending the desire for trained pilots (Fig. 4.17).

The technological characteristics may include the combination of fifth-generation aircraft capabilities with unmanned capability, unrefueled combat radius greater than 1000 nm, and directed-energy weapon. The latter is a subject of this article. One form of this energy is electronic bomb (E-bomb) [6].

Fig. 4.17 Sixth-generation aircraft—air forces to end the desire for pilots

This article aims to explore the technical aspects and potential capabilities of this type of bomb, target measurements, and its comparison with other form of electromagnetic weaponry.

Military action involves the use of directed-energy weapons, devices, and countermeasures to either cause direct damage or destruction of enemy equipment, facilities, and personnel or to determine, exploit, reduce, or prevent hostile use of the electromagnetic spectrum through damage, destruction, and disruption. The defensive part of electronic warfare includes the offensive actions such as preventing the enemy's use of the electromagnetic spectrum through countermeasures such as damaging, disrupting, or destructing the enemy's electromagnetic capability. Such weaponry (DEW) is an evolving addition to the EW [7, 8].

In a comparison between four recent joint service programs (F-35 Joint Service Fighter, Beechcraft T-6A Texan II (Fig. 4.18) Joint Primary Aircraft Training System as illustrated in Fig. 4.19, E-8 Joint Surveillance Target Attack Radar System (JSTARS) (Fig. 4.20), and V-22 Osprey (Fig. 4.21)) and four recent single-service programs (C-17 Globemaster III, F/A-18E/F Super Hornet, F-22 Raptor, and T-45 Goshawk (Fig. 4.22)), costs for joint programs rose to 65%, 9 years after a Milestone B decision to move into engineering and manufacturing development compared to 24% for independent programs during the same time span.

Note that the Beechcraft T-6 Texan II is a single-engine turboprop aircraft built by the Raytheon Aircraft Company (which became Hawker Beechcraft and later Beechcraft Defense Company and was bought by Textron Aviation in 2014). A trainer aircraft based on the Pilatus PC-9, the T-6, has replaced the Air Force's Cessna T-37B Tweet and the Navy's T-34C Turbo Mentor. The T-6A is used by the United States Air Force for basic pilot training and Combat Systems Officer (CSO) training, the United States Navy and United States Marine Corps for primary and intermediate Naval Flight Officer (NFO) training, and the Royal Canadian Air Force (CT-156 Harvard II designation), Greek Air Force, Israeli Air Force (with the

Fig. 4.18 A USAF T-6A Texan II image

Fig. 4.19 A US Air Force E-8C Joint STARS, in flight

"Efroni" nickname), and Iraqi Air Force for basic flight training. The T-6B is the primary trainer for US Student Naval Aviators (SNAs). The T-6C is used for training by the Mexican Air Force, Royal Air Force, Royal Moroccan Air Force, and the Royal New Zealand.

Note that the Northrop Grumman E-8 Joint Surveillance Target Attack Radar System (Joint STARS) is a United States Air Force airborne ground surveillance, battle management, and command and control aircraft. It tracks ground vehicles and

Fig. 4.20 An MV-22

Fig. 4.21 An image of a C-17 Globemaster III flying test

some aircraft, collects imagery, and relays tactical pictures to ground and air theater commanders. The aircraft is operated by both active duty Air Force and Air National Guard units and also carries specially trained US Army personnel as additional flight crew.

Note that the Bell Boeing V-22 Osprey is an American multi-mission, tiltrotor military aircraft with both vertical takeoff and landing (VTOL) and short takeoff and landing (STOL) capabilities. It is designed to combine the functionality of a

Fig. 4.22 T-45A Goshawk image

conventional helicopter with the long-range, high-speed cruise performance of a turboprop aircraft.

Note that the Boeing C-17 Globemaster III is a large military transport aircraft. It was developed for the United States Air Force (USAF) from the 1980s to the early 1990s by McDonnell Douglas. The C-17 carries forward the name of two previous piston-engine military cargo aircraft, the Douglas C-74 Globemaster and the Douglas C-124 Globemaster II. The C-17 commonly performs tactical and strategic airlift missions, transporting troops and cargo throughout the world; additional roles include medical evacuation and airdrop duties. It was designed to replace the Lockheed C-141 Starlifter and also fulfill some of the duties of the Lockheed C-5 Galaxy, freeing the C-5 fleet for outsize cargo.

Note that the McDonnell Douglas (now Boeing) T-45 Goshawk is a highly modified version of the British Aerospace (BAE) Systems Hawk land-based training jet aircraft. Manufactured by McDonnell Douglas (now Boeing) and British Aerospace (now BAE Systems), the T-45 is used by the United States Navy as an aircraft carrier-capable trainer.

Now that we have gathered basic information and built our knowledge around aircraft generation, we will end this section with a historical perspective of each generation of these aircraft, depicted in Figs. 4.23 and 4.24.

Considering that both Figs. 4.22 and 4.23 are illustrations that start with the first generation, the first flying jet which is World War II German ME 262 made by Messerschmitt factory is considered as zeroth generation and sometime gets included as part of the first generation (1944–1955), while the second generation are the jets flying between 1950–1960 and 1965, the third generation flew between

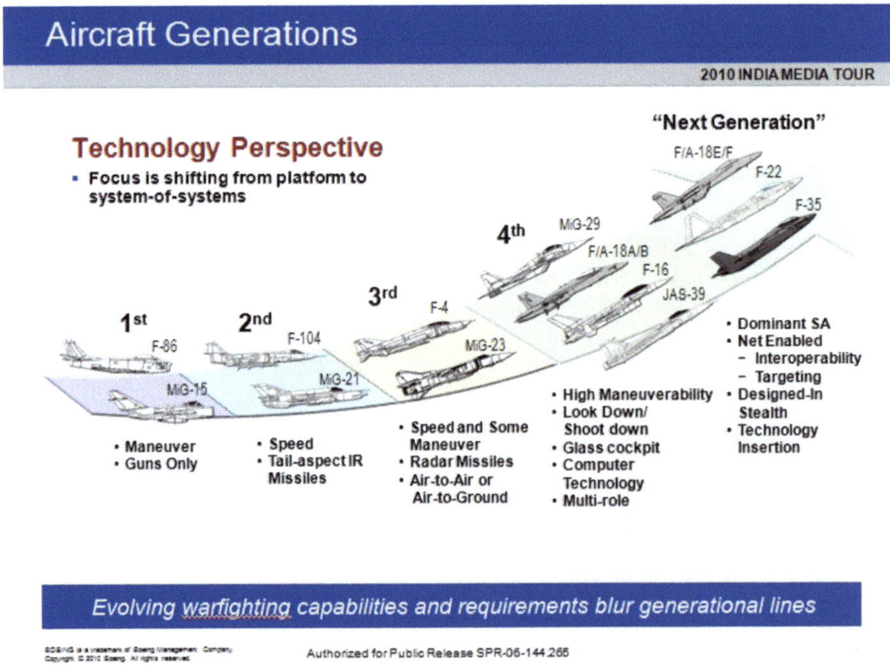

Fig. 4.23 Aircraft generation

time period 1960–1965 and 1970–1975, and the fourth generation flew during the period 1970–1994.

4.4 Science and History of Stealth Technology

The stealth technology being stealthy is nothing new and is not really was a methodology as today's engineers and scientist of modern radar and aircraft designer relied on it for hiding and evading detection.

From the ancient times that human beings discovered a tool known as a weapon system to defend themselves against adversary and going after conquering territories to rule, they were fighting in two-dimensional battlefields of those days; they were utilizing techniques known to us as camouflage. In the two-dimensional battlefields that existed in the "Before Aviation" era, stealth and surprise were achieved through positioning some forces in areas where the enemy would be unable to visually observe and learn of their location due to inadequate line of sight and thereafter using these forces to achieve a final strategic blow or tactical surprise on adversary forces followed by a victory over such enemy force.

Both adversaries facing each other in two-dimensional battlefield in past from ancient time up to recent time as World War-I (WW-I) and World War-II (WW-II),

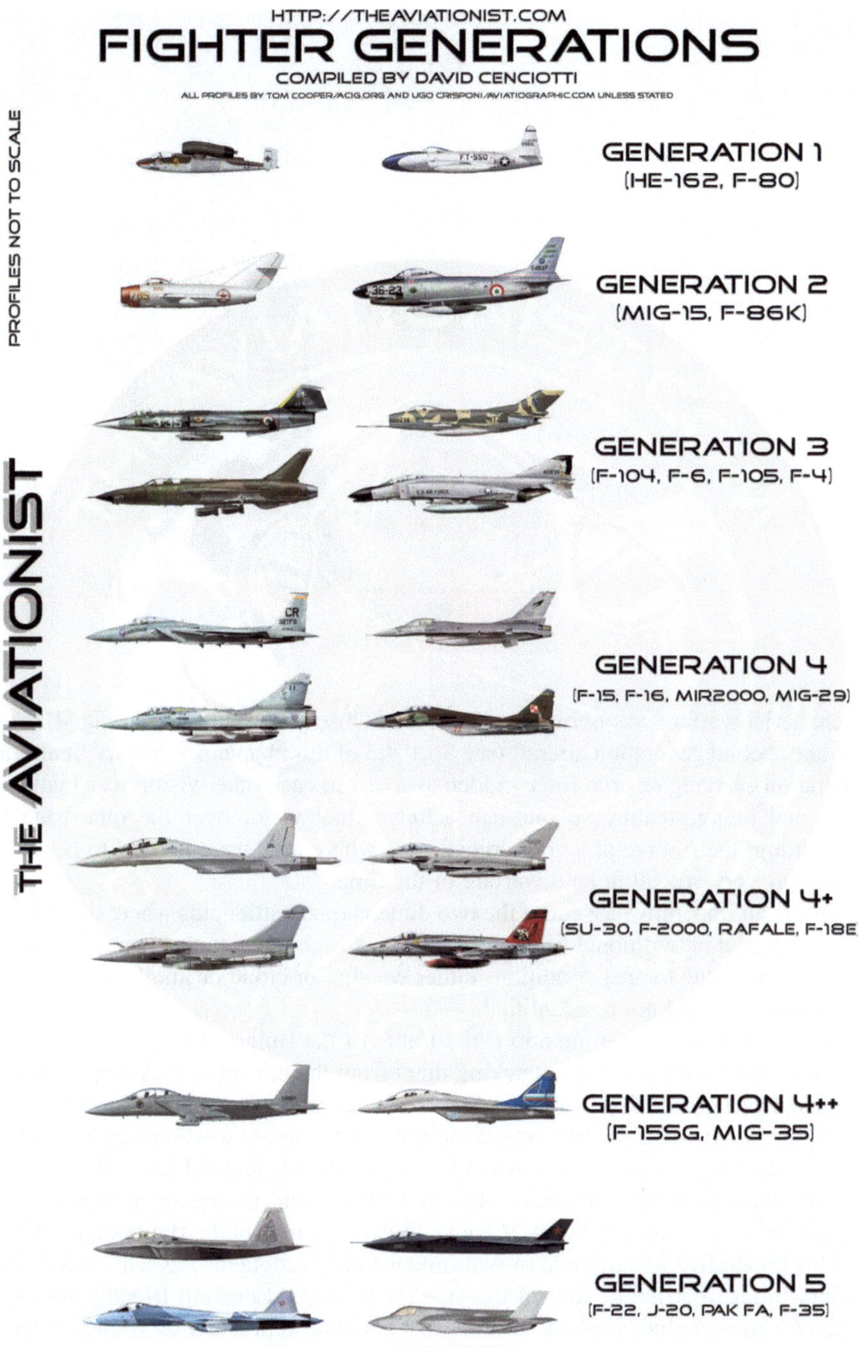

Fig. 4.24 Fighter generations

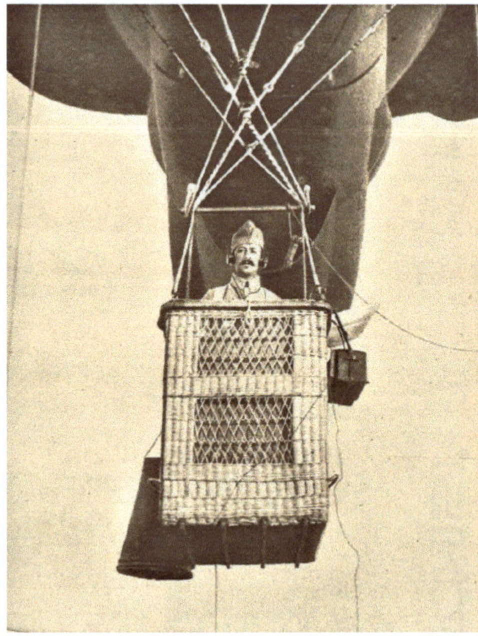

Fig. 4.25 Image of the typical observing balloon of World War II

where aerial warfare was born either in term of Observing Balloon (i.e. Fig. 4.25) or first and second generation aircraft (see Sect. 4.3 of this chapter for history of aircraft generation), having reserve force hidden away from each other vision was matter of tactic and being stealthy so one can achieve final victor over the other one by committing, the reserve at a very crucial time where they were needed to be called upon, was very crucial in land warfare of the time.

This is an imperative aspect of the two dimensional battlefields where the warfare was conducted in traditional lining forces against each other even during the time of WW-II where due to area conditions either weather or cloud or jungle coverage the enemy forces could not be identified.

As an example, we can mention "the Battle of the Bulge" where German forces surprised the Allied forces by attacking them from the coverage of Ardennes jungle that acted as a camouflage for German panzers and infantry personnel on ground as the Allied Air Force could not get off the ground due to the weather condition. This was the last occasion in the war when Hitler, an inveterate gambler, still possessed enough chips to double his stake. It was a bold plan, sweeping in concept and impossible to execute. See Fig. 4.26 for an illustration map of the Battle of the Bulge.

This imperative requirement of two-dimensional battlefields has led armies since ancient times to strive to control the "higher ground." Locating friendly forces at higher locations helped expand the areas that could be kept under surveillance due to the higher or longer line of sight available from elevated position. Of course, the longer line of sight available from higher locations helped detect and track the stealthy or hidden deployment of enemy forces [9].

Fig. 4.26 Map of Ardennes and the Battle of the Bulge. (Source: www.wikipedia.com)

Fig. 4.27 A soldier camouflaged with the surrounding

Furthermore, as part of augmentation stealth technology past time and a more tactical level ground forces have used camouflage through modifications of their equipment to reduce their detectability since armies first seriously applied their minds towards delaying detection of their troops in order to gain tactical and/or strategic advantage over the enemy. See Fig. 4.27, where 95% of people cannot spot these soldiers in full camouflage.

Surprise was achieved when forces the enemy had not seen earlier unexpectedly entered combat. Means employed have included strapping or tying freshly cut vegetation (grass and leafy twigs) to soldier's bodies, choosing the colors of uniforms and other equipment to match with the prevailing background (green and brown in jungle areas, khaki or sand brown in desert terrain, and white in snowbound arctic areas and in mountains), and breaking down of the shapes of personnel and equipment through use of camouflage patterns comprising two or more colors to break the distinctive outline of soldiers and their equipment. See Fig. 4.28 as illustration of vegetation and other means of camouflage, where a soldier has melted into his background environment.

War paint, applied directly on the skin especially on faces, has been used since very early times to serve a similar purpose. See Fig. 4.29.

Notable exceptions to the use of camouflage by ground troops of armies have been when there was a perceived advantage, usually psychological, to be gained through displaying one's own superbly armed, equipped, and trained troops in large numbers and all splendor to intimidate the enemy and to assist in cohesion and control of friendly forces. Here note the bright red uniforms favored by the British in the 1500s and 1600s. See Fig. 4.30 for an illustration of the British Red Coat Soldier.

At a much more basic level in the animal kingdom predators such as the big cats (lions, tigers, panthers, etc.), they conceal their approach from prey by staying

Fig. 4.28 A soldier camouflaged with freshly cut vegetation. (Source: www.imfunny.net)

Fig. 4.29 Army face paint. (Source: http://livegreenhealthy.co/army-face-paint/)

downwind to hide their odor and use vegetation in the area coupled with their natural coloring and patterns to reduce visual detection until it is too late for the prey to escape (see Fig. 4.31).

All these measures were intended to delay the detectability of friendly forces by the enemy in order to surprise the enemy. Thus, stealth as a basic concept is not new to war fighting or in the animal kingdom for that matter. In both instances above, of human land forces and predators of the animal kingdom, a stealthy deployment or stealthy movement of friendly forces has been aimed at achieving surprise. In earlier

Fig. 4.30 British Red Coat Soldier with musket illustration. (Source: www.wikipedia.com)

Fig. 4.31 A lion in hunting position

times with warfare limited to surface forces, the methods of achieving stealth were quite simple and rudimentary as was the general level of technology available to military forces for fighting.

Note: Initially man fought with handheld sharp-edged weapons like the sword, lance, and spear. These gave way in time to the smoothbore muzzle-loading musket which itself was replaced by the breech-loading bolt-action rifle.

As technology advanced, more complex equipment became available to war fighters. The introduction of heavier-than-air aircraft to the battlefield heralded a

Fig. 4.32 Early World War I reconnaissance and observation airplane. (Source: www.wikipedia.com)

major increment in the technology available for war fighting. Early aircraft with their distinctive shape, slow by today's standard speed and other signatures, were relatively easy to spot in the air. See Fig. 4.32, which is photo of an early reconnaissance and observation plane during World War I.

World War I involved a lot of territory and a quickly changing battle, and the lack of information and the element of surprise contributed to some German successes early in the war. The Allies quickly took advantage of aerial reconnaissance and learned how to accurately map and monitor troop movements. The value of information from aerial reconnaissance became of vital importance, and being able to stop your enemy's aerial capabilities was paramount to success, and thus the aerial "dogfight" was born. The use of aerial reconnaissance in World War I changed the nature of war forever.

As visual means were all that were initially available for spotting of aircraft, visual acquisition was the most prevalent; it was supplemented a while later by equipment meant to detect, amplify, and locate in azimuth and elevation the distinctive acoustic signature of aircraft engines. Over time increasingly advanced special equipment was developed to detect and identify the azimuth and elevation of aircraft through use of visual and acoustic sensors [10].

4.5 Stealth Technology

Today's battlefields are now fought in three-dimensional form, where aircraft plays the third dimension of the battlefield. Stealth or low observability (LO) as it is scientifically known is one of the most misunderstood and misinterpreted concepts in military aviation by the common man.

Aircraft are a relatively new entrant on the battlefield. The first heavier-than-air flying machine to take part in military operations did so as recently as in the first decade of the twentieth century. Air warfare, more than any other form of modern warfare, has been driven largely by technology. As was the case in surface warfare, in aerial combat also, surprise came to occupy a central position for advantage to be gained over the enemy. This surprise could be achieved through different means.

If there is one dimension in the air attack-air defense succession that is riding high on the wings of enabling edge technologies, it is the use of stealth, both in the offensive and defensive spheres. The air defense combatants are engaged in fielding high-technology and high-definition sensors in active, passive, and electromagnetic (EM) based anti-stealth domains to challenge stealth attack. The cause-effect duel, thus, continues undyingly.

Stealth aircraft are considered as invisible aircraft, which dominate the skies. With an additional boost from Hollywood action movies, stealth is today termed as the concept invincibility rather than invisibility. Though the debate still continues on whether stealth technology can make an aircraft invincible, it was found that stealth aircraft are detectable by radar. The motive behind incorporating stealth technology in an aircraft is not just to avoid missiles being fired at it but also to give total deniability to covert operations. This is very much useful to strike targets where it is impossible to reach. Thus, we can clearly say that the job of a stealth aircraft pilot is not to let others know that he was ever there.

In simple terms, stealth technology allows an aircraft to be partially invisible to radar or any other means of detection. This does not allow the aircraft to be fully invisible on radar. Stealth technology cannot make the aircraft invisible to enemy or friendly radar. All it can do is to reduce the detection range or an aircraft. This is similar to the camouflage tactics used by soldiers in jungle warfare. Unless the soldier comes near you, you can't see him. Though this gives a clear and safe striking distance for the aircraft, there is still a threat from radar systems, which can detect stealth aircraft.

Stealth is all about using technology to defeat detection systems that operate using the electromagnetic (EM) spectrum. Ability to defeat such detection systems would deliver the benefit of surprising the enemy (as the enemy would be unaware of the presence and location of stealth-enabled weapon systems). Such surprise, by achieving a high level of stealth, would be delivered primarily through application of technology. To be considered stealthy in practice, an aircraft should have minimal signatures in the following areas of the EM spectrum:

- Radar
- Infrared

- Acoustics
- Visual

These four factors, from the point of view of the methods of reducing detectability in these parts of the EM spectrum, will now be discussed briefly here:

1. *Radar*: For understanding stealth technology and how it works against radar, it is essential to understand the basic and simplified working principle radar, which we covered in Chap. 1 following with other chapters as well. We presented the history of radar which was actively deployed during the World War II and became the most important sensor and advanced warning system in aerial warfare and has continued up to now.

   Radar also gives very precise information on target parameters including azimuth/elevation, range, target vector, etc., thus enabling effective engagement of the target.
2. *Infrared*: Other sensors have also been experimented with over the years. A few such sensors have used optical and infrared (IR) as the primary means of detection and tracking. IR sensors were effective in determination of azimuth and elevation fairly accurately. These IR sensors, however, were severely affected by weather, basically by atmospheric transparency, and, moreover, these required to be coupled with radar or other means such as lasers for range determination.

   However, IR sensors provided a passive means of detecting and tracking, and when coupled with lasers for ranging, they could provide required target information often without the target being warned that it was being tracked. Hence, these found widespread applications despite the limitation of the variable atmospheric transparency affects their performance. Radar is thus the most used and most potent sensor to detect and track aircraft and hence the greatest threat to an aircraft. However, optical sensors proved inadequate due to difficulties in reliably detecting targets especially at large ranges, low accuracy in determination of position, as well as lack of range and lack of other parameter extraction capabilities.
3. *Acoustic*: Another type of sensor that is utilized for detection is the acoustic technique, and in case of airplane detection, aircraft noise is another means of its detection. Furthermore, it was found that acoustic methods could give only rough indications of the azimuth and elevation of an aircraft without range and any other parameters.

   In case of airplane, the aircraft engines are the primary source of this signature. Application of civilian technologies, driven by noise pollution laws, to military aircraft may serve to reduce the acoustic signature. This is already underway as seen in the shift in military jet engines from pure turbojets to turbofans with ever-increasing bypass ratios. Other design changes to reduce noise are also finding their way from civil engines to military engines. Jet engine noise is caused primarily by the high-speed movement of high-pressure air. Attempts to shape the engine's airflow passages to reduce this noise are underway. Control of the pressure distribution inside the engine as well as the redesign of the jet nozzles is reportedly to be a promising way forward [11].

Note: The Naval Research Laboratory (NRL) physics-accurate Large Eddy Simulation (LES) tool JENRE will be used to run numerous high-fidelity fluid flow simulations of full-scale nozzle modifications on DoD High-Performance Computing (HPC) resources. This approach is not currently utilized by the industry in developing designs for practical jet engines. However, this effort will result in far more accurate predictions of jet turbulence and resulting jet noise than the Reynolds-averaged Navier Stokes (RANS) or unsteady RANS (URANS) approaches used by the industry as the current state of the art. A joint effort will be conducted to experimentally optimize and refine several new noise reduction concepts with application to a GE F404 nozzle. Finally, the investigators will settle on the top two or three passive noise reduction concepts for further maturation. A set of full-scale/production F404 nozzle seals will be modified to implement each concept. Then a series of uninstalled GE F404 engine noise measurements will be performed in order to confirm or deny the predicted noise reduction at full scale. This multilevel approach is expected to rapidly mature up to three new concepts, each of which have been shown to be effective and promising in preliminary laboratory testing [12].

As benefits of this program, this effort will result in the rapid maturation from below Technology Readiness Level (TRL) 4 to at least TRL 5 of several promising passive jet noise reduction technologies. The project focuses on those concepts that are potentially applicable to retrofit on the legacy of Department of the Navy (DON) supersonic aircraft engines. The joint effort between the modern full-physics LES tools and laboratory-scale tests to screen, optimize, and confirm the effectiveness of these approaches in more realistic conditions ensures a high likelihood of success. The final evaluation at full scale, on an engine at realistic operating conditions, will confirm a jump to TRL5+. This rapid maturation timetable has the potential to save substantial DoD resources and allow transition to the fleet far faster than any other approach. The concepts being considered include [12]:

(a) Nozzle seal mounted micro vortex generators.
(b) TRL 6 proven chevron nozzle seals, modified to vary azimuthally so as to focus noise reduction below and to the sides, while reducing thrust impact.
(c) Sweeping jet actuators. They are easy to retrofit in an existing engine and have a relatively high likelihood of providing substantial improvement over the current state-of-the-art chevron solution.

Experiments in the United States have led to the discovery that making the exhaust nozzle lip in a sawtooth shape, referred to as cutting chevrons in it, contributes to jet engine noise reduction [13].

Another source of noise is the sonic footprint of an aircraft flying at supersonic speeds. The "sonic bang" is the noise heard as the pressure discontinuity which is the shock wave caused by supersonic flight crosses over the observer's position. With adequate listening posts deployed, it could be possible to track an aircraft in

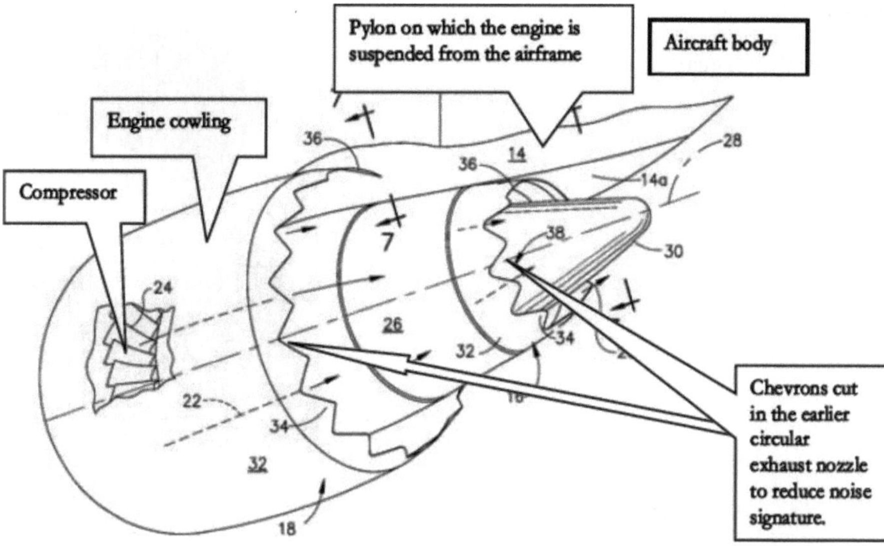

Fig. 4.33 Noise reduction engine design concept

supersonic flight through this noise signature alone, especially as the sonic shock waves are directional.

A stealthy aircraft would usually operate at subsonic speeds. Its supersonic capability, if any, would be reserved for the escape or die/kill situations where survival or the need to secure a victory in combat over an opponent outweighs the need for stealth. Both the B-2 "Spirit" and F-117A "Nighthawk," the two longest in service stealth aircraft, are definitely subsonic. Newer designs such as the F-22, F-35, J-20, J-31, and PAK FA have supersonic capability (this supersonic capability is expected to be coupled with very high agility for success in tactical engagements). High-performance stealth aircraft have the ability to use supersonic flight only when considered necessary, thus retaining the desired level of stealth in the acoustic domain for the mission concerned [9].

Figure 4.33 is a depiction of a jet exhaust designed to reduce noise by cutting chevrons in from the exhaust nozzle.

4. *Visual Signature*: Visual detection was all that was available to detect aircraft in the infancy of the military use of aircraft in World War I and that is why famous German Luftwaffe Ace Manfred von Richthofen (May 2, 1892–April 21, 1918, Fig. 4.34), also known as Red Baron, had his airplane painted in red; thus it could transfer some means of fear to enemy eyes.

   Richthofen painted his aircraft red, and this combined with his title led to him being called the "Red Baron" (Fig. 4.35), both inside and outside of Germany.[1] During his lifetime, he was more frequently described in German as *Der Rote*

[1]https://en.wikipedia.org/wiki/manfred_von_richthofen#cite_note-Kilduff6-1.

Fig. 4.34 Manfred Albrecht Freiherr von Richthofen (Red Baron)

Fig. 4.35 Replica of Richthofen's Fokker Dr. I triplane, at the Berlin Air Show in 2006. (Source: www.wikipedia.com)

Kampfflieger, variously translated as "The Red Battle Flyer" or "The Red Fighter Pilot."

Detection of an aircraft by the human eye still remains important, especially at the close ranges typical of within visual range (WVR) aerial combat and for use of lightweight man-portable air defense systems (MANPADS) such as the "Stinger," RBS-70, Mistral, and SAM-16 "Igla" anti-aircraft missile systems. This is because of the possibility that even a beyond visual range (BVR) air combat may terminate in a WVR engagement. Additionally, in case sensors such as radar and IR fail to pick up an opposing aircraft, it may, at closer ranges, be acquired and engaged visually. Several anti-aircraft weapon systems also use optical tracking.

The main factors in the visual signature of an aircraft are:

- Size and shape.
- Camouflage paint.
- Active camouflage.
- Contrails and exhaust smoke.

Each of the above factors is very self-explanatory due to their names. However, bear in mind that smoke is present in the exhaust due to incomplete or inefficient combustion of fuel in the combustion chamber. Any power plant considered for use in a stealth aircraft would have to be smokeless. This can be achieved by electronic air-fuel mixture control and good combustion chamber designed for efficient combustion [14].

4.6 More About Stealth Technology

The concept behind the stealth technology is very simple. As a matter of fact, it is totally the principle of reflection and absorption that makes aircraft "stealthy."

Deflecting the incoming radar waves into another direction, it thus reduces the number of waves, which returns to the radar. Another concept that is followed is to absorb the incoming radar waves totally and redirect the absorbed electromagnetic energy in another direction. Whatever may be the method used, the level of stealth an aircraft can achieve depends totally on the design and the substance with which it is made of. The idea is for the radar antenna to send out a burst of radio energy, which is then reflected back by any object it happens to encounter. The radar antenna measures the time it takes for the reflection to arrive, and with that information, one can tell how far away the object is.

The metal body of an airplane is very good at reflecting radar signals, and this makes it easy to find and track airplanes with radar equipment. The goal of stealth technology is to make an airplane invisible to radar. There are two different ways to create invisibility:

(a) The airplane can be shaped so that any radar signals it reflects are reflected away from the radar equipment.
(b) The airplane can be covered in materials that absorb radar signals.

Most conventional aircraft have a rounded shape as illustrated in Fig. 4.36. This shape makes them aerodynamic, but it also creates a very efficient radar reflector. The round shape means that no matter where the radar signal hits the plane, some of the signal gets reflected back.

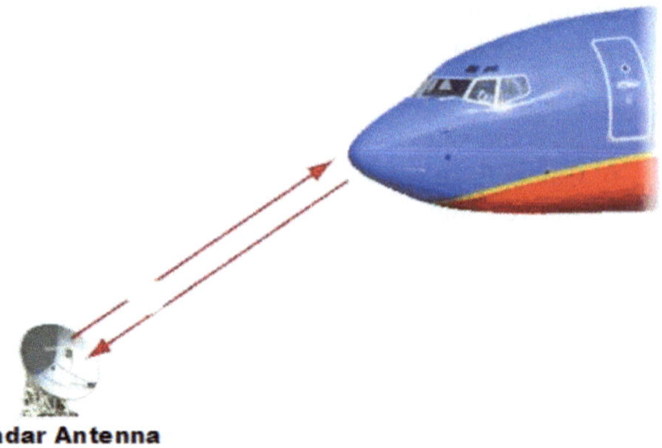

Fig. 4.36 A conventional round-shaped airplane

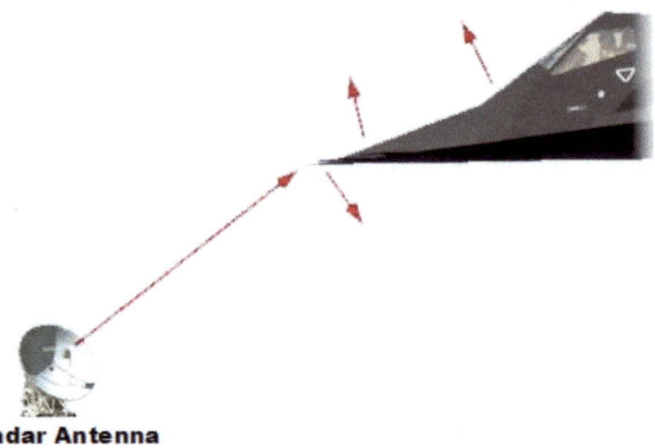

Fig. 4.37 Stealth aircraft shape

A stealth aircraft, on the other hand, is made up of completely flat surfaces and very sharp edges. When a radar signal hits a stealth plane, the signal reflects away at an angle, like the illustration in Fig. 4.37.

In addition, surfaces on a stealth aircraft can be treated so they absorb radar energy as well. The overall result is that a stealth aircraft like an F-117A can have the radar signature or radar cross section (CRS) of a small bird rather than an airplane (see Table 3.3 or Fig. 3.16). The only exception is when the plane banks—there will often be a moment when one of the panels of the plane will perfectly reflect a burst of radar energy back to the antenna.

4.6.1 Further Radar Cross Section (RCS) Discussion

In Chap. 3, Sect. 3.4 of this book, we discussed the nature of radar cross section (RCS) to some details, and here in this section we expand upon it. The reflections from a radar reflective target that are illuminated by radar are not proportional just to the size of the target. The material of which the target is made plays a major role in

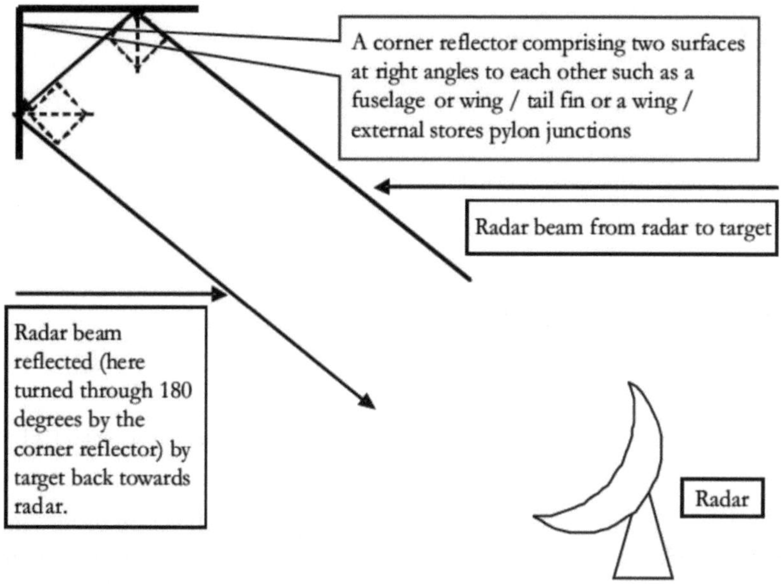

Fig. 4.38 A corner reflector [9]

determining the reflection of radar energy from an object. Metals are very good reflectors, while wood and several plastics are much worse at reflection of radar energy. Hence, a large object made of, say, wood would reflect much less efficiently or reflect less energy than a similar object made of metal. If two targets made of the same material but of different sizes were to be illuminated by the same radar at the same range, it is still possible for the physically smaller target to have a larger signature on the radar due to the influence of shape of the target on radar reflections from it. Hence, the term RCS is used in place of mere physical size as a measure of the radar detectability of a target. RCS equates the returned radar energy from the target to the size of a reflective sphere that would have returned the same amount of energy.

The projected area of this reflective sphere or the area of a disc of the same diameter placed normal to the path of the incident radiation is the "RCS number" itself [15]. A small efficient reflector such as a flat metal plate of area 1.0 m^2, normal to the radar beam, illuminated by a radar operating at 3 GHz, would have an RCS of about 12 m^2. For radar operating at 10 GHz, the RCS of the same plate would have increased to about 150 m^2. The RCS is thus seen to be a function of the physical size and shape of the target and also the frequency, or wavelength (as frequency is equal to *c*/wavelength), of the illuminating radar. The aspect or incident angle of illumination also plays a part in deciding the RCS at that instant. The effect of shape can be clearly understood by examining the issue of corner reflectors.

A typical corner reflector is depicted diagrammatically at Fig. 4.38. A "corner reflector" comprises two or more flat plates at right angles to each other [16]. If EM energy falls upon one plate of the corner reflector such that it is turned through 90°

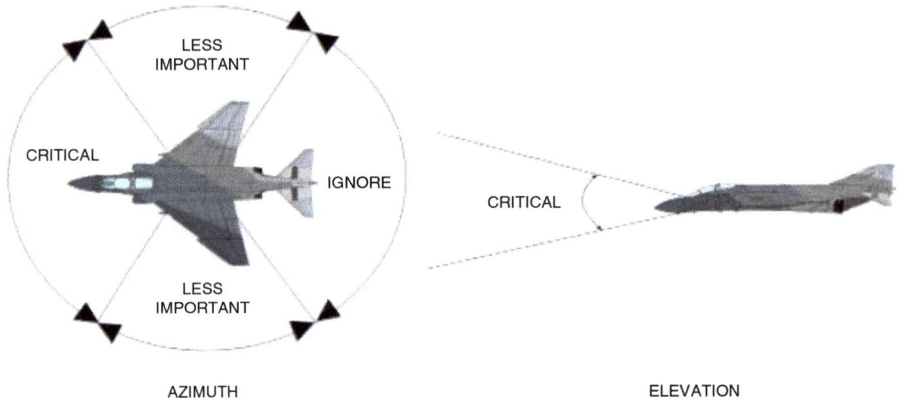

Fig. 4.39 Radio-frequency RCS importance according to the expected threat angles [16]

toward the other plate, it will again go through a change of direction by 90° and thus will be sent back toward its source with full strength. A corner reflector can turn an incident radar beam through 180° and thus can return the full strength of the original energy toward the radar showing a signature equal to a sphere of very large diameter and so have a very large RCS [17].

The angle of interception between the incident radar beam and the target aspect displayed in the direction of approach of the radar beam dictates the presentation of corner reflectors and reflective surface to the beam. The frequency of the radar beam dictates which parts of the aircraft will resonate and thus strengthen the reflected energy [15].

The design parameters of all radars are based on the ability to pick up a target of a specified RCS at a given range. Variation in the RCS of the target would therefore affect a radar's target detection range appreciably [18]. Radar has one more limitation touched upon earlier: it works best against metal than any other material. The design features required for stealth against radar are discussed in the next paragraphs.

The biggest effort in reducing the RCS is given to the forward aspects of the aircraft as illustrated in Fig. 4.39. However, in this case, greater returns for the other aspects or at least some angles are inevitable. This trade-off promises some advantage to countermeasures of stealth such as well-designed bi-static radar networks. Secondly, though shaping is the first principle in reducing RCS and must be carefully considered in the design of low observables, long wavelengths are less affected by the shape of the airframe and its details. Details of shaping are discussed in next section.

The RCS of the airframe can be reduced by geometrically controlling the incoming signals' reflection (directionally) and scattering. The first way to accomplish this is to use flat surfaces and rectilinear surfaces all around the aircraft fuselage, which are oblique to the radar signals. The F-117 Nighthawk, shown in Fig. 4.40, is a very good example of this kind of RCS reduction technique with

Fig. 4.40 Nighthawk's RCS by scattering the incoming signals nearly every direction [19]

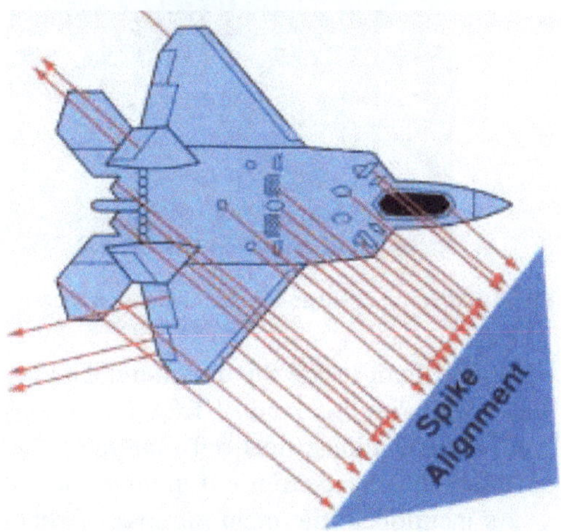

Fig. 4.41 F-22 Raptor's RCS reduction technology by shaping [19]

shaping. F-117 Nighthawk uses careful faceting technique to reduce RCS by scattering the incoming signals in nearly every direction [16].

The second reduction method is similar and involves reflecting the incoming signals in a limited number of directions rather than scattering them in all directions. So, a monostatic receiver never gets the transmitted signal back, unless the radar signal reflects with two 90° angles from a surface, which is improbable when extreme look-down angles are not present. If a bistatic system is considered, its receiver can only get the radiated beam when the spatial geometry is perfect [19].

In this technique, every straight line on the entire airframe should be designed carefully; shape of the aircraft, from main aircraft components such as wings, vertical and horizontal stabilizers, engine inlets, and rudders to all other moving parts such as rudders, elevators, ailerons, weapon bays, landing gear doors, canopy fasteners, etc., should be aligned in the direction of the few selected spikes (to reflect the incoming signal toward only these specific directions), as shown in Fig. 4.41. Using serrated (sawtooth shape shown in Fig. 4.42) parts on surfaces may also help to achieve the desired results [19].

The third method is modeling the aircraft with a compact, smoothly blended external geometry [21] which has changing curves. These curves do not have regular reflection characteristics, and they usually diminish the radar signal's energy by

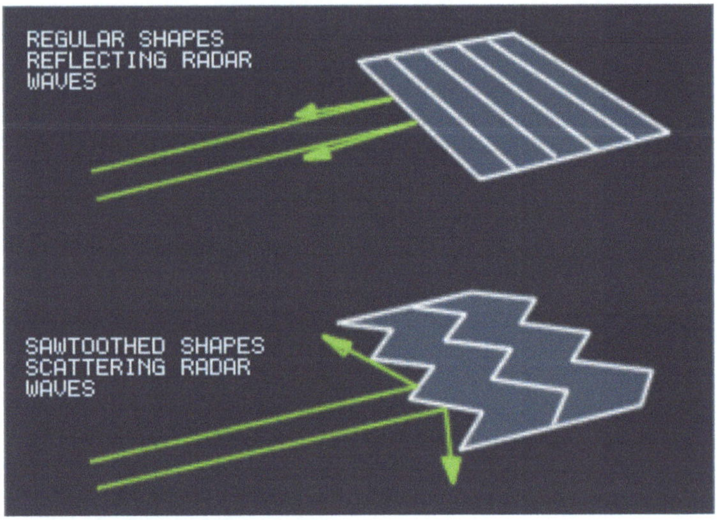

Fig. 4.42 Serrated shape for RCS reduction measures [20]

capturing them inside the curvature. The B-2 Spirit, especially its engine nacelles, was made with this kind of RCS technology. However, this method requires very precise calculations; thus only the latest (after the 1980s) low observable aircraft have had the chance to use it in their computer-based designs.

As mentioned, the main purpose of shaping is reducing or, ideally, eliminating the major RCS contributors. However, shaping measures for low RCS has some trade-offs, such as poor aerodynamic performance, increased costs, more maintenance requirements, or less ordnance capacity. Despite these drawbacks, which will be discussed in the next sections, the gains in RCS reduction compensate for the diminished qualities for the purpose of improving aircraft survivability during operations.

4.6.2 Minimizing Radar Cross Section (RCS)

There are two broad aspects of RCS minimization techniques. One falls under the effort to shape the airframe and covers the geometric design considerations that are taken into account when aiming for a low RCS as illustrated in Fig. 4.43a–c.

Note: The study for a Tactical High Altitude Penetrator (THAP) aircraft was prepared by the USAF's Aeronautical Systems Division and was released in 1980. It is a flying triangle with two buried turbofan engines and a deep layer of RAM (comprising non-conducting skins and foam cores) extending around its entire perimeter. The canted vertical fins provide pitch, roll, and yaw control in cruising flight. It is interesting to note the similarity of the design with the so-called TR-3

Black Manta, a supposedly advanced stealthy tactical/operational reconnaissance aircraft (Source: USAF via Interavia).

The other principle is referred to as "radar-absorbent materials (RAMs)" and is concerned with the materials that help to reduce the reflectivity of the airframe, as well as the structures that will support these materials and integrate them into the airframe (often referred to as "radar-absorbent structures"). These two axes are of course not taken in isolation during the design; trade-offs often have to be made between them.

a

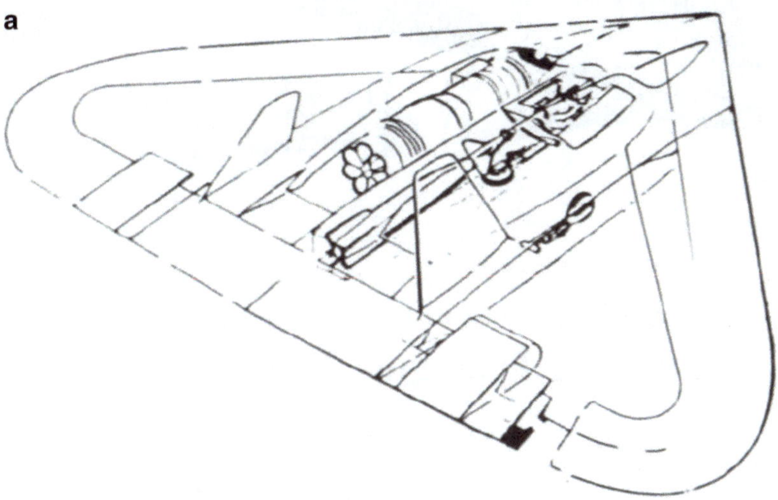

b

Fig. 4.43 (**a**) TR-3 Black Manta conceptual design configuration. (**b**) Possible USAF Top Secret Nuclear Powered Flying Triangle—the TR-3B. (Source: VSkylabs). (**c**) Possible USAF Top Secret Nuclear Powered Flying Triangle—the TR-3b top view [22]. (Source: www.techblog.com)

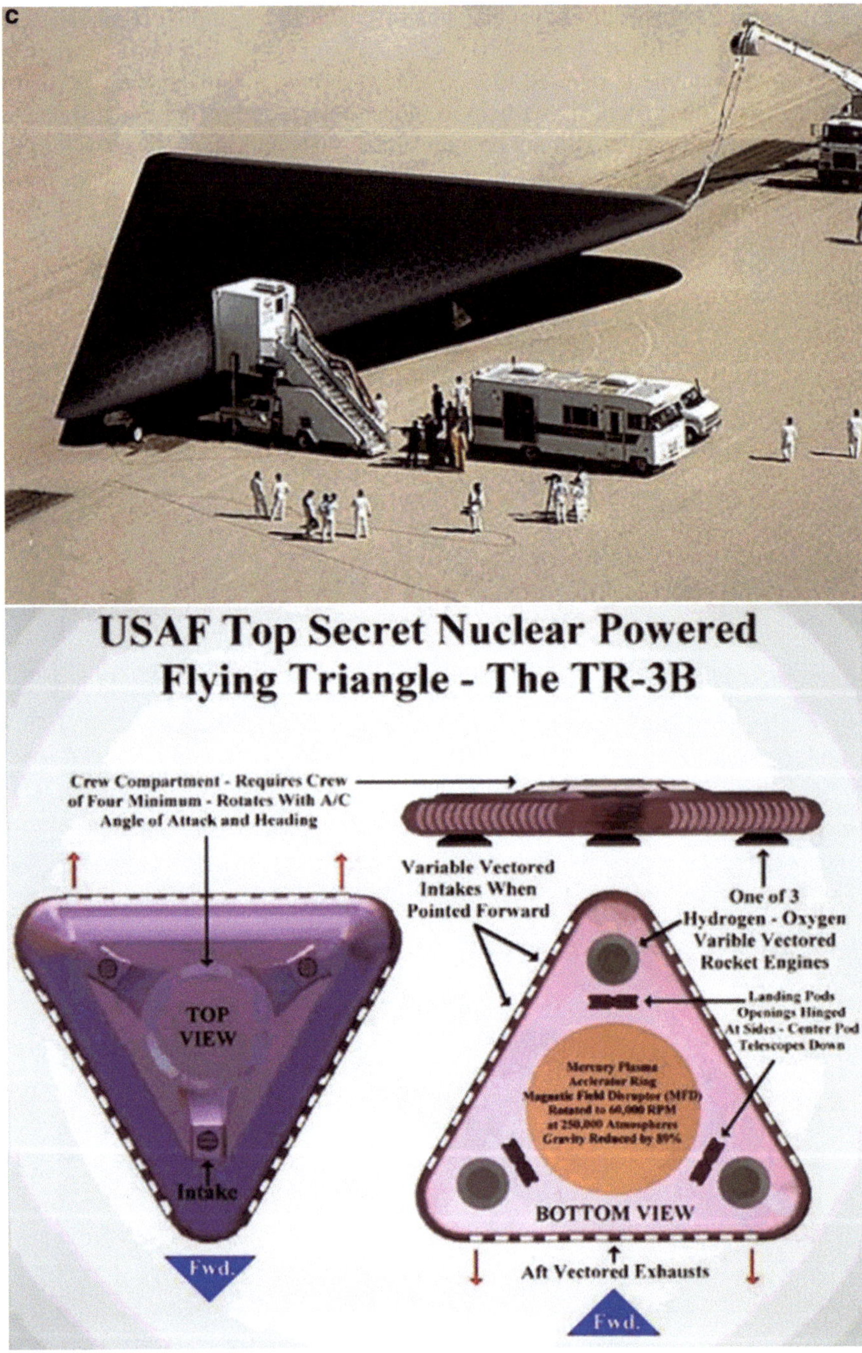

Fig. 4.43 (continued)

4.6.3 *Shape of the Airplane*

As most radar waves impinge on an aircraft at near horizontal angles, vertical surfaces on the aircraft have to be eliminated completely or at least kept to the minimum in order to reduce radar reflections. Furthermore, even high flying aircraft fly at altitudes of about 15–20 km above the surface, while radars have ranges of several hundred kilometers; thus the radar energy incident on aircraft from ground-based radars arrives at quite shallow angles. Thus, vertical surfaces such as vertical stabilizers efficiently reflect this back toward the radar. This should be avoided for reduced RCS [9].

This dictates the elimination or canting inward/outward of the dorsal fin (or vertical stabilizer), elimination of external pylons for weapons carriage with these stores moving to internal weapon bays, and elimination of corner reflectors such as conventional wing-fuselage junctions [23].

The last leads to extensive wing-body blending such as on the F-16 and Rafale so as to reflect the incident radar energy away from the radar. It may be noted, however, that at least in the F-16 when it was initially designed, wing-body blending was undertaken not primarily for stealth but for aerodynamic and structural reasons. Once it was realized that this blending also helped in radar cross section reduction (RCSR), this became a bonus spin-off.

The ultimate in this direction of aircraft shaping is the elimination of distinction between the fuselage and wings giving rise to a flying wing design such as of the Horten Ho-IX referred to earlier and the B-2 "Spirit" Stealth Bomber [24].

The design is not quite so simple. Any uniformly curved surface would act as part of a sphere and reflect energy randomly, some of it toward the radar site we are trying to avoid. Therefore, the curved surfaces on a stealth aircraft have to be such that they form the surface of a sphere of ever changing radius, the radii tailored to reflect the incident energy away from the radar site such a design would require powerful supercomputers to design the spread and magnitude of the ever changing radii to ensure that reflected radar energy is directed as desired). Such a design process could be expected to be and actually is very complex and costly. Stealthy aircraft currently in squadron service include the B-2 "Spirit," F-22 "Raptor," and the F-35 "Lightning II." Each F-22 "Raptor" is claimed to cost, including development and production spending, an enormous $412 million [9, 24].

The USAF fleet of 21 B-2 bombers cost as much as $2.1 billion each [25]. Operating costs are also high for these advanced aircraft. In 2010 the F-22 ad B-2 cost the USAF $55,000 and $135,000 to operate per flying hour, respectively [26].

Another approach to use shaping to reduce RCS is to make the aircraft body of a number of flat plates inclined to reflect energy away from its origin (the radar location), as on the US F-117 stealth fighter, which has actually been used for strike or bombing missions and never in the fighter, or air-to-air, role due to the severe limitations on its maneuverability and other performance parameters required for air-to-air engagements caused by its unique stealth design [16].

Furthermore, to achieve radar cross section reduction (RCSR), four approaches are used. The first one applies shaping features. In a conventional radar configuration, the transmitter and receiver are collocated, so the stealth platform is shaped to reflect the incoming radar signal in a direction other than directly back to the radar. The second approach seeks to absorb, cancel, or scatter the incoming radar transmitter signals so as not to reflect them to the radar receiver(s). This is accomplished by the application of special coatings to the platform's body or using special composites or materials in platform construction. The third technique implements passive cancelation. Cancelation is achieved by adding a skin to the surface of the platform which acts as a secondary scatter means and cancels the reflected field from the primary target [27]. The fourth technique implements active cancelation of incoming radar signals.

Technologies, including the use of platform-mounted active transmitters, are employed that mask and cancel out these signals. One additional approach involves the absorption of radio-frequency (RF) signals using a plasma layer, formed with ionized and conductive gas particles. There are not many applications of this technique; however, some scientists consider it promising for future low observable designs.

Furthermore, as part of studying the shaping factor, the first radar cross section reduction (RCSR) principle, we need to analyze the major RCS contributors of an aircraft that can be very beneficial in gaining a better understanding of the subject. Radar cross section (RCS) so far has been discussed extensively in previous chapters. The complex shape of an ordinary aircraft reveals many surfaces that can reflect incoming signals back to the radar, including air inlets, compressor blades, vertical stabilizers, external payloads, all cockpit instruments, all cavities (discontinuities), and corners. Illustration of Fig. 4.44 shows these contributors. All these contributors must be worked on very precisely to get desired reductions in RCS values.

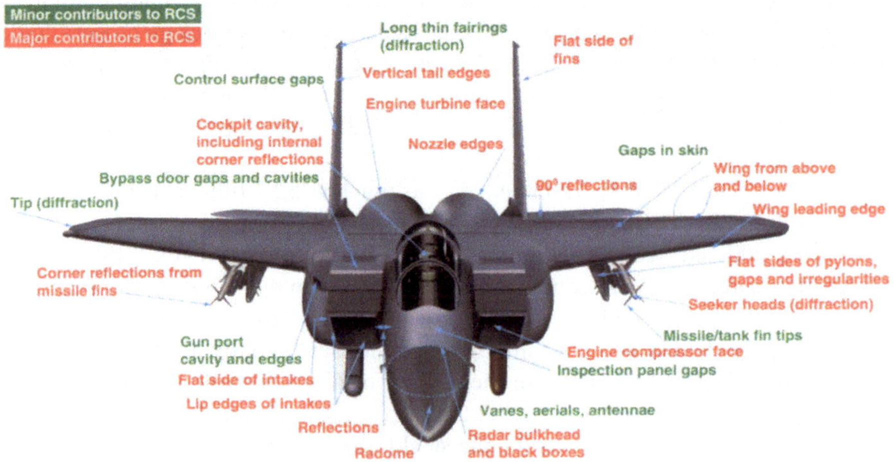

Fig. 4.44 The minor and major contributors to RCS of a fighter aircraft [28]

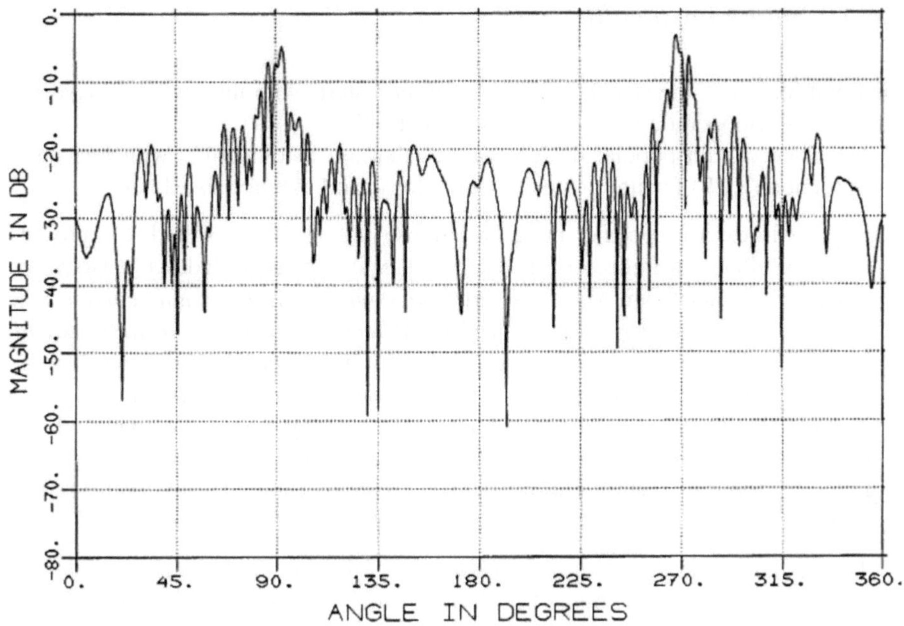

Fig. 4.45 Radar cross section (RCS) pattern

Composite or complex shapes can be even worse. Reflective surfaces at 90° to one another (as, e.g., the tail-mounted horizontal and vertical stabilizers of numerous aircraft) can turn a radar signal through two right angles and fire it back to the receiver in full intensity. Many modern aircraft are full of such reflectors, and the resulting RCS figures are almost staggering. Viewed from the side, a typical fighter, such as the F-15, may have a projected area of 25 m^2. Because of the aircraft's design, however, the broadside RCS may be 16 times as large, at 400 m^2, or the size of a very large house. Typical frontal-aspect RCS figures for modern aircraft run around 3–10 m^2 for fighters and up to 1000 m^2 for a bomber such as the B-52 or a transport aircraft like the Boeing 747 [16].

Other than these contributors, the angle of the incoming radar signals is also very important. This is because, as the normal of a surface to a signal changes, total reflected energy and the RCS also change. For example, an aircraft with a 25 m^2 head on RCS may have a 400 m^2 broadside RCS. Figure 4.45 illustrates a RCS pattern of a target reflecting a radar echo that is of relatively low frequency. The amplitude values for the pattern are on relative basis and don't represent a real aircraft.

The target is located in a plane where 0° represents the nose on position. To understand RCS value variation of an aircraft, in level flight, against radars at the same altitude but at different angles, the target is rotated in the yaw axis. Such patterns are used to analyze the ability of an aircraft to penetrate air defenses [28].

4.6.4 Shape for Stealth

First and foremost, the stealth designer's mission starts with the same words as the physician's Hippocratic Oath: "First, do no harm."

There are certain popular design features that are incompatible with low radar cross section (RCS) feature of stealth technology, and they are listed here:

- Engines in external pods or hung on pylons, such as those of the B-52, provide many excellent retroreflectors. Their first-stage compressor blades are also prime reflectors on their own. *Note*: *It is far from coincidental that many current new non-cooperative target recognition (NCTR) techniques are, to a large extend, based on the processing of strong radar returns from the first-stage engine compressor blades to determine the identity of the illuminated target.*
- Vertical stabilizers and slab-sided bodies (particularly when combined with the unavoidable horizontal wings) are ruled out.
- External stores are a strong no-no, as they create multiple hard-to-control reflections on their own.

The designers can, however, take advantage of the fact that the most threatening radar beams will illuminate his or her aircraft from a point that is much more distant horizontally than vertically. Most radar waves will impinge on the target from a narrow range of shallow angles. If as much as possible of the surface of the aircraft is highly oblique to those angles, the RCS will be low because most of the energy will be scattered. This can be accomplished by blending the airplane's bulky body into the wing.

Obviously, one aspect of being stealthy is aircraft shaping as a useful approach over a wide range of radar frequencies, but over a limited range of aspect angles. The forward cone is of greatest interest and hence, large returns can be shifted out of this sector into the broadside directions.

As we have stated before, engines produce strong radar reflections and have to be concealed in some way while permitting air to reach the engine efficiently. This tends to demand a long, complex inlet system, which takes up a great deal of internal space. The prohibition on external stores puts further pressure on internal volume.

There are a number of basic methods in using geometry to control the way the airframe will reflect and scatter a radar wave. One is to make the shape flat or rectilinear and at the same time oblique to the incoming waves, as already mentioned, so that reflection will never go toward the likely location of a receiver. This is the principle behind the "faceted" F-117A as it is illustrated as an ideal configuration in Fig. 4.40 [16].

Another trick, similar but antipodal to the first one in principle, is to shape the airframe in such a way that, instead of having the reflected energy scatter in all directions and thus a portion of it being always picked up by the enemy radar, it will bounce back on a very limited number of directions, maybe only one or two. This means that an enemy radar will get only one strong reflection (a spike) when the spatial geometry is "just perfect," but virtually no reflection at all in any other instance (see Fig. 4.41) [16].

Unless the radar beam makes two 90° angles to one of the surfaces (which is unlikely, except at extreme look-down angles), the aircraft may remain undetectable. A good example is the frontal wing surface of the B-2. A radar which illuminates the B-2 from anywhere in the front quadrant would produce only two strong "glint" reflections, one from each wing, and these two spikes are impossible to generate concurrently. This method is extensively used in numerous stealthy and semi-stealthy aircraft in order to minimize radar cross section (RCS). It does have the drawback that, in order to make a useful difference, pretty much every straight line on the entire airframe has to be aligned in the direction of the few selected spikes, thus posing extra headaches for the design of everything from landing gear doors to access panels to stabilizers to fasteners, etc. [16].

Another method is to use a compact, smoothly blended external geometry to achieve a continuously varying curvature. Most conventional aircraft have constant-radius curves for simplifying the design and manufacturing processes. However, a constant curve is isotopically scattered: it reflects energy equally in all directions, an effect which has been likened to the rear window of a Volkswagen Beetle car, gleaming in the sun regardless of the incoming angle. A varying curvature is similar to a seashell helix (i.e., Fig. 4.46).

Bear in mind to use a "double S"-shaped air intake duct on the Eurofighter. The vertical offset from the inlet almost achieves 100% line-of-sight blockage to engine compressor face. With RAM lining of the duct, the combination greatly reduces the frontal RCS due to backscattering. The duct is reminiscent of that used on the F-16, but the latter is a simple single S-shape and exposes about 60% of the compressor face. A straight duct has the largest RCS by far (Pete West/AIR International) [16].

The curves have an ever-changing circle radius, as though they are sections of a spiral rather than arcs of a circle, and thus do not reflect energy in the usual

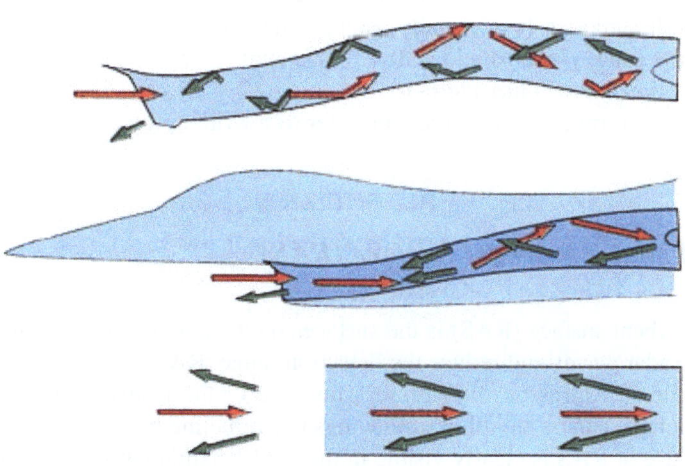

Fig. 4.46 Use of a "double S"-shaped intake [16]

Fig. 4.47 Primary French multirole fourth-generation fighter aircraft Rafale

predictable way. Rather, they tend to absorb the energy as it scatters toward the interior of the curve itself (in a fashion similar to the manner in which hi-fi sound speakers absorb superfluous sound in their internal helix structures). This careful shaping technique can be observed in the overwing engine nacelles of the B-2, as well as the basic fuselage cross section of the French Dassault's multirole fourth−generation fighter jet, namely, Rafale (see Fig. 4.47).

This method, however, requires far greater predictive ability and enormously increased computational capacity over the much simpler faceting. It is thus barely surprising that the F-117, an aircraft almost completely based on faceting, has been operational since the early 1980s, while more complex designs were significantly later in the pipeline [16].

Eliminating the radar reflections of the cockpit also results in a useful RCS reduction. Techniques here usually include the application of several absorbent layers on the canopy/windshield walls. This is applicable both on stealthy airframes and conventional assets like the F-16.

For more information, reader should refer to a write-up by Dranidis [16].

4.6.5 Radar-Absorbent Surface (RAS)

Radar-absorbent surface (RAS) is the surfaces on the aircraft, which can deflect the incoming radar waves and reduce the detection range. RAS works due to the angles at which the structures on the aircraft's fuselage or the fuselage itself are placed. These structures can be anything from wings to a refueling boom on the aircraft. The extensive use of RAS is clearly visible in the F-117 "Nighthawk." Due to the facets (as they are called) on the fuselage, most of the incoming radar waves are reflected to another direction. Due to these facets on the fuselage, the F-117 is a very unstable aircraft.

The concept behind the RAS is that of reflecting a light beam from a torch with a mirror. The angle at which the reflection takes place is also more important. When we consider a mirror being rotated from 0° to 90°, the amount of light that is reflected in the direction of the light beam is more. At 90°, maximum amount of light is reflected back to the same direction to the light beam's source. On the other hand, when the mirror is tilted above 90° and as it proceeds to 180°, the amount of light reflected in the same direction decreases drastically. This makes the aircraft like F-117 stealthy.

4.6.6 Radar-Absorbent (or Absorbing) Materials (RAMs)

Radar-absorbent surfaces absorb the incoming radar waves rather than deflecting it in another direction. Radar-absorbent surface (RAS) totally depends on the material with which the surface of the aircraft is made. Though the composition of this material is a top secret, the F-117 extensively uses radar-absorbent materials (RAMs) to reduce its radar signature or its radar cross section (RCS).

The radar-absorbent surface (RAS) is believed to be silicon-based inorganic compound. This is assumed by the information that the RAM coating on the B-2 is not waterproof. This is just a supposition and may not be true. What we know is that the RAM coating over the B-2 is placed like wrapping a cloth over the plane. When radar sends a beam in the direction of the B-2, the radar waves are absorbed by the plane's surface and are redirected to another direction after it is absorbed. This reduces the radar signature of the aircraft.

The concept behind the radar-absorbent material (RAM) is that of reflecting a light beam from a torch with a mirror. The angle at which the reflection takes place is also more important. When we consider a mirror being rotated from 0° to 90°, the amount of light that is reflected in the direction of the light beam is more. At 90°, maximum amount of light is reflected back to the same direction as the light beam's source. On the other hand, when the mirror is tilted above 90° and as it proceeds to 180°, the amount of light reflected in the same direction decreases drastically.

Today's highly developed technologies include dielectric composites and metal fibers containing ferrite isotopes. Paint comprises of depositing pyramid like colonies on the reflecting superficies with the gaps filled with ferrite-based RAM. The pyramidal structure deflects the incident radar energy in the maze of RAM. Ablative paints, as the name suggests, are paints that do not absorb radiation but conduct it over the skin tending to cool down any electromagnetic (EM) hot spots on the airframe. A commonly used material is known as "iron ball paint." Flight service stations (FSS) are planar periodic structures that behave like filters to electromagnetic energy. The considered frequency-selective surfaces are composed of conducting patch elements pasted on the ferrite layer. FSS are used for filtration

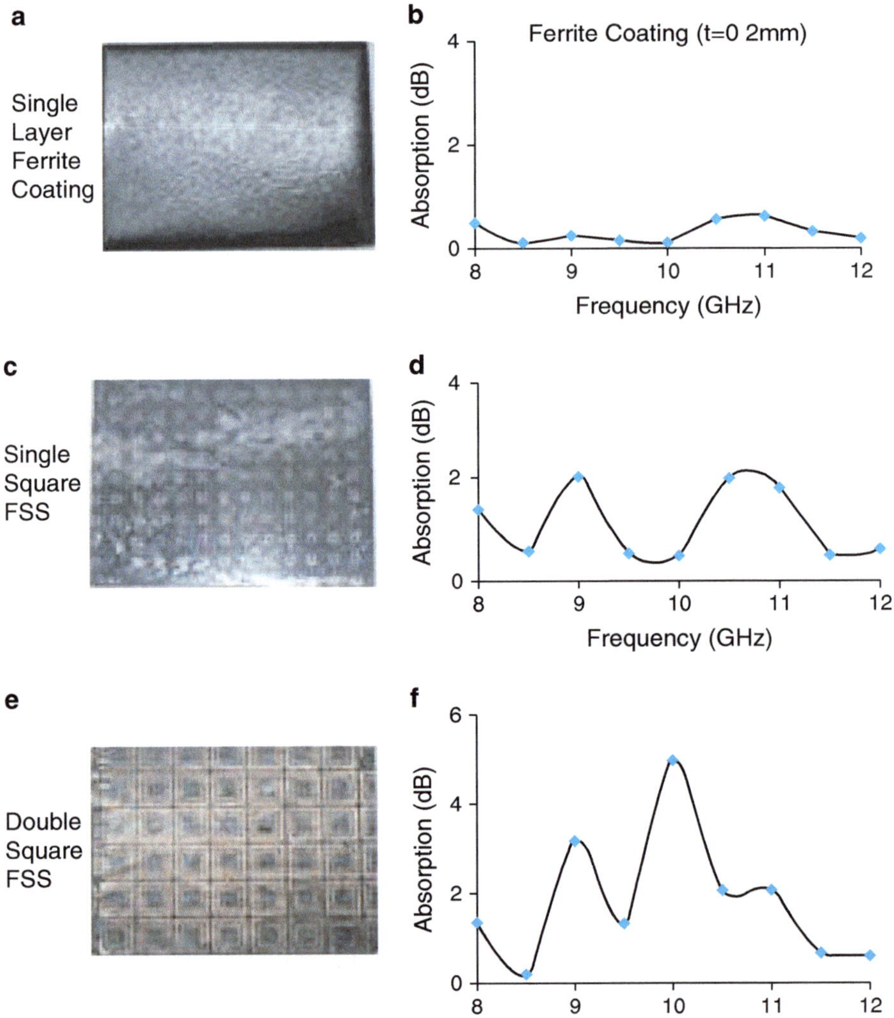

Fig. 4.48 (**a**) Effect of different FSS on microwave absorption on normal electromagnetic incident angle. (**b**) Effect of different FSS on microwave absorption on normal electromagnetic incident angle

and microwave absorption. The available results in Fig. 4.48a, b show that FSS can modify and improve the absorbing performances of RAM.

A flight service station (FSS) is an air traffic facility that provides information and services to aircraft pilots before, during, and after flights, but, unlike air traffic control (ATC), is not responsible for giving instructions or clearances or providing separation [29].

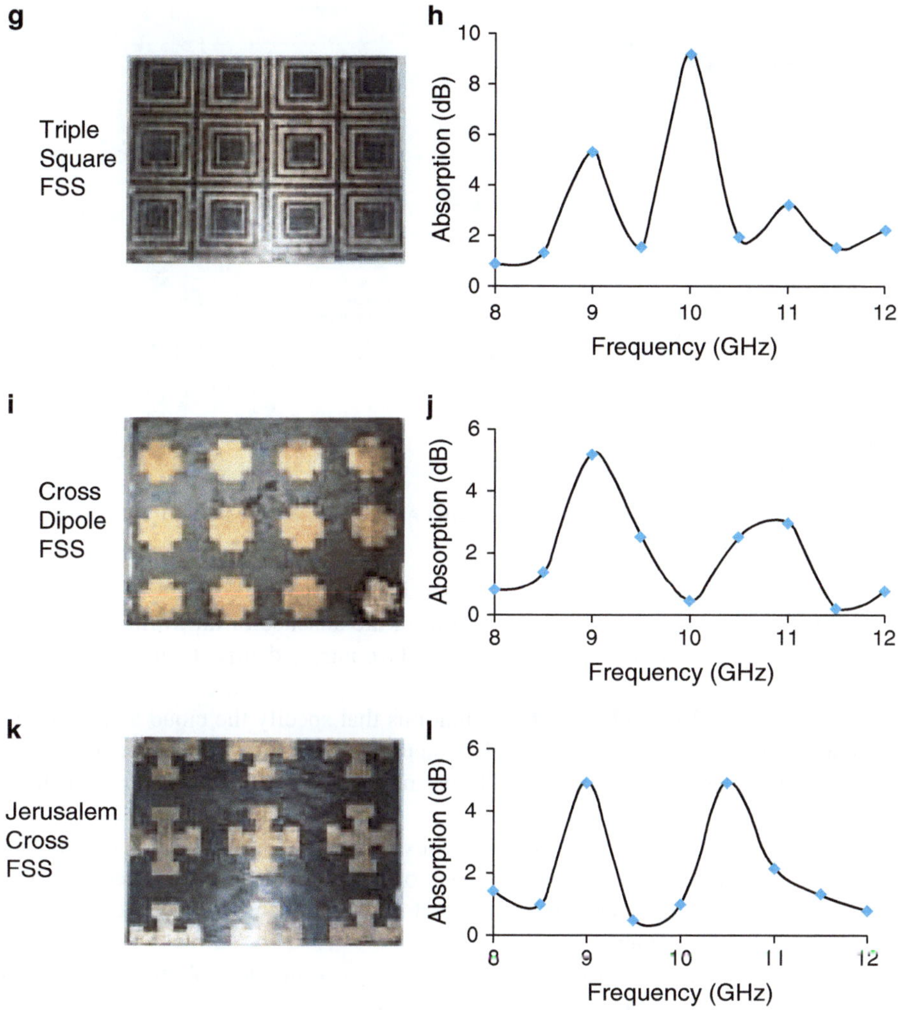

Fig. 4.48 (continued)

The precise services offered by stations vary by country, but typical FSS services may include providing preflight briefings including weather and notices to airmen (NOTAMs); filing, opening, and closing flight plans; monitoring navigational aids (NAVAIDs); collecting and disseminating pilot reports (PIREPs) and airport surface weather observations; offering traffic advisories to aircraft on the ground or in flight; relaying instructions or clearances from air traffic control; relaying information from or about airborne aircraft to their home bases, military bases, or homeland security; providing weather advisories to aircraft in flight; initiating search and rescue on missing visual flight rules (VFR) aircraft (i.e., Fig. 4.49); and providing assistance in

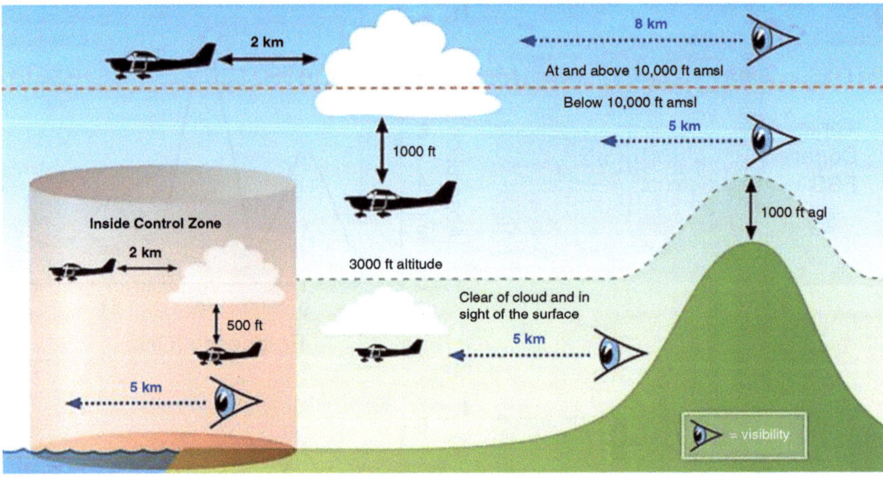

Fig. 4.49 Visual flight rules (VFR). (Source: Civil Aviation Authority, n.d.)

an emergency. In many countries, flight service stations also operate at mandatory frequency airports to help coordinate traffic in the absence of air traffic controllers and may take over a control tower frequency at a controlled airport when the tower is closed [29].

Visual flight rules (VFR) are the regulations that specify the cloud and visibility limitations for aircraft operating with visual reference to terrain. For a pilot to continue flight under VFR, the conditions must be equal to or greater than those specified by the governing body.

The basic premise of VFR is that the pilot will be able to navigate and manipulate the aircraft with reference to external cues only. Pilots are also required to avoid other aircraft using the "see-and-avoid" technique. To achieve this, the following requirements shall be met.

For military aircraft, especially for stealth assets, visual low observable capabilities are essential in deceiving opponents. Camouflage blends the aircraft with its environment. However, because the aircraft environment is susceptible to aspect changes and the relative position of the observer can vary, camouflage should be chosen very carefully. In cases where the observer is below the aircraft, blending the aircraft with the sky background should be considered. However, this is dependent upon altitude, general weather conditions, and time of day. In cases where the observer is above the aircraft, blending the aircraft with the terrain becomes the best approach, as depicted in Fig. 4.50.

The operational task and the type of aircraft under consideration are also very important. Special terrain tones or mixed colors are chosen according to an aircraft's operational area. That kind of color schemes are dependent on local flora and terrain features, like sand, and are applied to the upper sides of aircraft designed for low altitude operations. The lighter blue or gray tones applied to the lower sides of an aircraft are intended to match the sky. These countershading effects reduce the

Fig. 4.50 The Kingdom of Jordan's F-16 with the first advanced visual mitigation method application in the world [20]

visibility from threats located below. Night missions or very high altitude operations require matte and dark colors. Low observable aircraft, such as the F-117 and B-2, usually have black or dark gray hues because they typically operate at night. Moreover, reflections from cockpit glass or other smooth surfaces can be minimized with special coatings.

Visual low observability in daylight is of concern to modern air forces. Earlier attempts, during the World War II and later Vietnam, to decrease daytime visibility of aircraft proved successful during experimental programs. One of the basic principles that affect the ability to see an object is its luminance difference from its background or the amount of light scattered from it. For example, if an aircraft flies at high altitude, the reflected light from its underside increases, while the luminance of the sky decreases. Thus, a black- or dark-toned U-2 spy plane which flies at more than 70,000 ft appears white to an observer below the aircraft. In daylight, because the background of the sky is clear, dark tones can be detected more easily compared to light ones. When this contrast difference is eliminated, it is possible to hinder visual detection until at very close ranges [18].

Part of the aircraft camouflage in order to be factored in terms of stealth augmentation requires a complicated design. The design of camouflage for aircraft is complicated by the fact that the appearance of the aircraft's background varies widely, depending on the location of the observer (above or below) and the nature of the background. Many aircraft camouflage schemes of the past used countershading, where a light color was used underneath and darker colors above.

Other camouflage schemes acknowledge that the aircraft will be twisting and turning while in combat, and the camouflage pattern is applied to the entire aircraft.

Neutral and dull colors are preferred, and two or three shades are selected, depending on the size of the aircraft.

Though air-to-air combat is often initialized outside of visual range, at medium distances, camouflage can make an enemy pilot hesitate until certain of the attitude, distance, and maneuver of the camouflaged aircraft.

The higher speeds of modern aircraft and the reliance on radar and missiles in air combat have reduced the value of visual camouflage while increasing the value of electronic "stealth" measures. Modern paint is designed to absorb electromagnetic radiation used by the radar, reducing the signature of the aircraft, and to limit the emission of infrared light used by heat-seeking missiles to detect their target. Further advances in aircraft camouflage are being investigated in the field of active camouflage.

The purpose of vehicle and equipment camouflage differs from personal camouflage in that the primary threat is aerial reconnaissance. The goal is to disrupt the characteristic shape of the vehicle, reduce shine, and make the vehicle difficult to identify even if it is spotted. See Fig. 4.51a, b.

Methods to accomplish this include paint, nets, ghillie-type synthetic attachments, and natural materials. Paint is the least effective measure but forms a basis for other techniques. Military vehicles often become so dirty that pattern-painted camouflage is not visible. Patterns are designed to make it more difficult to interpret shadows and shapes; matte colors are used to reduce shine, but a wet vehicle can still be very shiny, especially when viewed from above. Nets can be highly effective at defeating visual observation but are useful mostly for stationary vehicles. They also take a lot of time to set up and take down. Nets are occasionally fixed in place around gun tubes or turrets and, if adequately attached, can remain in place while the tank is moving. Nets are far less effective in defeating radar and thermal sensors.

(a): Vehicle Camouflage (b) Visby –Ship Camouflage

Fig. 4.51 Military camouflaged augmentation. (**a**) Vehicle camouflage. (**b**) Visby—ship camouflage

Synthetic attachments, analogous to ghillie-suit attachments, are sometimes used to break up shape. These are prone to loss as armored fighting vehicles (AFVs) move across terrain but can be effective. Natural materials, such as tree branches, bundles of leaves, piles of hay, or small bits of urban wreckage, can be highly effective when the vehicle is in a defensive position.

Furthermore, until the twentieth century, naval weapons had a very short range, so camouflage was unimportant for ships or the men on board them. Paint schemes were selected on the basis of ease of maintenance or aesthetics, typically buff upper works (with polished brass fittings) and white or black hulls. At the turn of the century, the increasing range of naval engagements, as demonstrated by the Battle of Tsushima, prompted the introduction of the first camouflage, in the form of some solid shade of gray overall, in the hope that ships would fade into the mist.

4.6.7 Infrared (IR)

In Sect. 4.5 of this chapter, we did mention the technique of infrared (IR) as part of stealth technology implementation and augmentation as well, and another important factor that influences the stealth capability of an aircraft is the IR signature given out by the plane.

Usually planes are visible in thermal imaging systems because of the high temperature exhaust they give out. This is a great disadvantage to stealth aircraft as missiles also have IR guidance system. The IR signatures of stealth aircraft are minute when compared to the signature of a conventional fighter or any other military aircraft. If reducing the radar signature of an aircraft is tough, then reducing the IR signature of the aircraft is tougher. It will be like flying a plane with no engines. The reduced IR signature totally depends on the engine and where the engine is placed in an aircraft.

Engines for stealth aircraft are specifically built to have a very low IR signature. The technology behind this is top secret like others in stealth aircraft. Another main aspect that reduces the IR signature of a stealth aircraft is to place the engines deep into the fuselage. This is done in stealth aircraft like the B-2, F-22, and the Joint Service Fighter (JSF). The IR reduction scheme used in F-117 is very much different from the others. The engines are placed deep within the aircraft like any stealth aircraft, and at the outlet, a section of the fuselage deflects the exhaust to another direction.

This is useful for deflecting the hot exhaust gases in another direction, thus, to defeat any heating-seeking missile to go after the aircraft to bring them down. Moreover, the IR topcoat reduces the IR signature, along with ensuring the radar and infrared signatures are balanced. Early low observable programs made extensive use of RAM and RAS, which resulted in weight and manufacturing problems.

4.6.8 Infrared Signature and Infrared Stealth

All substances with a temperature above absolute zero (0 K, or −273.15 °C, or −459.67 °F) emit electromagnetic waves. The heat content of a material produces molecular vibrations which cause electron oscillations. These oscillations provide electromagnetic coupling that produces an emission of energy. This emission is called infrared radiation (IR). IR has a wavelength spectrum of 0.7–14 μm, and the amount of radiation emitted is primarily dependent on the physical temperature of the associated object (proportionally). The emissivity characteristics of an object are related to the material's molecular structure and the surface conditions of the object. IR energy that comes from another body is either absorbed or reradiated by the object according to its emissivity properties [22].

As with visible light, IR energy also travels in a straight line at speed of light. Similarly, IR energy is either reflected or absorbed and converted to heat when it hits the surface of an object. These absorption and reflection qualities change with material specifications. For example, polished surfaces reflect more IR energy but also have a much lower emissivity than matte surfaces [30, 31].

IR energy considerations are important to stealth designers, because IR detectors, also known as infrared homing devices, such as passive missile guidance systems, can use IR emissions from a target to track it. Detector systems, especially missile-guiding seekers, which detect the radiated infrared signals of their target, are often referred to as "heat-seekers." If unaided by IR countermeasures, aircraft are vulnerable to detection by such systems by means of the strongly radiated energy from their hot bodies. Some precautions to mitigate such detection include reducing or suppressing an aircraft's IR signature and adding some noise, deploying decoys or flares, and jamming the sensor by emitting high-power signals toward the detector.

For an asset designed to remain undetected, one of the most important measures is reducing or suppressing the aircraft's IR emissions. Thus, sources, surfaces, or components which produce and/or conserve heat are of great concern to low observables. Moreover, the IR detection capability of the new IR search and track (IRST) systems, such as shown in Fig. 4.52, and electro-optic (EO) systems deployed on the SU-27, Eurofighter Typhoon, and F-35 Lightning II reveal the importance of IR signature reduction.

These EO detectors absorb electromagnetic radiation and output an electrical signal that is useful for tracking and targeting their target. Another major advantage of these systems is that they are passive systems in which a target never knows that there is a threat trying to detect it. Further consideration for IR detection is revealed by the efforts required to increase combat effectiveness of stealth aircraft. When radar detection range is minimized by RCS reduction methods, other signatures such as IR, visual and acoustic, become more pronounced, especially for close-range engagements.

IR signal reduction is focused on engine exhausts. The back side of an engine is the major source of IR radiation in an aircraft, and when the afterburner is applied, the heat increases significantly, by nearly 50 times, since IR energy emitted from the

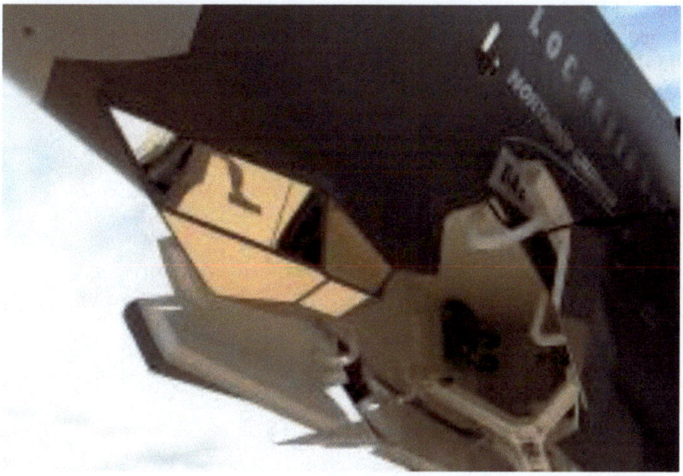

Fig. 4.52 IRST sensor of the F-35 Lightning II [30]

engines is proportional to the fourth power of absolute temperature [32]. Thus, the second-generation stealth F-117 Nighthawk and the third-generation strategic stealth bomber B-2 Spirit have non-afterburning engines. On the other hand, the fourth-generation stealth F-22 Raptor has the ability to cruise at supersonic speeds, but without afterburner. Being dependent to high Mach numbers for operation survivability, the first-generation stealth SR-71 Blackbird is also an exception, with its high-power afterburner engines [30].

One method to decrease the IR signature of the engines is to use exhaust masking. This is accomplished by placing the engines on top of the body and the wings. This is the reason the F-117A and B-2 exhausts cannot be seen from below, which is shown in Figs. 4.53 and 4.54, respectively. Over the rear conical sector of the aircraft, the hottest parts of the tailpipe can be easily detected by IR seekers. While outside of this sector, sensors can only detect the hot parts of the nozzle surface.

Another technique to decrease the IR signature is using the aircraft's aft fuselage and vertical surfaces to shield the jet pipes from view over as large a part of this rear sector as possible [32].

Another method to decrease IR signature is the shaping of exhaust geometry. Exhausts that are shaped flat and wide, as shown in Fig. 4.55, are effective in this regard. This increases the perimeter of the plum compared to conventional round nozzles and results in an increased mixing rate of exhaust gases, cooling them with air. This reduces probability of detection, but thrust efficiency is decreased with flat and wide designs.

High-bypass engines also benefit from the mixing of air with exhaust for exhaust nozzle temperature reduction purposes. Masking the hot turbine stages with curved jet pipes and concealing the forward emissions of the engine with curved air intakes are other measures to reduce IR signature.

Fig. 4.53 The body of the F-117 is designed to mask IR emission [33]

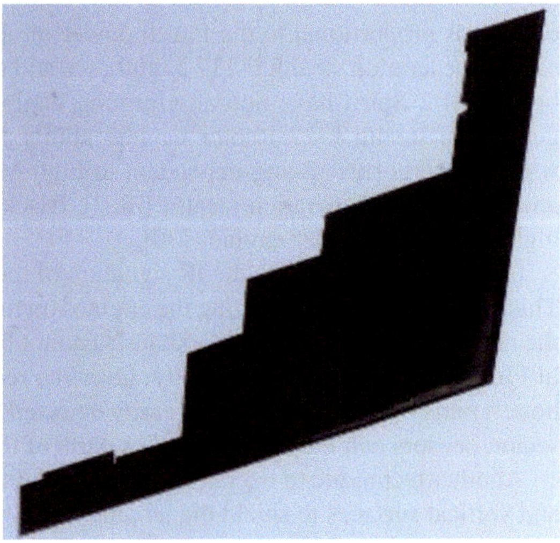

Fig. 4.54 The engine nozzles of B-2 are concealed to be seen from below [34]

After engine heat, kinetic heating of the aircraft body is the second major source of IR radiation. Some closed-loop cooling systems and special materials, such as IR signal-absorbent material, can be used to dissipate the heat from the body as well as the engine and exhaust parts. However, this method has some disadvantages, such as increased weight and special maintenance requirements, similar to RCS reduction-oriented RAMs. Dumping the heat into the fuel is another technique to reduce kinetic heating and was first used in the SR-71 Blackbird. However, at high Mach numbers, the high temperature from kinetic heating is inevitable. In general, limiting aircraft to relatively low speeds is required to minimize this source of IR radiation [30].

Fig. 4.55 F-22 Raptor's sawtooth-, wide-, and flat-shaped nozzles to reduce both radar and IR signatures [35]

4.6.9 Plasma Stealth

The principle of plasma stealth is to generate an ionized "layer" surrounding the aircraft to reduce radar cross section (RCS). It is a quasi-active system in which dangerous radar signatures are received and absorbed/scattered by plasma capable of absorbing/spreading a wide range of radar frequencies, angles, polarizations, and power densities. The use of plasmas to control the reflected electromagnetic radiation from an object (plasma stealth) is feasible *stealth technology and counter-stealth radars*, at higher frequency where the conductivity of the plasma allows it to interact strongly with the incoming radio wave, but the wave can be absorbed and converted into thermal energy rather than reflected.

Plasma stealth technology is what can be called as "active stealth technology" in scientific terms. This technology was first developed by the Russians. It is a milestone in the field of stealth technology. The technology behind this was not at all new. The plasma thrust technology was used in the Soviet/Russian space program. Later the same engine was used to power the American Deep Space 1 probe. See image of this probe as illustrated in Fig. 4.56.

Note: Deep Space 1, originally designed to test a dozen new technologies including the use of an ion engine for spacecraft propulsion, far outstripped its primary mission goals by also successfully flying by the asteroid 9969 Braille and comet Borrelly. The flybys produced what are still considered some of the best images and data ever collected from an upclose encounter with an asteroid or comet.

The success of Deep Space 1 set the stage for future ion-propelled spacecraft missions, especially those making the technically difficult journey to asteroids or comets, such as NASA's Dawn mission.

Fig. 4.56 Artistic image of American Deep Space 1 probe. (Source: National Aeronautics and Space Administration (NASA))

Fig. 4.57 Mikoyan MiG-35 image in full details. (Source: www.wikipedia.com)

In plasma stealth, the aircraft injects a stream of plasma in front of the aircraft. The plasma will cover the entire body of the fighter and will absorb most of the electromagnetic energy of the radar waves, thus making the aircraft difficult to detect. The same method is used in magnetohydrodynamics (MHD).

Using magnetohydrodynamics (see Appendix C as well), an aircraft can propel itself to great speeds. Plasma stealth will be incorporated in the MiG-35 "Super Fulcrum/Raptor Killer" as illustrated in Fig. 4.57.

Fig. 4.58 Su-57 inflight left banking position

This is a fighter which is an advanced derivative of the MiG-29. Initial trials have been conducted on this technology, but most of the results have proved to be productive.

Moreover, Russia claims its Su-57 (Fig. 4.58) fighter is even stealthier than the previous versions of Sukhoi aircraft that were manufactured. This is possible due to an indium-tin-oxide canopy that is slightly stealthier.

Although that barely matters when Russia might operate just one squadron of the planes due to cost of each plane, Russian firm Rostec has built new canopies for Su-57 fighters, Tu-160 bombers, and other warplanes, state-run TASS News Agency reported on January 11, 2018. The canopies include "a new composite material with enhanced radar-wave absorbing properties," TASS explained.

With or without its indium-tin-oxide canopy treatment, the Su-57 is too expensive for Russia to buy in meaningful numbers. The Russian Air Force acquired just 10 of the stealth fighters in the 8 years following the type's first flight in 2010 [27].

Plasma stealth can be considered as a specific stealth method employed for Ariel stealth. Couple of things to keep in mind: plasma is ionized gas particles. Therefore, plasma flow is a flow of ionized gas particles. Ion is an electrically charged particle or group of atoms. Plasma cloud is a quasi-neutral (total electrical charge is zero) collection of free charged particles. The vast majority of matter in the universe exists in plasma state. Near the Earth plasma can be found in the form of solar wind, magnetosphere, and ionosphere.

The main property of plasma (for our purposes) is its frequency, which is equal to a square root of a ratio of 4 $*$ Pi $*$ square of ion charge $*$ concentration of ions to the mass of ion and mathematically written as following equation:

$$\omega_{\mathrm{p}} = \sqrt{\frac{4\pi ne^2}{m}} \tag{4.1}$$

where e is electron or ion charge, r is concentration of ions per volume of plasma, and m is mass of ion.

There are several types of oscillations in plasma: low frequency (ion-sound waves), high frequency (oscillations of electrons relative to ions), spiral waves (in the presence of a magnetic field—"magnetosound"), and cross waves propagating along a magnetic field. A device for generating plasma is called plasmatron. This device generates the so-called low-temperature plasma.

This is truly unbelievable, but even this theoretically and technologically is perfectly possible. It is not known whether the plasma stealth system developed by the Russians employs a plasma laser or some other method for creating a plasma field.

However, our opinion is that it has nothing to do with a plasma laser (which is a very large and very power-hungry device).

Plasma physics was given priority in Russia many years ago, which resulted in a number of breakthroughs in theory as well as practical applications of plasma. Perhaps one of the most interesting and promising applications of plasma is the so-called ion thruster, used to propel spacecraft. This technology was first developed in Russia (mainly by Keldysh Research Center) and recently successfully used on an American satellite.

Plasma layers around aircraft have been considered for purposes other than stealth. There are many research papers on the use of plasma to reduce aerodynamic drag. In particular, electrohydrodynamic coupling can be used to accelerate air flow near an aerodynamic surface. One paper considers the use of a plasma panel for boundary layer control on a wing in a low-speed wind tunnel.

This demonstrates that it is possible to produce plasma on the skin of an aircraft. Xenon nuclear poison isotopes when successfully suspended in generated plasma layers or doped into vehicle hulls may be utilized in order for a reduction in radar cross section. If tunable this could shield against high-powered microwave (HPM)/ electromagnetic pulse (EMP) and high-energy radio-frequency (HERF) weaponry or act as optical radiation pressure actuators.

As we stated, plasma is a partially ionized and electrically conductive gas by means of the ability of the positive and negative charges to move somewhat independently [36]. Its free electrons make plasma respond strongly to electromagnetic fields. Thus, using plasma, which is sometimes considered an active cancelation technique, has been studied and proposed as a possible method of radar cross section reduction (RCSR). The inspiration for this method emerged in the late 1950s after spacecraft with a natural plasma layer over their airframes experienced communication interruption incidents while traveling through the ionosphere. Basically, radar waves (actually all electromagnetic waves of certain frequencies) traveling through this conductive plasma cause electrons to exchange their places, ending up with the electromagnetic waves losing their energy and transforming it to other forms, such as heat. Interaction between plasma and electromagnetic radiation is strongly dependent on the physical properties and parameters of the plasma [37]. The most dominating of these properties are the temperature and the density of the plasma.

Another important issue is frequency of the incident radar beam. Radar waves, below a specific frequency, are reflected by plasma layer. Plasma layer's physical properties have significant effect on this process. Long-distance communications with high-frequency (HF) signals by means of ionosphere scattering and reflection are a good example of these same phenomena.

Thus, RCS reduction plasma devices should also control and dynamically adjust the plasma properties, such as density, temperature, and composition, for effective radar absorption results.

Plasma stealth technology has some drawbacks from a low observable perspective. Some of these include emitting own electromagnetic radiation with a visible glow, existence of a plasma trail of ionized air behind the aircraft [38] before dissipation by the atmosphere, and difficulty in producing a radar-absorbent plasma around an entire aircraft traveling at high speed [38]. However, some Russian scientists have declared achieving a hundredfold RCS reduction with plasma technology, and this result, if real, is sufficient enough to focus on this method for further research and success in the stealth world [37].

Another application of plasma is utilizing this technology to deploy antenna surfaces to generate low observability characteristics. While metal antenna poles are reflective parts, a hollow glass tube filled with low pressure plasma can provide an entirely radar transparent surface when not in use [37].

Although there are some problems in the operational processes associated with plasma, such as the high-energy requirement in long-interval applications and the necessity of holes in the plasma fields for aircraft onboard radar activation, Russian plasma stealth research teams have announced the development of a plasma generator which weighs 100 kg and is thus feasible for a tactical air platform. This critical technology may be available on the Su-27 versions (such as Su-34 and Su-35), MIG-35 fighters, and also the MIG 1.44 prototype as illustrated in Fig. 4.59, see Fig. 4.45, according to recent claims by Russian officials [16, 37].

In conclusion, plasma stealth is a proposed process to use ionized gas (plasma) to reduce the radar cross section (RCS) of an aircraft. Interactions between electromagnetic radiation and ionized gas have been extensively studied for many purposes, including concealing aircraft from radar as stealth technology. Various methods might plausibly be able to form a layer or cloud of plasma around a vehicle to deflect or absorb radar, from simpler electrostatic or radio-frequency (RF) discharges to more complex laser discharges. It is theoretically possible to reduce RCS in this way, but it may be very difficult to do so in practice. Some Russian systems,

Fig. 4.59 MIG 1.44 with possible plasma stealth capabilities

Fig. 4.60 3M22 Zircon/SS-N-33 hypersonic maneuverable anti-ship missile. (Courtesy of Katehon.com)

e.g., the 3M22 Zircon (SS-N-33) (Fig. 4.60) missile that is also known as the Russian hypersonic missile, have been reported to make use of plasma stealth.

Note: The news that Russia's Zircon missile has attained unprecedented speed provoked alarming headlines in the western media. RBTH summarizes what is known about one of Russia's most classified military programs.

This month, Russia's Zircon reached the highest speed for any cruise missile in history. Citing sources close to the military, TASS news said the missile, during trials, was able to fly at a speed that is eight times faster than speed of sound—the so-called Mach 8—which is approximately equal to 9800 km/h.

With a possible range of around 400 km, it will be able to cover the entire distance in just 2.5 min. Western media outlets fear that this will render much of NATO's naval equipment obsolete.

In summary, as we stated in above, plasma layers around aircraft have been considered for purposes other than stealth. There are many research papers on the use of plasma to reduce aerodynamic drag. In particular, electrohydrodynamic (EHD) coupling can be used to accelerate airflow near an aerodynamic surface. One paper [21] considers the use of a plasma panel for boundary layer control on a wing in a low-speed wind tunnel. This demonstrates that it is possible to produce plasma on the skin of an aircraft. Xenon nuclear poison isotopes when successfully suspended in the plasma layers or vehicle hull can be utilized to reduce radar cross section and will shield against HMP/EMP and HERF weaponry.

Note: Electrohydrodynamic (EHD), also known as electro-fluid-dynamics (EFD) or electrokinetics, is the study of the dynamics of electrically charged fluids [39]. It is the study of the motions of ionized particles or molecules and their interactions with electric fields and the surrounding fluid. The term may be considered to be synonymous with the rather elaborate electrostrictive hydrodynamics. ESHD covers the following types of particle and fluid transport mechanisms: electrophoresis, electrokinesis, dielectrophoretic, electro-osmosis, and electrorotation. In general, the phenomena relate to the direct conversion of electrical energy into kinetic energy and vice versa [39].

Furthermore, plasma is a *quasineutral* (total electrical charge is close to zero) mix of ions (atoms which have been ionized, and therefore possess a net charge), electrons, and neutral particles (possibly including un-ionized atoms). Not all plasmas are fully ionized. Almost all the matter in the universe is plasma: solids, liquids, and gases are uncommon away from planetary bodies. Plasmas have many technological applications, from fluorescent lighting to plasma processing for semiconductor manufacture.

Plasmas can interact strongly with electromagnetic radiation: this is why plasmas might plausibly be used to modify an object's radar signature. Interaction between plasma and electromagnetic radiation is strongly dependent on the physical properties and parameters of the plasma, most notably, the temperature and density of the plasma. Plasmas can have a wide range of values in both temperature and density; plasma temperatures range from close to absolute zero and to well beyond 10^9 K (for comparison, tungsten melts at 3700 K), and plasma may contain less than one particle per cubic meter or be denser than lead. For a wide range of parameters and frequencies, plasma is electrically conductive, and its response to low-frequency electromagnetic waves is similar to that of a metal: plasma simply reflects incident low-frequency radiation. The use of plasmas to control the reflected electromagnetic radiation from an object (plasma stealth) is feasible at higher frequency where the conductivity of the plasma allows it to interact strongly with the incoming radio wave, but the wave can be absorbed and converted into thermal energy rather than reflected.

Plasmas support a wide range of waves, but for unmagnetized plasmas, the most relevant are the Langmuir waves, corresponding to a dynamic compression of the electrons. For magnetized plasmas, many different wave modes can be excited which might interact with radiation at radar frequencies.

4.6.9.1 Absorption of Electromagnetic Radiation by Plasma Stealth

When electromagnetic waves (EMWs), such as radar signals, propagate into a conductive plasma, ions and electrons are displaced as a result of the time-varying electric and magnetic fields. The wave field gives energy to the particles. The particles generally return some fraction of the energy they have gained to the wave, but some energy may be permanently absorbed as heat by processes like scattering or resonant acceleration or transferred into other wave types by mode conversion or nonlinear effects. Plasma can, at least in principle, absorb all the energy in an incoming wave, and this is the key to plasma stealth. However, plasma stealth implies a substantial reduction of an aircraft's RCS, making it more difficult (but not necessarily impossible) to detect. The mere fact of detection of an aircraft by a radar does not guarantee an accurate targeting solution needed to intercept the aircraft or to engage it with missiles. A reduction in RCS also results in a proportional reduction in detection range, allowing an aircraft to get closer to the radar before being detected.

The central issue here is frequency of the incoming signal. A plasma will simply reflect radio waves below a certain frequency (characteristic electron plasma frequency). This is the basic principle of shortwave radios and long-range communications, because low-frequency radio signals bounce between the Earth and the ionosphere and may therefore travel long distances. Early-warning over-the-horizon radars utilize such low-frequency radio waves (typically lower than 50 MHz). Most military airborne and air defense radars, however, operate in VHF, UHF, and microwave band, which have frequencies higher than the characteristic plasma frequency of ionosphere; therefore microwave can penetrate the ionosphere, and communication between the ground and communication satellites is possible. (*Some* frequencies can penetrate the ionosphere.)

Plasma surrounding an aircraft might be able to absorb incoming radiation and therefore reduces signal reflection from the metal parts of the aircraft: the aircraft would then be effectively invisible to radar at long range due to weak signals received. A plasma might also be used to modify the reflected waves to confuse the opponent's radar system: for example, frequency shifting the reflected radiation would frustrate Doppler filtering and might make the reflected radiation more difficult to distinguish from noise.

Control of plasma properties like density and temperature is important for a functioning plasma stealth device, and it may be necessary to dynamically adjust the plasma density, temperature, or combinations, or the magnetic field, in order to effectively defeat different types of radar systems. The great advantage plasma stealth possesses over traditional radio-frequency stealth techniques like shape morphing into low observing (LO) geometry and use of radar-absorbent materials (RAM) is that plasma is tunable and wideband. When faced with frequency hopping radar, it is possible, at least in principle, to change the plasma temperature and density to deal with the situation. The greatest challenge is to generate a large area or volume of plasma with good energy efficiency.

Plasma stealth technology also faces various technical problems. For example, the plasma itself emits EM radiation, although it is usually weak and noise-like in spectrum. Also, it takes some time for plasma to be re-absorbed by the atmosphere, and a trail of ionized air would be created behind the moving aircraft, but at present there is no method to detect this kind of plasma trail at long distance. Thirdly, plasmas (like glow discharges or fluorescent lights) tend to emit a visible glow: this is not compatible with overall low observability concept. However, present optical detection devices like forward-looking infrared (FLIR) has a shorter range than radar, so plasma stealth still has an operational range space. Last but not least, it is extremely difficult to produce a radar-absorbent plasma around an entire aircraft traveling at high speed; the electrical power needed is tremendous. However, a substantial reduction of an aircraft's RCS maybe still be achieved by generating radar-absorbent plasma around the most reflective surfaces of the aircraft, such as the turbojet engine fan blades, engine air intakes, vertical stabilizers, and airborne radar antenna.

There have been several computational studies on plasma-based radar cross section reduction technique using three-dimensional finite-difference time-domain simulations. Chaudhury et al. studied the electromagnetic wave attenuation of an Epstein profile plasma using this method. Chung studied the radar cross change of a metal cone when it is covered with plasma, a phenomenon that occurs during reentry into the atmosphere [40]. Chung simulated the radar cross section of a generic satellite and also the radar cross section when it is covered with artificially generated plasma cones.

4.7 Advantages of Stealth Technology

The benefits of stealth apply not only to platforms but to a lot of weapons as well. Anti-surface munitions like the JSOW, JASSM, Apache/SCALP/Storm Shadow, Taurus/KEPD, and many others are specifically shaped and treated to minimize their radar and IR signatures. This has two useful payoffs: on the one hand, the weapon itself becomes less vulnerable to enemy defensive systems, which means that fewer of the weapons launched will be shot down before reaching their target(s). This in turn means that fewer weapons and their parent platforms need to be allocated to any given mission, and finally the end result is that a greater number of targets can be confidently engaged with a given force. The other benefit is the advantage of surprise and its effect in cases where shrinking the enemy's available reaction time is of the essence.

A good example of such a situation is a typical OCA strike against an airfield. If non-stealthy strike aircraft or standoff weapons are used, it is quite likely that they will be detected far enough out that the enemy will have some time available (even just 4–5 min will do) to get many of his ready-to-fly aircraft in the air and fly them somewhere else to preserve them. If the aircraft being flushed include armed hot-pad alert fighters (a common protective measure), these can immediately and actively contribute to the basic defense against the incoming attack. Contrast this with a situation where, as a result of using stealthy weapons and/or platforms, the base is caught virtually napping, and the attack is detected so perilously close that the enemy has no time to get anything in the air but instead can only rely on his ground-based terminal defenses. This can mean the difference between the base suffering little or no damage and being virtually obliterated.

1. A smaller number of stealth vehicles may replace fleet of conventional attacks vehicles with the same or increased combat efficiency. Possibly resulting in longer-term savings in the military budget.
2. A stealth vehicle strike capability may deter potential enemies from taking action and keep them in constant fear of strikes, since they can never know if the attack vehicles are already underway.
3. The production of a stealth combat vehicles design may force an opponent to pursue the same aim, possibly resulting in significant weakening of the economically inferior party.
4. Stationing stealth vehicles in a friendly country is a powerful diplomatic gesture as stealth vehicles incorporate high technology and military secrets.
5. Decreasing causality rates of the pilots and crew members.
6. Weapons and/or platforms, the base is caught virtually napping, and the attack is detected so perilously close that the enemy has no time to get anything in the air but instead can only rely on his ground-based terminal defenses. This can mean the difference between the base suffering little or no damage and being virtually obliterated.

Moreover, as a benefit of stealth technology, a smaller number of stealth aircraft may replace fleet of conventional attack jets with the same or increased combat efficiency, possibly resulting in longer-term savings in the military budget.

A stealth aircraft strike capability may deter potential enemies from taking action and keep them in constant fear of strikes, since they can never know if the attack planes are already underway. The production of a stealth combat aircraft design may force an opponent to pursue the same aim, possibly resulting in significant weakening of the economically inferior party. Stationing stealth aircraft in a friendly country is a powerful diplomatic gesture as stealth planes incorporate high technology and military secrets.

The goal of stealth technology is to make an airplane invisible to radar. There are two different ways to create invisibility. The airplane can be shaped so that any radar signals it reflects are reflected away from the radar equipment. The airplane can be covered in materials that absorb radar signals. Most conventional aircraft have a rounded shape. This shape makes them aerodynamic, but it also creates a very efficient radar reflector. The round shape means that no matter where the radar signal hits the plane, some of the signal gets reflected back: a stealth aircraft, on the other hand, is made up of completely flat surfaces and very sharp edges. When a radar signal hits a stealth plane, the signal reflects away at an angle, like this. In addition, surfaces on a stealth aircraft can be treated so they absorb radar energy as well. The overall result is that a stealth aircraft can have the radar signature of a small bird rather than an airplane. The only exception is when the plane banks, there will often be a moment when one of the panels of the plane will perfectly reflect a burst of radar energy back to the antenna.

4.8 Disadvantages of Stealth Technology

Stealth technology has its own disadvantages like other technologies. Stealth aircraft cannot fly as fast or is not maneuverable like conventional aircraft. The F-22 and the aircraft of its category proved this wrong up to an extent. Though the F-22 may be fast or maneuverable, it can't go beyond Mach 2 and cannot make turns like the Su-37. Another serious disadvantage with the stealth aircraft is the reduced amount of payload it can carry. As most of the payload is carried internally in a stealth aircraft to reduce the radar signature, weapons can only occupy a less amount of space internally. On the other hand, a conventional aircraft can carry much more payload than any stealth aircraft of its class.

Whatever may be the disadvantage a stealth aircraft can have, the biggest of all disadvantages that it faces is its sheer cost. Stealth aircraft literally costs its weight in gold. Fighters in service and in development for the USAF like the B-2 ($2 billion), F-117 ($70 million), and the F-22 ($100 million) are the costliest planes in the world. After the cold war, the number of B-2 bombers was reduced sharply because of its staggering price tag and maintenance charges.

There is a possible solution for this problem. In the recent past, the Russian design firms Sukhoi and Mikoyan Gurevich (MiG) have developed fighters which will have a price tag similar to that of the Su-30MKI. This can be a positive step to make stealth technology affordable for third-world countries.

Moreover, the B-2 Spirit carries a large bomb load, but it has relatively slow speed, resulting in 18–24 h long missions when it flies halfway around the globe to attack overseas targets. Therefore, advance planning and receiving intelligence in a timely manner is of paramount importance.

Stealth aircraft are vulnerable to detection immediately before, during, and after using their weaponry, due to the nature of reduced RCS and cruise. Missiles are yet not available; all armament must be carried internally to avoid increasing the radar cross section (RCS). As soon as the bomb bay doors opened, the RCS will be multiplied.

As we stated so far, stealth technology also termed as low observable (LO) technology is a sub-discipline of military tactics and passive electronic countermeasures, which cover a range of techniques used with personnel, aircraft, ships, submarines, and missiles, to make them less visible (ideally invisible) to radar, infrared, sonar, and other detection methods. It corresponds to camouflage for these parts of the electromagnetic spectrum. Quantum stealth is a material that renders the target completely invisible by bending light waves around the target. The material removes not only your visual, infrared (night vision), and thermal signatures but also the target's shadow. Thus, it paves way to the development of increasingly sophisticated technologies that help in evading the enemy's ever vigilant eyes. These types of materials are known as "metamaterials" which are artificial materials engineered to have properties that may not be found in nature.

They are assemblies of multiple individual elements fashioned from conventional microscopic materials such as metals or plastics, but the materials are usually arranged in periodic patterns. Metamaterials gain their properties not from their composition, but from their exactingly designed structures. These materials show the negative index, and their response is usually linked to the resonant behavior of the unit cells. Perhaps the epitome of stealth was the cloaking device, which used selective bending of light and emission dampening to render a ship totally invisible and undetectable.

Again, stealth technology is clearly the future of defense service. In the future, as air defense systems grow more accurate and deadly, stealth technology can be a factor for a decisive by a country over the other. In the future, stealth technology will not only be incorporated in fighters and bombers but also in ships, helicopters, tanks and transport planes, and army uniforms.

However, as we expressed in above, as part of disadvantages of quantum stealth per its definition, we can list the following functionality of stealth technology as follows;

1. Quantum stealth has its own disadvantages like other technologies. Stealth aircraft cannot fly as fast or is not maneuverable like conventional aircraft. The F-22 and the aircraft of its category proved this wrong up to an extent. Though the F-22 may be fast or maneuverable or fast, it can't go beyond Mach 2 and cannot make turns like the Su-37.
2. Another serious disadvantage with the stealth aircraft is the reduced amount of payload it can carry. As most of the payload is carried internally in a stealth aircraft to reduce the radar signature, weapons can only occupy a less amount of space internally. On the other hand, a conventional aircraft can carry much more payload than any stealth aircraft of its class.
3. Whatever may be the disadvantage, a stealth vehicle can have the biggest of all disadvantages that it faces, which is its sheer cost. Stealth aircraft literally costs its weight in gold. Fighters in service and in development for the USAF like the B-2 ($2 billion), F-117 ($70 million), and the F-22 ($100 million) are the costliest planes in the world. After the Cold War, the number of B-2 bombers was reduced sharply because of its staggering price tag and maintenance charges.

One of the disadvantages or drawbacks of stealth airplane design is the poor aerodynamic properties common to stealth airframe as it is illustrated in Figs. 4.36 and 4.37 and the difference between their nose shape.

Rather than aerodynamic perfection, stealth aircraft are designed according to requirements for RCS reduction, and in general this results in handling difficulties. Most modern aircraft are made unstable at one axis for greater maneuverability; however, stealth aircraft are usually unstable in all axes. Unlike other modern fighters, stealth assets require highly redundant, fly-by-wire systems for flight safety, which increase the cost and add extra weight to the airframe. During training and experimental flights, there were many failures of these flight control systems, some of which resulted in crashes; one known B-2 crash, one of seven F-117 crashes, and both F-22 crashes were related to flight control unit malfunctions.

Moreover, most stealth aircraft do not have engines with afterburners; thus they do not have high speed performance and are not suitable for dogfighting. The F-22 Raptor is an exception and may be a future solution to this problem. It is both an agile and stealthy air superiority fighter, and that is why its shape is more conventional than other stealth assets.

The second disadvantage of stealth aircraft is the requirement to either restrict electromagnetic emissions completely or emit them in a very careful manner, such as via low probability of intercept (LPI) radars (see Appendix D). Fully autonomous systems and applications using different systems, other than radar, reduce this risk; however, these systems have many constraints that limit the operational capability of the aircraft. LPI is a potential remedy and is a property of radar that, because of its low power, wide bandwidth, frequency variability, or other design attributes, makes it difficult for it to be detected by means of a passive intercept receiver [41].

Thus, radars and radio and data connection methods, based on the same principle, are realistic solutions for remaining stealthy. LPI technology is more necessary to low observables than any other asset. LPI can be used to support systems, such as altimeters, tactical airborne targeting, surveillance, and navigation [41], while it also matches with other stealthy qualifications. However, such sophisticated LPI systems, which require continuous development to counter new receiver designs, result in very high costs and deployment of complex electronically instrumentation and software [30].

Another drawback is the high maintenance costs associated with stealth. To remain low observable, an aircraft's surfaces must sustain their faultlessness. Surfaces must be examined very carefully, considering the fact that even an improperly tightened screw might degrade the stealthiness of an aircraft. All RAM-coated parts and special paintings must be treated before each mission. Moreover, this kind of maintenance requires special shelters, such as the B-2's climate-controlled hangars as it is shown in Fig. 4.61. After each sortie, B-2 Spirit has to be maintained for nearly 119 h with experienced staff and high-tech automated devices. It is preferable to deploy these aircraft on missions from their home bases only where they can be prepared for flight [30].

The issue is that long-range sorties conducted from the homeland against overseas targets still place a serious economic burden on stealth aircraft operators [43].

The fourth disadvantage is that stealth aircraft are limited by the amount of ordnance they can carry. This is because in full stealth mode, aircraft are required to carry all of their ordnance internally, at least until the time when stealth weapons

Fig. 4.61 Special climate control maintenance shelters of B-2 Spirit [42]

become operational. Thus, pre-operational intelligence is critical, and the judicious use of ordnance is important, as reattack of targets is limited by inventory. Furthermore, when the weapon bays are opened, the RCS increases which raises an enemy's probability of detection.

Another drawback of stealth aircraft is their visual signatures. Although decreased by paintings, night missions (dependency on nights and weather conditions is another drawback), and other camouflage tactics, stealth aircraft are still visible to the naked eye. Currently, experiments are being conducted to develop approaches for total cancelation of visual illumination; however, there are no known applications of such a system on operational stealth aircraft at this time [43].

The sixth disadvantage is the negative reaction of the public to aircraft failures. Based on mission experience during various wars, stealth aircraft have proven to be extremely successful. However, there are several known failures that have had a negative influence on public opinion. Incidents include the shoot-down of an F-117, and there are speculations that more than one F-117 took severe damage from enemy fire on March 27, 1999 during the Kosovo War. Other losses include shoot-downs of U-2 Dragon Lady and several low observable UAVs during the Cold War. Normally, such small numbers of shoot-down incidents over battlefields and other losses of military aircraft during training are neglected. But, the loss of such expensive military assets, which are thought to be impervious to enemy defenses, receives significant public interest. In addition to the shoot-down of the F-117 over Serbian airspace, eight F-117s, two F-22A Raptors, and one B-2A Spirit have been lost during training flights [30].

Table 4.1 The table shows that relatively small production numbers increase the project total cost per aircraft [44]

	F-117A Nighthawk	B-2A Spirit	F-22A Raptor
First projected production amount	89	132	750
Actual production	59	21	Continuing 127 of total 184 have been produced
Average procurement unit cost per aircraft	$42.6 million	$737 million	$185.4 million
Program unit acquisition cost per aircraft	$111.2 million	$2.13 billion	$353 million

The final and the most important con of stealth technology is the cost. Cost is affected by three factors. The first factor is the level of effort required to achieve a perfect low observable capability. Though perfection has not been provided, gained capabilities have taken a very long time to achieve and have come at a high cost. These efforts have been effective, but designers have worked hard to find methods of defeating radars and other sensor systems.

The second cost factor is the total cost of improving operational effectiveness of stealth assets using other technologies, such as complex fly-by-wire systems, high-tech computer, and control units, special super cruise engines, low probability of intercept (LPI) radars, navigation, precision targeting systems, and stealth armaments, which are under development. These factors require spending exorbitant amounts of money. Moreover, production of all three currently operational stealth aircraft reveals that total program expense, together with sunk costs of these projects per aircraft, is extremely high. Projected production amount, actual production amount, average procurement unit cost per aircraft, and program unit acquisition cost per aircraft with sunk costs are presented in Table 4.1 [44–46].

The table exposes that relatively small production numbers increase the project total cost per aircraft. The reason for this is the increase in single airframe cost, when projected production amounts are decreased to relatively small numbers due to cost growth associated with unexpected commitments or changes in requirements. Moreover, it is difficult to recover development costs through sales to other nations, a common practice for non-stealth weapons systems. Stealth assets are protected from foreign military sales due to security concerns. In this context, the US Congress has banned their sales by declaring their critical technology, even though these sales would likely to recover some of these costs [30].

The third cost factor concerns operational expenses. For example, while the B-2 Spirit can be deployed anywhere in the world within 12 h, "...it is operationally crippled by its exorbitant replacement cost and results in a challenging risk/benefit analysis when considering its deployment [55]." Table 4.2 compares the B-2A Spirit with other US strategic bombers, semi-stealth B-1B Lancer and the highly

Table 4.2 Comparison of the three US strategic bombers [47]

	B-2A "Spirit"	B-1B "Lancer"	B-52H "Stratofortress"
Date deployed	1993	1985	1955
Prime contractor	Northrop Grumman	Rockwell	Boeing
Cost per aircraft	−$2.2 billion	$200 million	$74 million
Number in inventory	21	95	85 (+9 reserve)
Weapons payload	40,000 mph	72,000 + pounds	70,000
JDAM payload	16	24	12
Speed	~600 mph (high subsonic)	900 mph (Mach 1.2)	650 mph (Mach 0.86)
Crew	2	4	5

conventional B-52H Stratofortress, which were also designed and produced for heavy bombing missions.

Despite all these drawbacks and challenges in producing stealth assets, stealth technology has fulfilled the air force requirements for battlefield survivability since its first applications. Thus, many assets have been developed and deployed. These airframes used stealth technology in favor of their tactical combat superiority and overwhelming dominance over an opponent. In this context, specially designed air defenses with new radar systems and tactics have been required to withstand against low observables. The next chapter will discuss counter-stealth technologies which focus to improve solutions for air defenses by means of exploiting the technological limitations of stealth technology.

4.9 The Future of Quantum Stealth or Stealth Technology

Quantum stealth technology is clearly the future of air combat. In the future, as air defense systems grow more accurate and deadly, stealth technology can be a decisive factor for a country over the other. In the future, stealth technology will not only be incorporated in fighters and bombers but also in ships, helicopters, tanks, and transport planes.

These are evident from the RAH-66 "Comanche" (see Fig. 4.62) and the Sea Shadow stealth ship. Sea Shadow (IX-529) (i.e., Fig. 4.63) is an experimental stealth ship built by Lockheed for the United States Navy to determine how a low radar profile might be achieved and to test high-stability full configurations which have been used in oceanographic ships.

Ever since the Wright brothers flew the first powered flight, the advancements in this particular field of technology have seen staggering heights. Stealth technology is just one of the advancements that we have seen. In due course of time, we can see

Fig. 4.62 The first Boeing-Sikorsky RAH-66 prototype during its maiden flight on January 4, 1996. (Source: www.wikipedia.com)

Fig. 4.63 Unclassified miscellaneous (IX) ship photo index. (Source: www.wikipedia.com)

many improvements in the field of military aviation which would one day even make quantum stealth technology obsolete.

This is not a new idea; in fact several military fiction writers have already come up with the idea, in one particular instance having the aircraft continually modified from

Fig. 4.64 Magical image of stealth airplane

top and bottom like a magician's mirror box making the aircraft totally invisible (see Fig. 4.64).

Furthermore, in conclusion, the detection of stealth technology has improved significantly, more advanced in the last 50 years or so. This trend is likely to continue as these two oppose each other.

Till date stealth aircraft have been used in several low- and moderate-intensity conflicts, including Operation Desert Storm, Operation Allied Force, and the 2003 invasion of Iraq. In each case they were employed to strike high-value targets which were either out of range of conventional aircraft or which were too heavily defended for conventional aircraft to strike without a high risk of loss. In addition, because the stealth aircraft aren't going to be dodging surface-to-air missiles and anti-aircraft artillery over the target, they can aim more carefully and, thus, are more likely to hit the high-value targets early in the campaign or, even for it, before other aircraft had the opportunity to degrade the opposing air defense.

However, given the increasing prevalence of excellent Russian-built surface-to-air missile (SAM) system on the open market, stealth aircraft are likely to be very important in a high-intensity conflict in order to gain and maintain air supremacy. Stealth technology, in the future, would be required for clearing the way for deeper strikes, which conventional aircraft would find very difficult .For example, China are licensed to build a wide range of SAM systems in quantity and would be able to heavily defend important strategic and tactical targets in the event of some kind of conflict .Even if anti-radiation weapons are used in an attempt to destroy the SAM radars of such systems, these SAMs are capable of shooting down weapons fired against them. The surprise of a stealth attack may become the only reasonable way of making a safe corridor for conventional bombers. It would then be possible for the less-stealth force with superior weaponry to suppress the remaining systems and gain air superiority.

The development and the deployment of Visby's first commissioned stealth ships have raised new threats in the maritime boundaries. The sudden appearance of sea clutters on the radar at a region may be these ships.

The plasma stealth technology raises new hopes of engineering brilliance. As plasma is said to absorb all electromagnetic radiation, the development of a counter-stealth technology to such a mechanism will be a strenuous task.

Well to conclude, the current scenario appears something similar to the Cold War as both sides are accumulating weapons to counter each other, and each side can be termed as "stealth technology" and the other as "anti-stealth technology." It is an arm race except it is not between specific countries. "It is a fight between technologies."

As we have seen so far, stealth appears to gradually follow the timeless pattern of novel war principles, a cycle that has been repeated in the past with concepts such as the airplane, the armored warship, the tank, the submarine, the nuclear weapon, etc. Initially the "new way" is met with resounding success, as there is virtually no counter for it in place and is frequently hailed as the precursor of a revolution in military affairs (said revolution sometimes indeed happening, and sometimes not). Subsequently, as the lessons of its initial uses sink in, solutions to dealing with it are explored and at the same time its operational use is refined. Eventually, the new principle finds its true niche within the art of war and becomes one more arrow in a full quiver, rather than the silver bullet as originally envisioned.

An interesting shift in counter-stealth research in the last few years is the visibly increased Western attention in the field. This is hardly a surprise when one considers that, until quite recently, the West held a decisive advantage in Suppression of Enemy Air Defense (SEAD), very low observable (VLO), and cruise missile technologies, all resulting from its superiority in electronics and miniaturization. As this gap however tends to shrink, Western military branches increasingly find themselves faced with potential threats that may employ such technologies against them. Little wonder, then, that technologies such as passive/covert radar systems or advanced long-range IR sensors are being generously funded. Hard details on VLO programs in the East are usually hard to come by, but what is known is enough to cause interest—and in some cases unrest. Technologies such as plasma-stealth in augmentation with hypersonic velocity vehicles (see Appendix C and Fig. 4.65) and active cancelation, prototypes with a clear LO inclination such as the S-37, MiG-1.42, and even the still-shady J-10 and high-precision strike systems like the latest generation of Russian, Chinese, and Indian missiles are a clear indication of things to come [48].

At the same time, technologies previously reserved for high-value or silver bullet forces (primarily due to cost and complexity) are gradually trickling down to even wider portions of the air forces. Fitting phased array radars to light combat aircraft or advanced trainers was an absurd idea a decade ago, for example, yet it is actively considered nowadays. Similarly, stealth will likely find its way into such aircraft classes as multi-mission and C4ISR platforms (see Fig. 4.66), transports, utility craft,

Fig. 4.65 A possible view of the future as hyper-soar bomber [48]

Fig. 4.66 A C4ISR platform

and maybe even trainers. At this point, no doubt having lost much of its still-present glamor, it will have to compete with other principles that may yet be beyond our grasp. Interestingly, the next "darling" principle may not have to be something completely new, but rather a novel way of revisiting already established priorities. For example, the USAF is currently exploring the options for a future endo/exoatmospheric hypersonic bomber, perhaps an indirect admission that stealth alone will not cut it in the future. Going retro with the B-70/F-108 idea? Time will tell [48].

C4ISR, or command, control, communications, computers, intelligence, surveillance, and reconnaissance, brings together arguably the most important elements of the infrastructure of global national security into a single, memorable term. C4ISR can be defined as the web of platforms, payloads, sensors, and other systems that inform and connect warfighters and first responders.

Perhaps it was someone's dark sense of humor, but whoever created the acronym "C4ISR" surely did so with a wry smile. As an acronym, it almost single-handedly makes the case that defense and security agencies have an unhealthy appetite for unsightly abbreviations.

With such capability, "the value of knowing your next move is your most informed." As this author claims "Knowledge is Power in Four Dimension in Four Dimension" [49].

4.10 Stealth Aircraft of Yesterday, Today, and Tomorrow

Stealth technology is a concept that is not at all new. During the World War II, Allied aircraft used tin and aluminum foils in huge numbers to confuse German radar installations. This acted as a cover for Allied bombers to conduct air raids. This method was later used as chaffs by aircraft to dodge radar-guided missiles. The first stealth aircraft was the F-117 developed by Lockheed Martin. It was a top secret project developed by its Skunk Works unit. The F-117 was only revealed during the late 1980s and then saw action in the Persian Gulf. In due course of time, the B-2 was developed as a successor to the F-117. Though both of them serve different purposes, the B-2 went a step ahead of the F-117. The B-2 was developed to deliver nuclear weapons and other guided and unguided bombs. On the other hand, the F-117 was developed to deliver its precision laser-guided bombs. Another stealth aircraft, which made a lot of promises and in the end ended up in a trash can, was the A-12. It was a fighter that was designed to replace the F-14 and F-18 in the future. The capabilities of this aircraft were boasted to such an extent that the project ended up in a big mess. Billions of dollars were wasted for nothing.

Stealth technology became famous with the ATF contest. The Boeing-Lockheed YF-22 and the McDonnell Douglas-Grumman YF-23 fought for the multi-billion contracts to build the fighter that would take the USAF into the fifth-generation fighter era. The Boeing-Lockheed won the contract, and the F-22 was approved to be the replacement for the F-15 "Eagle" interceptor.

America now has a competitor, Russia, which decided to respond to the development of the F-22 by making the Su-47 (S-37) "Berkut" and the MiG-35 "Super Fulcrum/Raptor Killer." These fighters were developed by the two leading aviation firms in Russia, Sukhoi and Mikoyan Gurevich (MiG). The future of these projects totally depends on the funding which will be provided to the Russian defense sector. This time Boeing developed the X-32 and the Lockheed it's X-35. With the experience gained from developing the F-22, they were tasked with making a replacement for the F-16. This saw great technological advances, as they had to make the first operational supersonic VSOL aircraft. Lockheed Martin took the technical assistance of Russian scientists who developed the Yak-141. The Yak-141 is the first supersonic VSTOL aircraft. In the end the Lockheed team with its X-35 won the contract, and the fighter was re-designated as the F-35.

Many projects remain over the horizon that will use stealth technology as its primary capability. They come from some of the most unlikely contenders. These projects include the Euro JSF, which will be designed by the team that developed the EF-2000. Russia is stepping forward with its LFS project with the S-54 and other designs. Two new entries into this field will be India and China. India will be introducing its MCA, which is a twin-engine fighter without vertical stabilizers. This fighter will use thrust vectoring instead of rudders. China will be introducing the J-12 (F-12/XXJ) which is equivalent to the United States' fighter F-22.

These challenges reveal that stealth technology is an inevitable requirement for today's modern forces to dominate the battlefield. Its many advantages give the user tactical combat superiority and an overwhelming dominance over an opponent.

However, designing, manufacturing, operating, and maintaining stealth assets have some cons. The use of the terms cons, disadvantages, or drawbacks here does not intend to thwart advances in this sophisticated military technology, but it implies that there are some challenges to deploying these technologies. These challenges must be balanced by designers and users.

4.11 Stealth Technology Versus Electronic Warfare

In April 21, 2014 Dave Majumdar [50] published an article in USNI News under title of "Stealth Vs. Electronic." He went on to say that United States Navy is in need of stealth technology integrated with electronic warfare (EW) capabilities as a combined system in order to defeat Russian-made Advanced Anti-Access/Area Denial (A2/AD) (i.e., see Sect. 3.11 of Chap. 3 of this book for more details) threats as countermeasure against such measure in the near future as well as a long-term solution. Such requirements and need arise from the fact that we have seen in Chap. 3 of this book so far that the stealthy generation of new warplanes (i.e., fifth generation) is not really stealthy and new passive radar operating within frequency range of cellular phone can detect stealthy plane such as F-35 Lightning Fighter as pictured in Fig. 4.67 up to 100 miles away.

Chief of Naval Operation (CNO) Admiral Jonathan Greenert in April 16, 2014, in his speech at the US Naval Institute Annual Meeting in Washington, DC, stated that:

"*Stealth* is needed for what we have in the future for at least ten years out there and there is nothing magic about that decade," Greenert said. "But I think we need to look beyond that. So, the thought is that it is a combination of having aircraft that have stealth but also aircraft that can suppress other forms of radio frequency electromagnetic emissions so that we can get in."

Fig. 4.67 Fifth-generation F-35C Lightning stealth fighter

Electronic attack by itself will probably not be enough to enable US forces to penetrate enemy air defenses, according to Greenert and multiple US military and industry sources.

Admiral Greenert went on to say that:

"I doubt in the future we can just suppress everything, go rolling in until we do what we need to do and get out," Greenert said. "But we have the means for—way out in the future—with the Next Generation Jammer and what it'll bring, to be able to get in when we need to and get out."

Greenert's comments largely mirror a Boeing presentation last week at the Navy League's Sea-Air-Space Exposition where Mike Gibbons, the company's Vice President for the F/A-18E/F and EA-18G programs, had stated that stealth aircraft must be supported by airborne electronic attack capabilities.

"The point is anybody that goes in can't be good against any one frequency band because you will be seen by others, that's the key," Gibbons said. "The Growler, the Boeing EA-18G carrier-based as illustrated in Fig. 4.68, is the only aircraft that has that full spectrum sensor and jamming capability to take care of that for strikers."

The Boeing EA-18G Growler is an American carrier-based electronic warfare aircraft, a specialized version of the two-seat F/A-18F Super Hornet. The EA-18G replaced the Northrop Grumman EA-6B Prowlers (Fig. 4.69) in service with the United States Navy. The Growler's electronic warfare capability is primarily provided by Northrop Grumman. The EA-18G began production in 2007 and entered operational service with the US Navy in the late 2009. Australia has also purchased 12 EA-18Gs, which entered service with the Royal Australian Air Force in 2017.

The claim by Gibbons graphically is depicted in Fig. 4.70.

Fig. 4.68 A US Navy EA-18G carrier-based plane. (Source: www.wikipedia.com)

Fig. 4.69 Northrop Grumman EA-6B Prowler. (Source: www.wikipedia.com)

The Boeing presentation also reiterated the company's oft-stated position that low observable (LO) technologies are a "perishable" asset—particularly as potential enemies develop advanced low-frequency radars, and signal processors become ever more capable.

"Stealth is 'delayed detection' and that delay is getting shorter. SAM (Surface-to-Air Missile) radars are shifting their frequencies into lower frequency bands where

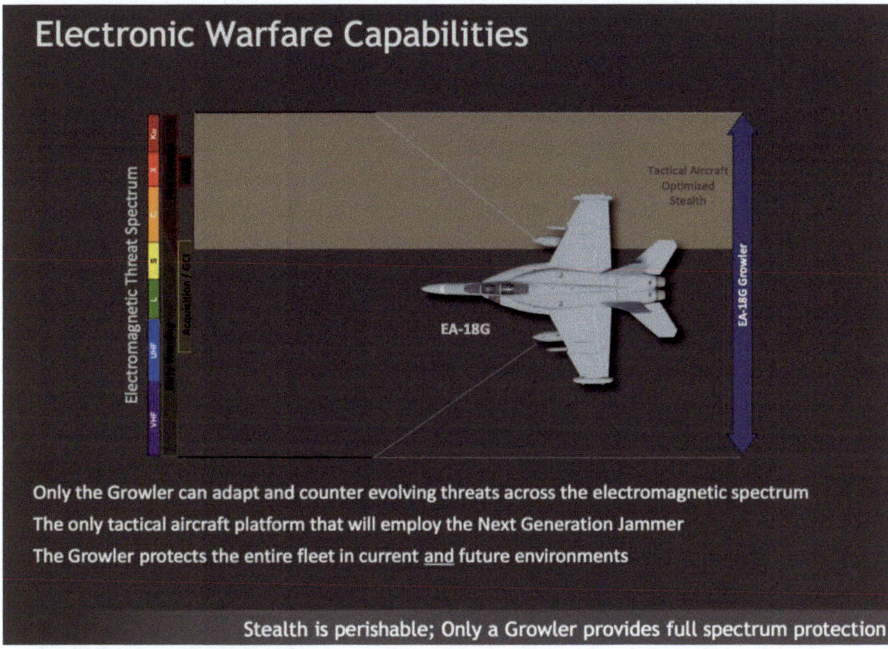

Fig. 4.70 EA-F-18G Growler electronic warfare capabilities

Fig. 4.71 The Navy's F/A-18 Super Hornet. (Source: www.wikipedia.com)

U.S. stealth is less effective," said Mark Gammon, Boeing's F/A-18E/F (Fig. 4.71) and EA-18G program manager for advanced capabilities, in an emailed statement. "Early warning radars are in the VHF spectrum where stealth has limited if any capability. These radars are networked into the SAM radars giving the SAM radars cued search. The threat is developing out of spectrum sensors like Infrared Search and Track (IRST) systems on their fighters. Stealth has no capability to delay an IRST detection and track."

While some military officials consulted by USNI News wholeheartedly concurred with Boeing's assessment, others dismissed the company's claims out of hand. Many others offered a more nuanced view.

"Boeing is in full-court press against the [Lockheed Martin] F-35 in this briefing. As such, when they describe the advantages of the Growler–which are accurate–they ignore the tradeoff for that advantage," said one US Air Force official. "The truth is that the Growler and Low Observable (LO) platforms complement each other extremely well."

Lockheed Martin officials, however, maintain that the F-35 is able to operate inside highly contested airspace without any support assets.

"By government contract specification, the airplane is required to be able to go into high threat anti-access environments, autonomously perform its mission and survive," said Eric Van Camp, Lockheed's domestic F-35 business development director. "The results of flight test indicate conclusively that the airplane will meet that contract specification."

While it is an indisputable fact that a tactical fighter-sized stealth aircraft must be optimized to defeat higher-frequency bands such the C-, X-, and Ku-bands as a simple matter of physics, in a real-world operational setting, there are often other factors involved that make detecting and tracking a stealth aircraft more difficult.

Industry, Air Force, and Navy officials agreed that there is a "step change" in an LO aircraft's signature once the frequency wavelength exceeds a certain threshold and causes a resonant effect.

Typically, that resonance occurs when a feature on an aircraft—such as a tail fin—is less than eight times the size of a particular frequency wavelength.

Effectively, small stealth aircraft that do not have the size or weight allowances for 2 ft or more of radar-absorbent material coatings on every surface are forced to make trades as to which frequency bands they are optimized for.

"You can't be everywhere at once on a fighter-sized aircraft," says another Air Force source.

What that means is that a radar operating at a lower-frequency band such as parts of the S- or L-band—like civilian air traffic control (ATC) radars—might be able to detect and possibly even track certain stealth aircraft to an extent.

However, a larger stealth aircraft like the Northrop Grumman B-2 Spirit, which lacks many of the features that cause a resonance effect, is much more effective against low-frequency radars than, for example, an F-35.

But those lower-frequency radars do not provide what Pentagon officials call a "weapons quality" track needed to guide a missile onto a target.

"Even if you can see a Low Observable (LO) strike aircraft with ATC radar, you can't kill it without a fire control system," an Air Force official said.

Meanwhile, Russia, China, and others are developing advanced UHF- and VHF-band early warning radars that use even longer wavelengths in an effort to cue their other sensors and give their fighters some idea of where an adversary stealth aircraft might be coming from.

But the problem with VHF- and UHF-band radars, as one US Navy official told USNI News, is that with long wavelengths come large radar resolution cells.

That means that contacts are not tracked with the required level of fidelity to guide a weapon onto a target.

"Does the mission require a cloaking device or is it Ok if the threat sees it but can't do anything about it?" the Navy official asked rhetorically.

Further, officials from the Air Force, Navy, and Marine Corps agreed that while aircraft like the F-35 or F-22 are not solely relying on low observables for survivability, stealth is an absolute requirement to survive in an A2/AD environment even with airborne electronic attack support.

As one Air Force official explained, stealth and electronic attack always have a synergistic relationship because detection is about the signal-to-noise ratio. Low observables reduce the signal, while electronic attack increases the noise. "Any big picture plan, looking forward, to deal with emerging A2/AD threats will address both sides of that equation," he said (see Fig. 4.72).

Air Force and Marine Corps officials took exception to Boeing pointing out that the F-35 only has X-band electronic attack coverage from the front. "Aft coverage may or may not be provided onboard any given fighter but is provided by the package overall—which will likely include EA-18s," one Air Force official pointed out.

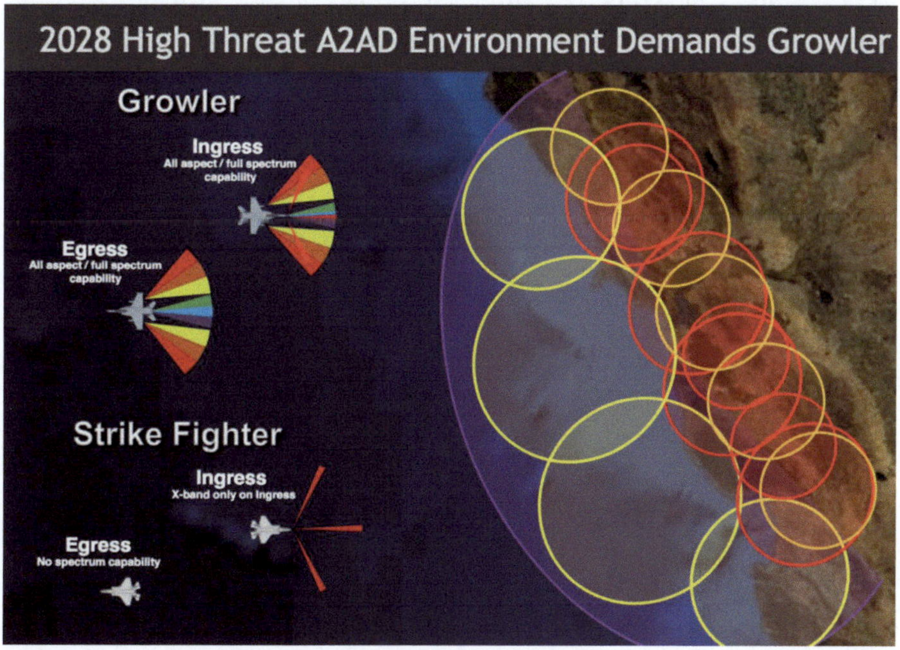

Fig. 4.72 Artistic Boeing presentation A2/AD threat

However, Air Force and Marine Corps officials said that the Growler may not be particularly useful against emerging threats and noted that there are electronic warfare upgrades planned for the F-35 in addition to its baseline capability.

"The Growler itself, while a very credible aircraft, has limited suitability in an advanced A2/AD area," one Air Force official said.

"While it is the state of the art for now, I don't know if it will be the appropriate jamming platform for the pictured environment."

Nonetheless, a number of Air Force officials expressed support for the Pentagon potentially increasing the size of its Growler buy. "The Growler is a great asset, we probably need more, and it is an important part of a strike package into an advanced Integrated Air Defense System (IADS)," one official said. "It is not as stand-alone as Boeing will claim."

However, those same officials pointed out that the Growler is not fully interoperable with joint forces.

"If there is a major enduring shortfall to the Growler, it's the degree and fidelity between it and other joint suppression platforms. The reasons for which could be as benign as joint interoperability [being seen] as an afterthought," one Air Force official said.

But "it's to Boeing's advantage to propagate a limited interoperability platform, especially one that doesn't communicate very well with competitor's platforms in the Suppression of Enemy Air Defenses/Destruction of Enemy Air Defenses (SEAD/DEAD) mission. But it doesn't make sense from a warfighter position."

An industry source agreed that the Growler still faces interoperability problems when operating with Air Force assets, but that is true of many platforms across the board. "There are interoperability issues across a lot of the platforms," the industry source said. For example, Lockheed F-22s (see Fig. 4.73) are only able to connect with other Raptors using the intra-flight data link (IFDL), while the F-35 uses a Joint

Fig. 4.73 Lockheed Martin F-22 Raptor. (Source: www.wikipedia.com)

Fig. 4.74 Northrop Grumman E-2D Advanced Hawkeye. (Source: www.wikipedia.com)

Service Fighter-only multifunction advanced data link (MADL). "This is one of the bigger issues the Air Sea Battle Office is attempting to resolve," the industry source said.

Gammon defended the EA-18G's ability to operate with other Pentagon assets. "Growlers have Link 16 which is compatible with [the] F/A-18 Super Hornet and F-35, E-2D (Fig. 4.74), F-15, F-16, and most of the bombers," he said. "The good news is the Growlers can stand-off from the threat, build the Electromagnetic (EM) picture, and pass weapons quality tracks to the other fighters via Link-16 (and soon TTNT [Tactical Targeting Network Technology])."

The industry source noted that while the F-35 will be fitted with the Link-16 data link, it would not be able to use that omnidirectional link inside a high threat environment because it could compromise the aircraft's position. "Aircraft such as the F-35 that might not want to transmit on their Link-16 can always receive Link-16 tracks from Growlers and employ weapons on those tracks," the industry source said.

Air Force officials conceded that the Pentagon probably needs more EA-18Gs.

"In truth, we never bought enough Growlers in the first place," one Air Force official said.

"They're worth their weight in gold and contribute immensely to the Electromagnetic Spectrum (ES) situational awareness and EA [electronic attack] communities.

But the Limiting Factor (LIMFAC) is, and always will be, the carrier deck park and cycle times."

A Navy official said that the carrier deck cycle would be a limiting factor only if the Growler was being used to launch missiles such as the AGM-88E Advanced Anti-Radiation Guided Missile (AARGM) or the High-speed Anti-Radiation Missile (HARM). The official noted that with aerial refueling, it is not unheard of for Navy fighters to remain airborne for more than 6 h at a time. "If the Growler was kinetic, launching all its HARM and then needing to reload... Yes, the deck cycle time would come into play here," the official said. "But, it's more realistic to provide standoff jamming than launch HARMs unless in a self-protect role." The official also pointed out that the need to land, refuel, swap crews, and perform maintenance at some regulated interval is a requirement for any aircraft.

Boeing also suggested in its presentation that the Growler could be used in the counter-air and strike roles. Gammon elaborated on how Boeing envisions the EA-18G might perform some of those missions—distancing the company's position somewhat from the diagrams shown in the presentation. "In the counter-air mission, the Growlers will use their ESM [electronic support measures] system to help the fighters detect, and just as important, ID the threat. They can do this from a stand-off position from the fighters and still contribute to the overall Situational Awareness (SA) and ID," Gammon said.

Gammon also clarified the company's position with regard to using the EA-18G in the air-to-ground strike role. "In the strike mission, the Growler is supporting by building that enemy EM order of battle, find, fix, track, and Identification (ID) those threat emitters and then quarterback the EM fight and determine which of those threat systems we are going to jam, attack, avoid. The Growler can employ weapons such as AGM-88E, the Advanced Anti-Radiation Guided Missile (AARGM) (Fig. 4.75, shows a launch from F-18 hornet) at those emitters as well as handing off that track to a strike fighter to engage."

The AGM-88E AARGM is a medium-range air-to-ground missile developed by Orbital ATK (previously Alliant Techsystems). The primary role of the missile is to target enemy air defenses. The missile can engage relocatable integrated air defense (IAD) targets and other targets equipped with shutdown capability.

The AGM-88E AARGM is a follow-on to the US Navy's AGM-88 High-Speed Anti-Radiation Missile (HARM). It was the only tactical extended-range,

Fig. 4.75 The AGM-88E Advanced Anti-Radiation Guided Missile. (Source: www.wikipedia.com)

supersonic, multi-role strike weapon in the US and Italian inventory which became operational in 2012.

As Navy officials had said that while the service might consider using the Growler as a battle manager, it is extremely unlikely the service would ever consider using the EA-18G in a direct strike role or the air superiority role where the jet would be the primary shooter.

An industry source conceded that while the Growler would likely never be used as an air superiority fighter or strike aircraft, it could play an important role in those missions. "I do agree that Growlers will not be bringing Joint Direct Attack Munition (JDAMs) to a target," the source said. "They will support the strike fighters as they fight their way into the target area."

Though there is broad support for purchasing additional Growlers, it is not a stand-alone solution for dealing with advanced A2/AD threats.

"Stealth has its flaws, as the brief points out; however, if a new pod on a fourth gen platform was a workable answer against the modern and future IADS, I'm about 100% certain that U.S. Air Force (USAF) would be trying to buy a pile of them as well," an Air Force official said.

"But the juice ain't worth the squeeze, as they say."

4.12 E-Bomb-Driven Directed-Energy Warfare

The rules of tomorrow's battlefield and defense system since the introduction of Strategic Defense Initiative (SDI) right around 1975–1980s and continuing to present have drastically changed.

There exits now, Directed Energy Weapons (DEWs) using either High Energy Laser (HEL) [8] or High Energy Beam (HEB) [6] or recent discussion on type of wave that is known as Scalar Longitudinal Wave (SLW) [7] that are fighting these new age battles and battlefield at speed of light or electron and with them come a new generation of warplane that are known as Sixth Generation stealth fighter and bomber that are flying two to threefold speed of sound and with the help of Artificial Intelligence they are pushing the age of flying to a pilotless mode as depicted in Fig. 4.76.

Even recently high-tech defense companies and adversary countries and their leaders are speaking of new generation of weapon systems that are able of either cruising or gliding to their designated target at 5–15 Mach speed, where today's radars either passive or active are having hard time to track them down, let alone to be able to shoot them with any existing air defense mechanism within superpower military arsenal. (See Appendix B for more details.) These days among scientist and engineers of stealth technology, they are considering speed in order to avoid radar detection as a new way of being stealthy. As we discussed previously under Sect. 4.4 of this chapter and has been shown in Appendix B, which is an article published by Zohuri and Moghaddam [4] indicates that by creating a weak plasma around the

Fig. 4.76 Pilotless sixth-generation airplane. (Source: Rodrigo Avella 2016)

supersonic flying object, not only you can evade radar detection, but you can increase speed to hypersonic edge, while reducing drag and friction.

In this section we discuss the usage of E-bomb as a driving element of directed-energy warfare (DEW), which is the idea behind the new generation of stealth fighter and bomber, where superpower air forces around the globe are trying to end the desire for pilots as presented in Fig. 4.77.

As we stated at the beginning of this section, the rules of battle have changed over the entirety of military history. Tools such as technology, strategy, tactics, and weapons have been the principal elements determining what kind of rules apply to the battlefield. What can constitute to a sixth-generation fighter jet—that is the question we can ask ourselves. Perhaps, it might be too early to think of these questions, when even planes like JSF, PAK-FA, or F-22 are not even fully operational.

The contemporary military rivalry is driven mostly by the ongoing military technical revolution. In particular, the weapons used on the future battlefield will play an important role in military affairs.

Then, which weapons can play a key role in the future?

Sixth-generation jet fighters are currently conceptual and expected to enter service in the United States Air Force and United States Navy in 2025–2030 timeframe. The technological characteristics may include the combination of fifth-generation aircraft capabilities with unmanned capability, unrefueled combat radius greater than 1000 nm, and directed-energy weapon. It is the latter which is a subject of this article. One form of this energy is electronic bomb (E-bomb). This section

Fig. 4.77 Russian Sukhoi 57 or PAK-FA stealth fighter

aims to explore the technical aspects and potential capabilities of this type of bomb, target measurements, and its comparison with other forms of electromagnetic (EM) weaponry.

An E-bomb (electromagnetic bomb) is a weapon that uses an intense electromagnetic field to create a brief pulse of energy that affects electronic circuitry without harming humans or buildings. At low levels, the pulse temporarily disables electronic systems; mid-range levels corrupt computer data. Very high levels completely destroy electronic circuitry, thus disabling any type of machine that uses electricity, including computers, radios, and ignition systems in vehicles. Although not directly lethal, an E-bomb would devastate any target that relies upon electricity: a category encompassing any potential military target and most civilian areas of the world as well. According to a CBS News report, the United States deployed an experimental E-bomb on March 24, 2003, to knock out Iraqi satellite television and disrupt the broadcast of propaganda [6].

In the United States, most E-bomb research has been carried out at the Air Force Research Laboratory at Kirtland Air Force Base in New Mexico, where researchers have been exploring the use of high-power microwaves (HPM). Although the devices themselves may be relatively uncomplicated to manufacture (popular mechanics illustrated a simple design in September 2001), their usage poses a number of problems. To create an effective E-bomb, developers must not only generate an extremely high-powered pulse of energy but must also find a way to control both the energy—which can behave in unpredictable ways—and the heat generated as its by-product. Furthermore, for non-nuclear E-bombs, the range is limited. According to most defense analysts' speculations, devices in development are likely to affect an area of only a few hundred yards [4–8].

The concept behind the E-bomb arose from nuclear weaponry research in the 1950s. When the US military tested hydrogen bombs over the Pacific Ocean, streetlights were blown out hundreds of miles away, and radio equipment was affected as far as away as Australia. At the time these effects were considered incidental since that time researchers have sought a means of focusing that energy [6, 8].

The E-bomb targets mission-essential electronic systems such as the computers used in data processing systems; communications systems; displays; industrial control applications, including road and rail signaling; and those embedded in military equipment, such as signal processors, electronic flight controls, and digital engine control systems. I must point out that when E-bomb outputs are too weak to destroy these systems but strong enough to disrupt their operations, system performance can be degraded. The relation between the altitude (shown below) where the E-bomb is detonated and a representation of the lethality range. Target information (to include location and vulnerability) becomes an important issue. See Fig. 4.78, where possibly a fifth- or sixth-generation warplane presents a conceptual delivery of E-weapon, although there are other methods of delivery of E-bomb to target can be entertained as well, and we discuss these options at the end of this section.

Research has shown that it is possible to develop such kind of device. Directed-energy (DE) research originated with research work done to determine the impact to important military systems operating in harsh electromagnetic environments. One of

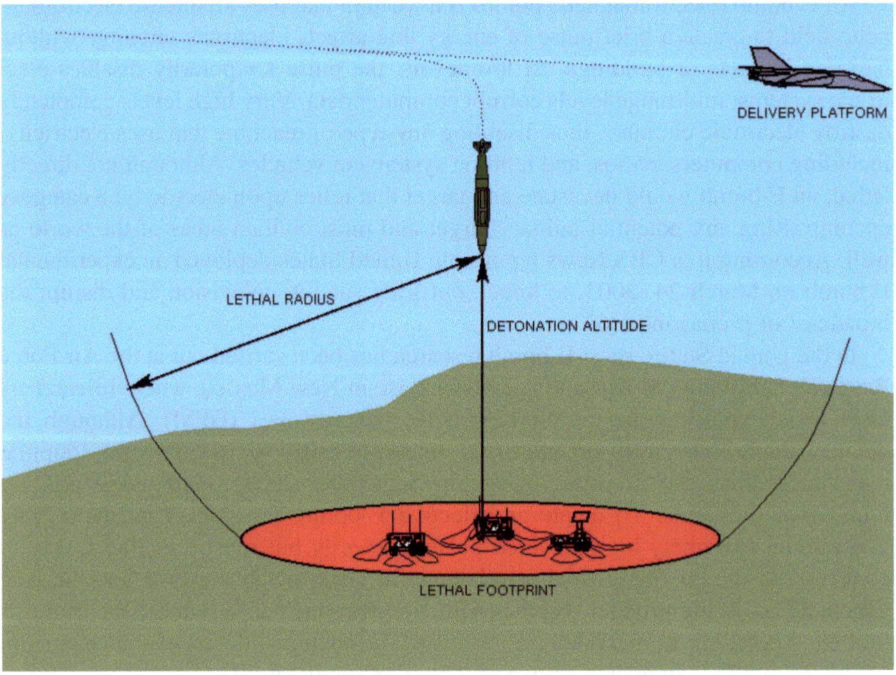

Fig. 4.78 Lethal footprint of low-frequency E-bomb in relation to altitude

the most threatening and pervasive of all electromagnetic threats is that due to electromagnetic pulse [6].

These pulses can burst of electromagnetic radiation that results from an explosion (usually from the detonation of a nuclear weapon) and/or a suddenly fluctuating magnetic field. However, it is not only the nuclear weapon that generates these pulses, non-nuclear electromagnetic pulse (NNEMP) is an electromagnetic pulse generated without use of nuclear weapons.

There are a number of devices that can achieve this objective, ranging from a large low-inductance capacitor bank discharged into a single-loop antenna or a microwave generator to an explosively pumped flux compression generator. To achieve the frequency characteristics of the pulse needed for optimal coupling into the target, wave-shaping circuits and/or microwave generators are added between the pulse source and the antenna. A vacuum tube particularly suitable for microwave conversion of high-energy pulses is the vircator. These high-altitude electromagnetic pulse (HEMP)-induced stresses can damage or severely disrupt some electronic systems, which are sensitive to transient disturbance. Significant potential damaging effects can occur at long ranges to virtually all systems located within line of sight of the detonation point.

Note: HEMP is a process where a large-scale EMP effect can be produced by a single nuclear explosion detonated high in the atmosphere. This method is referred to as high-altitude EMP (HEMP). A similar, smaller-scale EMP effect can be created using non-nuclear devices with powerful batteries or reactive chemicals. This method is called high-power microwave (HPM). Several nations, including reported sponsors of terrorism, may currently have a capability to use EMP as a weapon for cyber warfare or cyber terrorism to disrupt communications and other parts of the US critical infrastructure. Also, some equipment and weapons used by the US Military may be vulnerable to the effects of EMP.

Thus, it is feasible to say that NNEMP generators can be carried as a payload of bombs and cruise missiles, allowing construction of electromagnetic bombs with diminished mechanical, thermal, and ionizing radiation effects and without the political consequences of deploying nuclear weapons.

The fact that an electromagnetic pulse is produced by a nuclear explosion was known since the earliest days of nuclear weapons testing, but the magnitude of the EMP and the significance of its effects were not realized for some time. As a result of the test, a very short but extremely intense electromagnetic pulse was observed. This pulse propagated away from its source with a decreasing intensity, which is also to be expected according to the theory of electromagnetism.

According to the CBS reports dated March 2003, it stated the application of experimental EM pulse:

> *The U.S. Air Force hit Iraqi TV with an experimental electromagnetic pulse device called the "E-Bomb" in an attempt to knock it off the air and shut down Saddam Hussein's propaganda machine. The highly classified bomb created a brief pulse of microwaves powerful enough to fry computers, blind radar, silence radios, trigger crippling power outages and disable the electronic ignitions in vehicles and aircraft. Officially, the Pentagon does not acknowledge the weapon's existence.*

Fig. 4.79 Wedgetail Flares testbed

4.12.1 Directed-Energy Warfare

Military action involve the use of directed-energy weapons, devices, and countermeasures to either cause direct damage or destruction of enemy equipment, facilities, and personnel or to determine, exploit, reduce, or prevent hostile use of the electromagnetic spectrum through damage, destruction, and disruption. The defensive part of electronic warfare (EW) includes the offensive actions such as preventing the enemy's use of the electromagnetic spectrum through countermeasures such as damaging, disrupting, or destructing the enemy's electromagnetic capability. See Fig. 4.79, where Wedgetail Flares are tested.

Such weaponry (DEW) is an evolving addition to the EW.

4.12.2 Characteristic of Directed-Energy Weapons (DEWs)

The most common characteristics of the directed-energy weapons are that they attack at the speed of light. This poses some advantage over conventional weaponry. This helps in defeating targetssuch as theater and ballistic missiles before they can deploy defense-saturating sub-munitions. Another advantage of such weapons is that they can be used against multiple targets at the same time. The directed-energy weapons are classified into four categories: high-power microwave (HPM), charged particle beams (CPB), neutral particle beams (NPB), and high-energy laser (HEL). It is the latter which has high potential for military applications (both strategic and tactical missions) [6, 8].

However, for E-Bomb it is HPM is a base. But of course, when compared to laser technology, the microwave technology lags in terms of research. High-power microwave (HPM) use electromagnetic radiation to deliver heat, mechanical, or electrical energy to a target to cause various, sometimes very subtle, effects. When used against equipment, directed electromagnetic energy weapons can operate similar to omnidirectional electromagnetic pulse (EMP) devices, by inducing destructive voltage within electronic wiring. The difference is that they are directional and can be focused on a specific target using a parabolic reflector.

High-energy radio-frequency (HERF) weapons or high-power radiofrequency weapons (HPRF) use high-intensity radio waves to disrupt electronics. However, high and low power, pulsed microwave devices use low-frequency microwave radiation which can be made to closely mimic and interact with normal human brain waves having similar frequencies. Although it belongs to the same family of technology, the E-bomb deployment differs from that of HPM.

4.12.3 Potential for Aircraft Operations

Electronic warfare in the Information Age has defined the potential of these kinds of weapons for aircraft operations. DEWs have great potential for aircraft operations since crews can enhance their own survivability in the battlefield, where the aircraft are susceptible and vulnerable to missile threats, by protecting themselves with electromagnetic shields. In such environment, DEW systems may prevent the aircraft from threats by decreasing the detection and targeting capability of enemy. They may also aid in hit avoidance by deflecting, blinding, or causing the incoming missile to break lock and finally, where necessary, to destroy the missile itself before it reaches its target.

An additional approach might be to defeat the fusing system of the incoming missile. However, when deploying these bombs, getting the projectile successfully right is the key, such that useful damage can be produced. Further information about the deployment of these DEWs can be accessed from electronic warfare in the Information Age and utilization of artificial intelligence (AI), integrated with its components such as machine learning (ML) and deep learning (DL) as sub-system [49].

By this stage one difference between HPM and E-bomb is apparent, despite belonging to the same technological family, and this difference is their deployment. HEMP (high altitude electromagnetic pulse) is not a directed-energy weapon. The reason why HEMP is defined as an electromagnetic weapon is that it produces similar effects in electromagnetic spectrum and can cause similar impacts on electronic devices.

Fig. 4.80 Artistic AGM-154 Joint Standoff Weapon image

The potential effects of a designed HPM weapon strongly depend on the electromagnetic properties of the target. Since it is difficult to get the required intelligence, the complexity of real systems poses technical difficulties. A typical HPM weapon system basically includes a prime source that generates the intended power, an RF generator, a system that shapes and forms the wave into the intended form, a waveguide through which the generated wave travels, an antenna that propagated the wave, and the control unit that manages all the steps.

Delivery system considerations for E-bombs are very important. The massed application of such electromagnetic weapons in the opening phase of an electronic battle delivered at the proper instant or location can quickly lead the superiority in the electromagnetic spectrum.

This package might mean a major shift from physically lethal weaponry to electronically lethal attacks (via E-bombs) as a preferred mode of operation. Potential platforms for such weapons delivery systems are AGM-154 Joint Standoff Weapon (JSOW) glide bomb (Fig. 4.80) and the B-2 bomber (Fig. 4.81). The attractiveness of glide bombs delivering HPM warheads is that the weapon can be released from outside the effective radius of target air defenses, minimizing the risk to the launch aircraft, which can stay clear of the bomb's electromagnetic effects.

Another delivery method of E-bomb may be the use of UAVs. The technology of UAVs is still developing and partly immature; however, improvements can be expected in the next decade.

Whether E-bomb is science fiction or fact given today's technology is the question. So, can this hypothetical E-bomb be a significant weapon for the future

Fig. 4.81 B-2 bomber during refueling

battlefield? Theoretically, the military advantage obtainable with E-bombs is related mostly to their operational significance. Will future battlefields be won by the countries that best manage the revolution in military affairs or technological revolution? If latter is the case, then one has to remind himself that technology is not a winner on its own, but it has been, and it will continue to be, a critical enabler. If everything else is equal, the side with better technology will win.

Finally, can the country that first develops this new weapon have a significant and exploitable military advantage against other powers? Is it feasible for a nation to invest in this kind of bomb?—the debate continues.

4.13 Sixth-Generation Pilotless Driven Directed-Energy Warfare Delivery

Almost about 6 years ago for the first time, Aviation Week reported the existence of a large, classified pilotless aircraft developed by aviation companies such as Northrop Grumman along with the growing series of evidence that the stealthy aircraft as illustrated in Fig. 4.82 is now fully operational with the United States Air Force in a penetrating intelligence, surveillance, and reconnaissance (ISR) role.

From artistic illustration of Fig. 4.82, one can come to conclusion that B-2 (Fig. 4.80) bomber as just a wing shape plane has been in production and flying around as final production past few years ago and to be dubbed (i.e. an unofficial name or nickname) the RQ-180, the advanced design is believed to have been flying

Fig. 4.82 Dubbed RQ-180 artistic image

cross desert of California and Nevada since 2010 and under operational test and evaluation since 2014.

According to the new information provided to Aviation Week, the aircraft became operational with the recently reformed 427th Reconnaissance Squadron at Beale AFB, California, this year (2019). The Air Force declined to comment on the status of the program, although rumor on the street is that RQ-180 first flight believed to have occurred in 2010 and at least seven vehicles have been developed and are in operation by USAF.

Although images of the aircraft remain elusive, an assessment of new evidence enables a clearer picture to be drawn of the secret aircraft's progress through early flight testing, development, and initial deployment. New information from open sources backs up the first reports of its existence published in 2013 and fills in gaps in the program's earlier history as well as subsequent test and operational evaluation at sites mostly in and around California and Nevada.

Developed to conduct the penetrating ISR mission that has been left unaddressed since the retirement of the Lockheed SR-71 in 1999, the RQ-180 ultimately emerged from what was originally a large Unmanned Combat Air Vehicle (UCAV) design proposed by Northrop Grumman to the Air Force in 2005. At the time, Northrop was competing against Boeing with a smaller tailless design for the Air Force/US Navy Joint Unmanned Combat Air Systems (J-UCAS) program [51].

However, when J-UCAS was canceled in 2006 after the Pentagon's Quadrennial Defense Review opted to restructure the joint-service program into a Navy-only UCAV carrier suitability demonstration, funding was removed from the fiscal 2007 defense budget request. A total of $239 million was requested in lieu of the Pentagon funding to begin a US Navy carrier-based, long-endurance UCAV demonstration program.

At the same time, Air Force funds were transferred into a classified High-Altitude, Long-Endurance (HALE) program which, it is believed, led to a competition between Boeing, Lockheed Martin, and Northrop Grumman. Northrop also publicly discussed a range of longer-winged X-47C configurations around this time. The largest of these was a 172 ft-span design with two engines derived from General Electric's CF34 and capable of carrying a 10,000 lb weapon load.

Additional evidence now suggests the final configuration may be closer to the company's more familiar flying-wing designs, with a simpler trailing edge similar to that seen in the Air Force's official rendering of the B-21 Raider. Northrop Grumman originally crafted the same basic trailing edge configuration for the B-2 under the Advanced Tactical Bomber (ATB) program but changed it to the stronger load-carrying sawtooth design when the Air Force added the low-level penetration role.

The RQ-180 design also was likely strongly influenced by Northrop Grumman's work for the Air Force Research Laboratory's (AFRL) SensorCraft project, aimed at developing technologies for future stealthy, high-altitude unmanned surveillance platforms. In 2002, AFRL unveiled several SensorCraft vehicle studies, including a Northrop Grumman flying wing with a highly loaded airfoil capable of handling large aeroelastic deflections. Two years later, the company revealed it was partnering with AFRL to mature advanced conformal antenna integration technology for SensorCraft under a 5-year, $12 million effort called the Low-Band Structural Array (Lobstar) program. At the time, the company said Lobstar would "enhance the surveillance capabilities of aerial vehicles by embedding antennas in the primary load-bearing structures of composite aircraft wings."

Further signs of RQ-180 regular operations support activity are believed to have been indicated by the activation during 2018 and early 2019 of Detachment 5 of the 9th Operations Group at Beale to serve as the schoolhouse unit for the aircraft. Given the 9th Operations Group's role in training, planning, and execution of U-2 intelligence, surveillance, and reconnaissance (ISR) missions as well as training for RQ-4 flight crew members, this unit would be considered as a logical candidate to support and train RQ-180 operations.

Although the Air Force has made no reference to operations by the unit involving any particular aircraft type, the 427th Reconnaissance Squadron, together with Detachment 5 of the 9th Operations Group and Detachment 3 of the 605th Test and Evaluation Group, hosted the opening of a new Common Mission Control

Center at the base on April 23. The new center will "provide combatant commanders scalable, tailorable products and services for use in contested environments," the Air Force says. "Using software, hardware and human machines, the center will be able to manage C2 productivity, shorten the task execution chain, and reduce human-intensive communication." More details can be found in Ref. [51] of this chapter at the end under reference section.

4.14 Stealth in Strike Warfare

Dr. Carlo Kopp, a consultant to Air Power Australia in his published article on January 27, 2014 [52], under similar title, expresses that low observables (LO) or stealth is the most important paradigm in air warfare since the invention of the jet engine. Stealth technology aims to reduce the radar signature and infrared signature of an aircraft to the point, where detection ranges by hostile sensors and weapons are so small, as to render them tactically ineffective.

He goes on to say that the increasing capabilities of guided missiles and airborne radar during the late 1960s and early 1970s reached the level where the established methods of defense penetration, based upon a combination of maneuver and jamming, became increasingly less effective. The wide proliferation of pulse Doppler radar and IRS&T equipment, and improvements in missile performance and seeker technology, produced a situation where maneuver and low-altitude flight could not prevent engagements from being initiated, especially against bombers. Figure 4.83 is a presentation of F-117 getting in position of strike.

The increasing sophistication of radar and seeker technology caused significant and growing costs in electronic countermeasure (ECM) or jammer equipment, and

Fig. 4.83 Lockheed Nighthawk F-117 in air strike position

the increasing tempo of warfare meant that time would not be available to adapt existing in-service ECM equipment to hitherto unknown threat systems, before unacceptable combat losses were incurred.

Both maneuver and jamming are techniques which defeat specific weaknesses of an opponent's sensors and weapons. Without knowledge of these weaknesses, a priori, gained, for instance, through human intelligence operations, signals, and electronic intelligence, or capture of an opponent's equipment, it is extremely difficult and often impossible to develop particularly effective countermeasures.

The central philosophy behind Stealth is to defeat the basic physics underlying the opponent's sensors and weapons. By reducing the signatures of an aircraft down to an extremely low level, an opponent's sensor and weapons technology is denied any information about the aircraft. Very faint and fluctuating signatures will be extremely difficult to detect until the aircraft is very close to the threat system and will also be extremely difficult to track successfully.

A typical missile engagement requires that the aircraft be detected and tracked, its flightpath predicted, and missiles launched and guided to impact for the engagement to be successful. Should any of these phases of the engagement be disrupted or defeated successfully, the engagement will not be successful.

Extremely short detection ranges produce the further advantage of compressing the time available for the opponent and his automated equipment to react, thereby increasing the chances of the equipment not performing or the operators making mistakes.

Stealth restores the element of surprise at a tactical, operational, and strategic level and will place an opponent in a situation not unlike that which predated the invention of radar.

Stealth techniques are technologically demanding, since they require that designers address the necessary constraints inherent in signature reduction first and foremost, requiring significantly more complex tradeoffs in other areas of a design.

At this time only two operational types, the F-117A and B-2A, employ genuine stealth technology. The USAF's F-22A Raptor will be the next production aircraft to employ genuine stealth technology, which is also to be incorporated into the planned Joint Service Fighter.

An airplane of stealth capabilities of present fifth generation and future sixth generation involved in strike warfare mode needs to take a different approach as described as follows and as also demonstrated in artistic pattern of Fig. 4.84.

The established penetration technique for strike aircraft, pioneered by the F-111 design, involves flying into defended airspace at very low altitudes and high speeds and defeating hostile radar and weapon guidance by using jammers. For this purpose, conventional strike aircraft are equipped with terrain-following radar (TFR) or avoidance radars (TAR), thermal imagers, and typically comprehensive packages of radar warning and jamming equipment. In a situation where the opponent lacks pulse Doppler technology capable of detecting low-flying targets and uses relatively simple and unsophisticated radar and missile guidance equipment, low-level defense penetration can be very effective. Until recent times this has

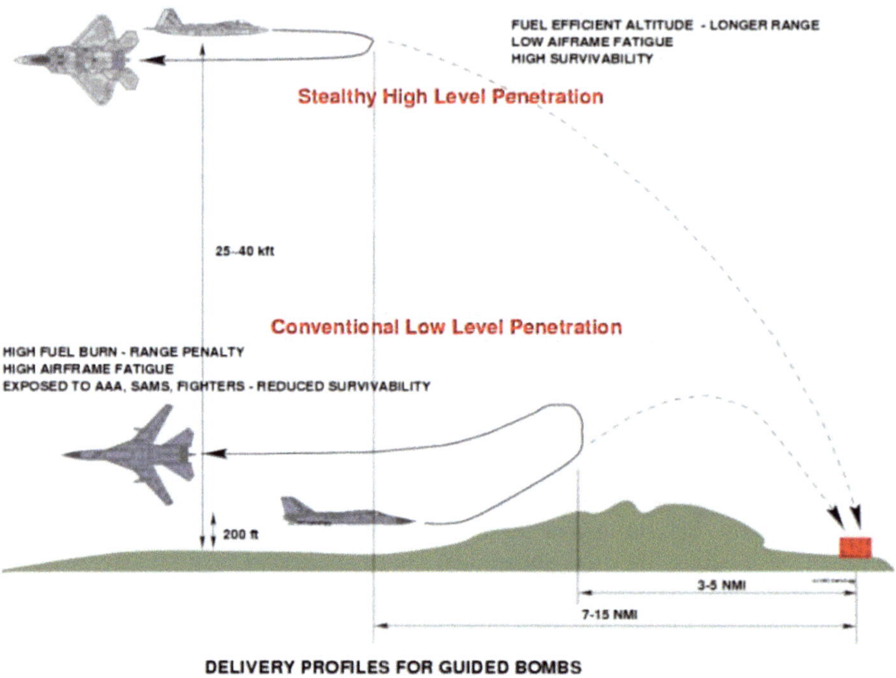

Fig. 4.84 Schematic of delivery profiles for guided bombs [52]

been true of the broader region, and thus the RAAF's F/RF-111C/G has been an effective penetrator.

Low-level penetration, while tactically effective in relatively benign threat environments, has some important limitations. The first is that it incurs a significant penalty in combat radius, since turbojet and turbofan-specific fuel consumption is poor at low altitudes and the higher air density requires higher thrusts be employed to achieve tactically useful airspeeds. Moreover, continuous maneuvers to clear terrain impose a significant fatigue load on the airframe, and the aircrew, thus limiting airframe life and aircrew endurance in combat (Fig. 4.85).

Often much effort is required in mission planning to select the lowest-risk ingress and egress routes, and in some instances supporting aircraft armed with anti-radiation missiles may be required, as well as fighter escorts. This technique is termed "strike packaging" and was pioneered during the Vietnam War. Its primary drawback is the costs incurred per damage inflicted, since the supporting assets typically outnumber the bombers [52].

At low and very low levels, aircraft will be exposed to fire from a wide range of weapons, including small arms, anti-aircraft artillery (AAA), short-range point-defense surface-to-air missiles (SAMs), and man-portable SAMs (MANPADS), collectively termed "trash fire." While not particularly effective on a per-firing basis, large numbers of firings will often yield a statistically significant outcome, and aircraft will be lost, as what happened with RAF Tornadoes during the early

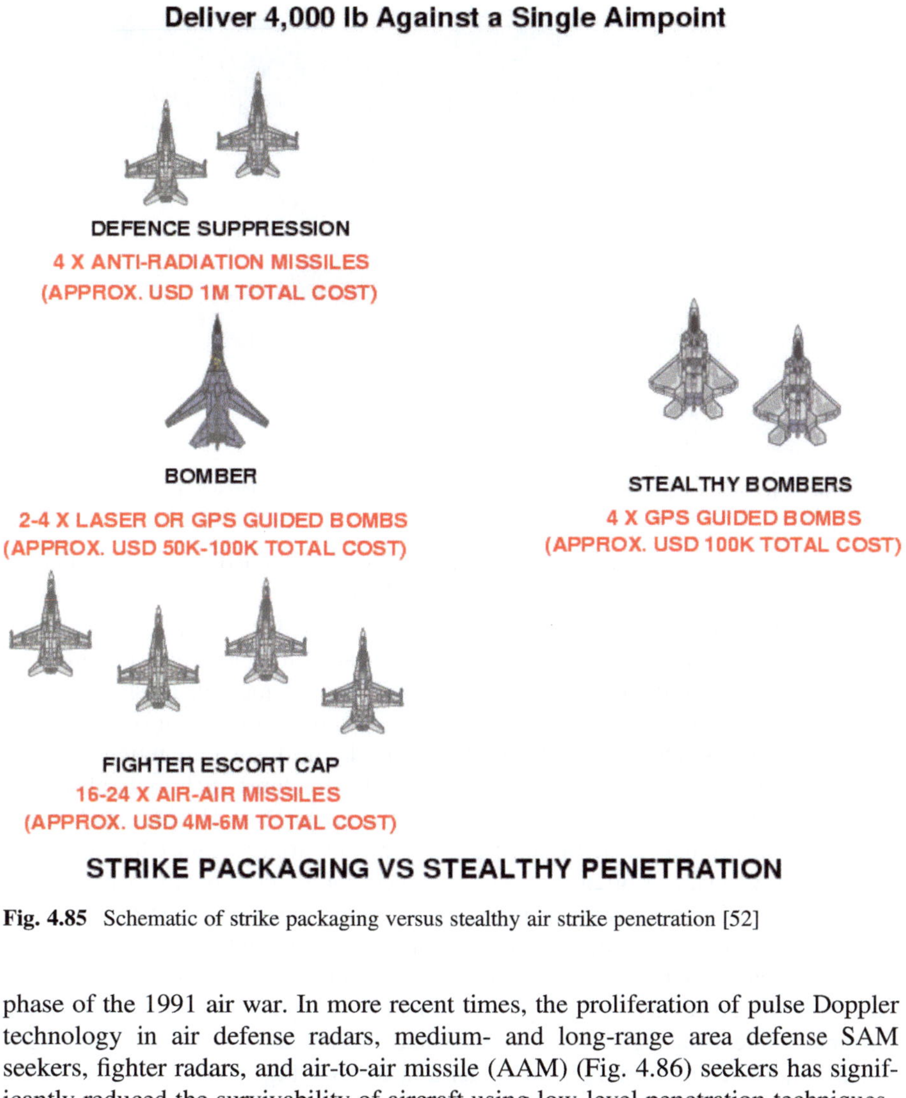

Fig. 4.85 Schematic of strike packaging versus stealthy air strike penetration [52]

phase of the 1991 air war. In more recent times, the proliferation of pulse Doppler technology in air defense radars, medium- and long-range area defense SAM seekers, fighter radars, and air-to-air missile (AAM) (Fig. 4.86) seekers has significantly reduced the survivability of aircraft using low-level penetration techniques.

Illustration in Fig. 4.87 is a presentation of Chinese and Russian air-launched weapons capabilities associated with their airplane of their deliveries.

As Fig. 4.88 illustration presents, the strategy recently adopted by users of conventional low-level penetration aircraft to defeat such defenses has been the adoption of standoff missiles and glide weapons, which may be launched from outside the effective range of the target's defenses. This technique can often be highly effective but incurs a major cost penalty since standoff weapons are typically 10–50 times more expensive than guided bombs.

Moreover, fighter aircraft can often engage bombers at ranges of hundreds of miles from the intended target. Defeating fighters requires standoff weapons such as

Fig. 4.86 Raytheon air-to-air missile AAM-A-1 Firebird. (Source: www.wikipedia.com)

medium- and long-range cruise missiles, which can be launched from safe distances. Such weapons are mostly very expensive, with costs in excess of a million dollars per round and carry relatively small warheads. Unless the conflict is very short, stocks of weapons may be expended before the desired military effect is achieved.

It is worth reiterating that cruise missiles and standoff missiles most often carry warheads of weights between 500 and 1000 lb. With the exception of the United Kingdom's Royal Ordnance BROACH warhead, most such munitions have a limited ability to defeat thick reinforced concrete structures such as bunkers and hardened aircraft shelters. It is worth noting the large numbers of Tomahawk cruise missiles typically expended by the United States in strikes against Iraq or, more recently, in Bosnia. It is often necessary to target four to eight rounds to achieve the same damage effects as produced by a pair of cheap guided bombs [52].

The use of stealth techniques avoids most of these difficulties. A stealth aircraft may penetrate at a high subsonic or low supersonic speed at medium or high altitudes, thus achieving the best possible fuel efficiency and combat radius for the airframe, while incurring minimal airframe and aircrew fatigue. Mission planning is much simplified since terrain is no longer a factor. See Fig. 4.87.

The target may be attacked with relatively cheap guided bombs, which provide very high lethality even against hardened targets. This will translate into a lesser number of sorties required to achieve the desired military effect, since the lethality per sortie is much increased. In terms of "bang per buck," stealthy penetration is significantly cheaper than either strike packaging or using standoff weapons.

This is most apparent in a sustained combat situation. If we make the arguably optimistic assumption that adequate standoff missile stocks are available for the duration of the conflict, we find that the USD 70–100M cost of a stealthy strike aircraft is equal to the cost of the standoff missiles expended after a mere 35–50 strike sorties flown against defended airspace. If we assume a turnaround time of 2 h per sortie, and a sortie duration of 4 h, i.e., four sorties per day (Fig. 4.89), then the

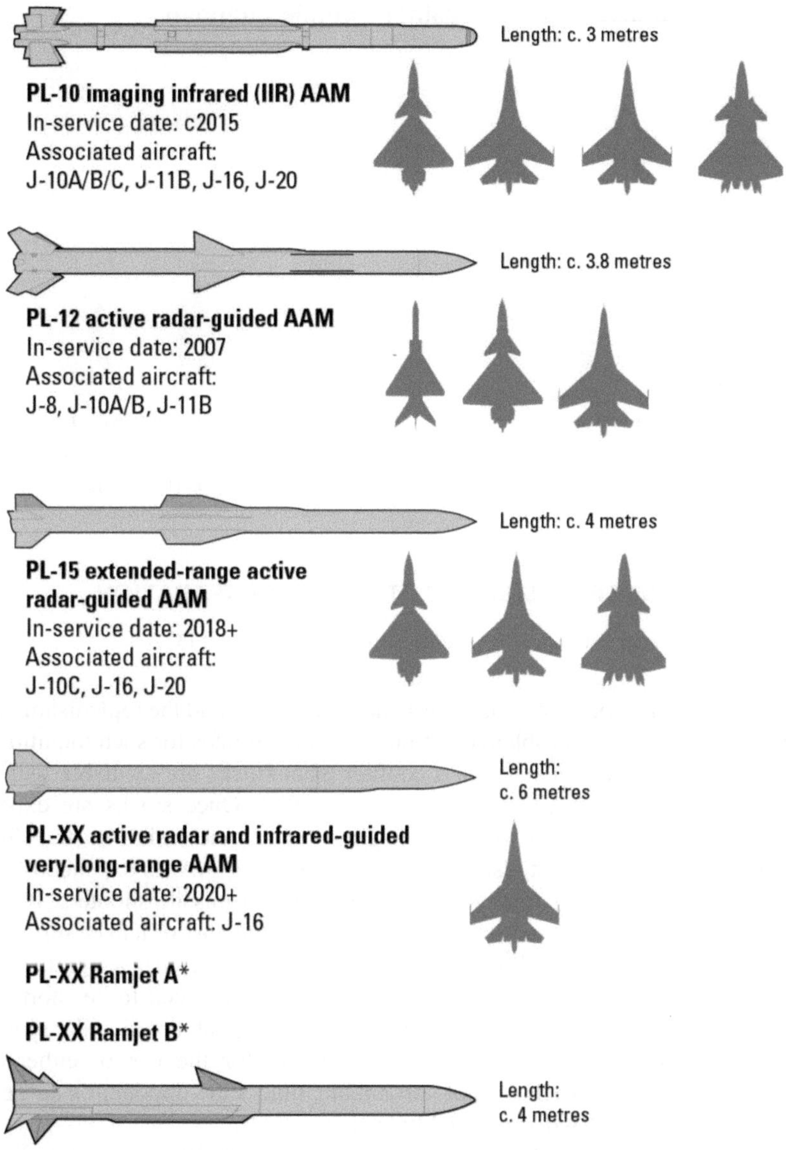

Fig. 4.87 Chinese and Russian air-launched weapons. *In development (Source: The Military Balance 2018)

cost of the stealthy strike aircraft is amortized in 9–12.5 days of sustained combat operations. For higher sortie rates at shorter ranges, this amortization rate is even higher. The case is even stronger should we consider using strike packaging rather than standoff weapons [52].

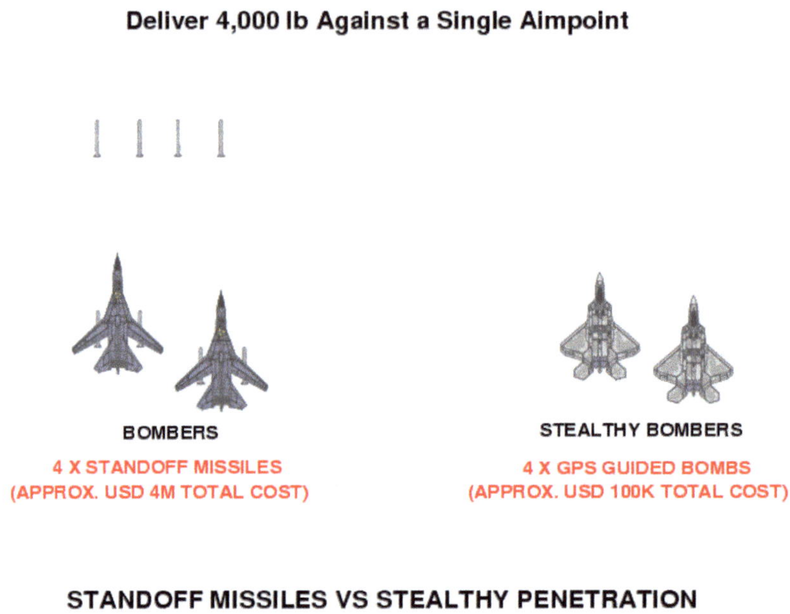

Fig. 4.88 Illustration of delivery of 4000 lb against a single aimpoint [52]

The issue of war stocks of expensive standoff weapons and the replenishment rate of these by production is problematic. Since production rates for such munitions are modest, due to their complexity, in a conflict what stocks are available will more than likely have to last the duration of the conflict. Once stocks are expended operations must fall back on strike packaging, further increasing costs. Where a fighter threat exists, we must also budget the costs of the AAMs expended and the costs of mounting fighter sorties to defend the standoff missile shooters. If we are operating beyond the CAP radius of the fighter, then the cost of tanker sorties must be included. Therefore, shooting standoff missiles may not confer a significant cost advantage unless the duration of the conflict can be guaranteed to be shorter than about 1 week. Recent historical experience suggests that conflict durations are typically of several weeks; therefore the argument for the use of either strike packaging or standoff missiles is not sustainable, unless the opponent's air defense capabilities can be defeated very quickly [52].

A scenario of regional relevance would be such where the Royal Australian Air Force (RAAF) is attempting to shut down several airfields, defended by fighters and SAMs. Given that an airfield basing one or two squadrons of aircraft will have a dozen or more critical aimpoints, and will most likely need to be reattacked to keep runways and taxiways closed, it is unlikely that the RAAF, or any air force of similar modest size, will be able to sortie enough aircraft to achieve a knockout blow in the first few days. Therefore, the opponent's air capability will have to be reduced by repeated strikes over a 1- or 2-week period until rendered operationally ineffective. See Fig. 4.90, an artistic illustration of stealth fifth-generation fighter jet.

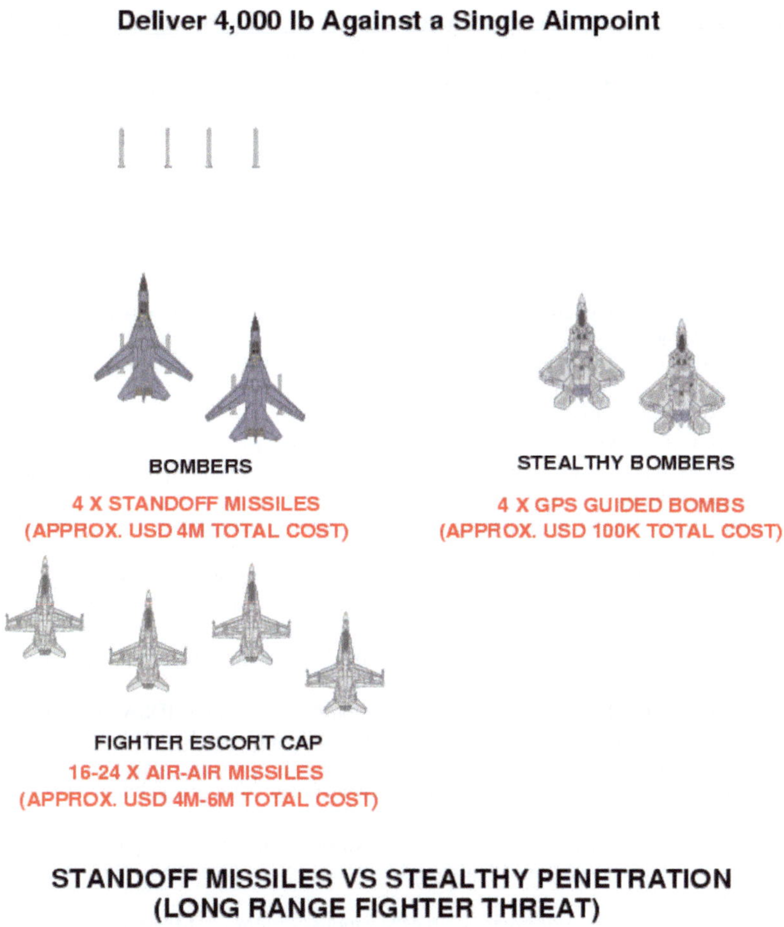

Fig. 4.89 Schematic of four sorties per day [52]

Fig. 4.90 Fifth-generation stealth jet fighter [52]

As a result, the expectation that air superiority can be achieved quickly and decisively is somewhat optimistic. Under such conditions, the cost advantages of stealthy strike over strike packaging or escorted standoff missile attacks are truly compelling [52].

References

1. https://www.dailymail.co.uk/sciencetech/article-7522413/amp/German-radar-tracked-two-U-S-F-35-stealth-jet-100-MILES-hiding-pony-farm.html
2. https://www.c4isrnet.com/intel-geoint/sensors/2019/09/30/stealthy-no-more-a-german-radar-vendor-says-it-tracked-the-f-35-jet-in-2018-from-a-pony-farm/
3. https://en.wikipedia.org/wiki/Mitsubishi_X-2_Shinshin
4. B. Zohuri and M. Moghaddam, "Neural Network Driven Artificial Intelligence: Decision Making Based on Fuzzy Logic (Computer Science, Technology and Applications: Mathematics Research Developments), Nova Publication July 24, 2017
5. B. Zohuri and Masoud Moghaddam, "Artificial Intelligence Driven by a General Neural Simulation System – Genesis" Nova Publication January 15, 2018
6. B. Zohuri, "Directed Energy Beam Weapons", Springer Publication Company July 2, 2019.
7. B. Zohuri, "Scalar Wave Driven Energy Applications", Springer Publication, September 28, 2019
8. B. Zohuri, "Directed Energy Weapons: Physics of High Energy Lasers" Springer Publications, August 30, 2016.
9. Vivek Kapur, "Stealth Technology and Its Effect on Aerial Warfare" IDSA Monograph Series, No. 33 March 2014, Institute For Defense Studies and Analyses, Rao Tula Ram Marg, Delhi Cantt, New Delhi, India.
10. "Searchlights and Sound Locators", http://www.antiaircraft.org/search.htm, (Accessed on September 14, 2013).
11. "Reduction of Advanced Military Aircraft Noise", http://www.serdp.org/Program-Areas/Weapons-Systems-and-Platforms/Noise-and-Emissions/Noise/WP-1583, Accessed September 16, 2013.
12. https://www.serdp-estcp.org/Program-Areas/Weapons-Systems-and-Platforms/Noise-and-Emissions/Noise/WP19-1125.
13. "Playing Skillfully With a Loud Noise", http://www.insidescience.org/content/playingskillfully-loud-noise/771, and Anlage A , "Noise Aspects of Future Jet Engines", http://www.mtu.de/en/technologies/engineering_news/others/Traub_Noise_aspects_en.pdf (Accessed October 04, 2013).
14. "Contrails", http://ww2010.atmos.uiuc.edu/(Gh)/guides/mtr/cld/cldtyp/oth/cntrl.rxml, (Accessed September 15, 2013). also see "Contrail Science", http://contrailscience.com/, (Accessed September 16, 2013), pp 85-87.
15. Allen E. Fuhs, and David C. Jenn, "Fundamentals of Stealth with Counter Stealth Radar Fundamentals: Applied to Radar, Laser, Infrared, Visible, Ultraviolet, & Acoustics," Naval Air Warfare Center Weapons Division China Lake, CA, Lecture Notes, 1999.
16. Dimitris V. Dranidis, "Airborne Stealth in a Nutshell-Part I," the Magazine of the Computer Harpoon Community http://www.harpoonhq.com/waypoint/, ().
17. Knight. op. cit., pp. 94-99, Air Chief Marshal Sir Michael Knight KCB, AFC, FRAeS. *Strategic Offensive Air Operations.,* Brassey's Defense Publishers Ltd, London, 1989.
18. Bill Sweetman, "Will Cost Kill Stealth?" Jane's International Defense Review 01.Oct.1996, www.janes.com, (Accessed January 2009).
19. F/A-22 Media Library, "Figure," http://f-22raptor.com/st_getstealthy.php, (Accessed February 2009).

20. Defense Update International Online Defense Magazine, "Visual Stealth for F-16? Hyper-Stealth Biotechnology Corp / Canada," "Figure," http://www.defenseupdate.com/products/f/f-16-camo.htm, (Accessed January 2009).
21. https://en.wikipedia.org/wiki/Plasma_stealth#cite_note-drag-5 (Last Accessed November 2019)
22. Ronald G. Driggers, Paul Cox, and Timothy Edwards, "Introduction to Infrared and Electro-Optical Systems," Artech House Norwood, MA, 1999
23. *The World's Great Stealth and Reconnaissance Aircraft*, Oriole Publishing Ltd, Hong Kong 1991, pp. 153-162.
24. Ralph Vartabedian and W.J. Hennigan, "F-22 program produces few planes, soaring costs", http://www.latimes.com/business/la-fi-advanced-fighter-woes-20130616-dto,0,7588480.htmlstory, Accessed October 13, 2013).
25. Winslow Wheeler, "Air Force Doesn't Know Aircraft Operations, Maintenance Costs; Audit Needed", http://breakingdefense.com/2011/09/21/air-force-doesnt-knowaircraft-operations-maintenance-costs-a/, (Accessed June 2013).
26. *The World's Great Stealth and Reconnaissance Aircraft*, Oriole Publishing Ltd, Hong Kong 1991, pp. 164-172.
27. Russian firm Rostec has built new canopies for Su-57 fighters, Tu-160 bombers and other warplanes, state-run TASS news agency reported on Jan. 11, 2018. The canopies include "a new composite material with enhanced radar-wave absorbing properties. (Accessed October 2019).
28. William E. Bahret, "The Beginnings of Stealth Technology," IEEE Transactions on Aerospace and Electronic Systems Vol.29, No 4 October 1993.
29. Aeronautical Information Manual (AIM) - Page 632". faraim.org. Retrieved 2 September 2015. FAA order 7110.10.
30. Serdar Cadirci, "RF Stealth (Or Low Observable) and Counter- RF Stealth Technologies: Implications of Counter- RF Stealth Solutions for Turkish Air Force", Naval Postgraduate School, Monterey, California, Thesis March 2009.
31. Robert P. Haffa Jr., and James H. Patton Jr., "Analogues of Stealth," Analysis Center Papers, Northrop Grumman Corporation, June 2002.
32. Doug Richardson, "Stealth Warplanes: Deception, Evasion, and Concealment in the Air," MBI Publishing Company, New York, 2001.
33. Global Security, "Figure," http://www.globalsecurity.org/wmd/systems/images/b-2_spirit_kitty_hawk092702.jpg, (Accessed February 2009).
34. Richard Seaman, "Figure," http://www.richardseaman.com/Aircraft/AirShows/Holloman2005/Highlights/F117AndT38Banking.jpg, (Accessed February 2009).
35. Latest Tech & Gadget News, "Figure" http://media.techeblog.com/images/f 22 5.jpg, (Accessed February 2009).
36. Coalition for Plasma Science, "What is Plasma", http://www.plasmacoalition.org/what.htm, (Accessed March 2009).
37. Writing by Tolip, "Russian Plasma Stealth Fighters," Military Heat 3 Oct 2007, http://www.military-heat.com/43/russian-plasma-stealth-fighters/, (Accessed February 2009).
38. T.R. Anderson and I. Alexeff, 2007 APS Division of Plasma Physics Annual Meeting November 12, 2007, Scientific Blogging Science 2.0, "Stealth Antenna Made of Gas Impervious to Jamming" http://www.scientificblogging.com/news_account/stealth_antenna_made_of_gas_impervious_to_jamming, (Accessed March 2009).
39. https://en.wikipedia.org/wiki/Electrohydrodynamics
40. Bill Sweetman, *Advanced Fighter Technology The Future of Cockpit Combat*, Airlife Publications Ltd, London,1988, pp. 105-106.
41. Phillip E. Pace, "Detecting and Classifying Low Probability of Intercept Radar, "Artech House Publishers, 2004.
42. Harrington Caitlin, "USAF looks for a more modest B-2 successor," Jane's Defense Weekly - October 10, 2007 www.janes.com, (Accessed January 2009).

43. Nation Master Encyclopedia, "Stealth Aircraft," http://www.nationmaster.com/encyclopedia/Stealth-aircraft, (Accessed February 2009).
44. F-117A the Black Jet, http://www.f-117a.com/Javaframe.html, (Accessed February 2009).
45. Christopher Bolkcom, "CRS Report for Congress F-22 A Raptor," June 12, 2007.
46. Carlo Kopp, "Lockheed F-117A Stealth Fighter" Australian Aviation, December 1990, http://www.ausairpower.net/Profile-F-117A.html, (Accessed February 2009).
47. Farhan Abdullah, Jeff Boyd, Mike Kowalkowski, Patricia Roman, Mandy Scott, and Joe Small, "B-2 Spirit," College of Engineering, http://aae.www.ecn.purdue.edu/~aae251/VOW_Presentations/Team_2_B_2.ppt, (Accessed February 2009).
48. Dimitris V. Dranidis, Airborne Stealth In A Nutshell Part II, Countering Stealth – Technology & Tactics,http://www.google.com/url?sa=t&rct=j&q=&esrc=s&source=web&cd=1&ved=2ahUKEwj8j56ch83lAhUYrZ4KHdffCRAQFjAAegQIAxAC&url=http%3A%2F%2Fwww.harpoonhq.com%2Fwaypoint%2Farticles%2FArticle_021.pdf&usg=AOvVaw3tvhiMWdEHOaeWmA-XnI8k (Accessed November 2019), *The magazine of the computer Harpoon community - http://www.harpoonhq.com/waypoint/*
49. Bahman Zohuri and Farhang Mossavar-Rahmani, "A Model to Forecast Future Paradigms: Volume 1: Introduction to Knowledge Is Power in Four Dimensions" Published by Apple Academic Press; 1 edition (January 2, 2020)
50. https://news.usni.org/2014/04/21/stealth-vs-electronic-attack
51. https://aviationweek.com/defense/usaf-unit-moves-reveal-clues-rq-180-ops-debut?utm_rid=CPEN1000002552854&utm_campaign=21863&utm_medium=email&elq2=b7408b78d4194c1e8548ef4f2e09143d (Accessed October 2013).
52. https://www.ausairpower.net/API-VLO-Strike.html

Appendix A: Luneburg Lens Radar Reflector

The Luneburg lens is a dielectric sphere with a permittivity varying with the distance from the center. This property allows a good focalization of the microwave beams in a focal point located at the peripheral of the lens. The lens can then be used as a reflector (with a metallization) or as emission or reception antenna with one or more feeds. The Luneburg lens is a passive radar augmentation device used to increase the radar reflectivity of a target without the use of additional energy.

A.1 Introduction

The lens reflector is a sphere in shape, usually composed of concentric dielectric shells. By the proper selection of dielectric constants for each shell, radar energy incident on one of the faces of the lens is focused at a point on the rear surface of the lens. The rear conductive surface reflects radar energy back to the source.

The physical characteristic of a Luneburg lens varies according to its application and the frequency at which it is required to operate in order to meet a variety of weapon system requirements; the company QinetiQ Target Systems integrates a variety of lens types into its targets. Generally, these are of 7.5 in. in diameter, but alternative sizes from 4 to 8.7 in. in diameter are available.

In general, radar reflector offers three models:

1. Luneburg reflectors
2. Trihedral reflectors
3. Active radar reflectors

The Luneburg lens is a passive radar enhancement device, that is, an augmentation device used to increase the radar reflectivity of a target without the use of additional energy. The lens reflector is a sphere in shape, usually composed of concentric dielectric shells as illustrated in Fig. A.1.

B. Zohuri, *Radar Energy Warfare and the Challenges of Stealth Technology*,
https://doi.org/10.1007/978-3-030-40619-6

Fig. A.1 The Luneburg lens reflector spherical shape. (Source: Meggitt Target Systems)

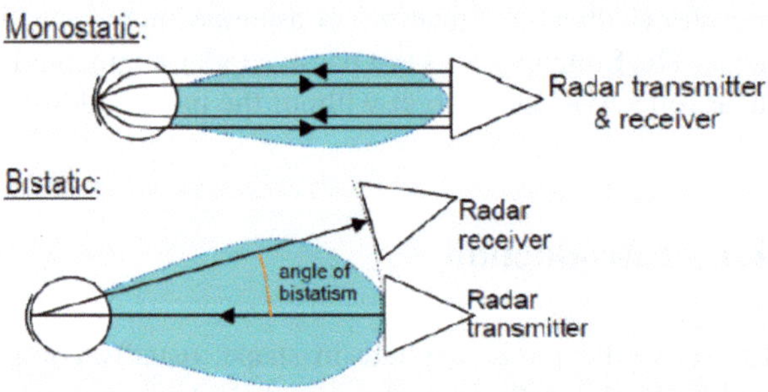

Fig. A.2 Fundamental Luneburg lens operation. (Source: Meggitt Target Systems)

By the proper selection of dielectric constants for each shell, radar energy incident on one of the faces of the lens is focused at a point on the rear surface of the lens. The rear conductive surface reflects radar energy back to the source. See Fig. A.2.

The physical characteristic of a Luneburg lens varies according to its application and the frequency at which it is required to operate. To meet a variety of weapon system requirements, Meggitt Target Systems integrates a variety of lens types into its targets. Generally, these are of 7.5 in. in diameter, but alternative sizes from 4 to 8.7 in. in diameter may be fitted to the Banshee and Snipe targets.

The radar cross section of a Luneburg lens is several hundred times that of a metallic sphere of the same size. Requiring no power supply or maintenance, the Luneburg lens is the most efficient, passive radar reflector available.

Lenses are generally of three types designed to fulfill different technical requirements.

(a) A monostatic unit where the radar source and the radar receiver are collocated. This type of lens is a retroreflector designed to operate with linear polarized radars. This is the most commonly used general-purpose reflector which has a broadband RF capability from S-band to Ku-band.

(b) A monostatic unit similar to that above but designed for use with radars that utilize circular polarization. These units, which look the same as the linear polarized radars, work differently and have a much narrower operating band. Therefore, they tend to be frequency specific.

(c) A bistatic unit designed for use where the radar source and receiver are located independent to each other, for example, where a radar is used to illuminate a target so that it can be acquired and identified by a missile's active radar seeker head. This unit is generally used for linear polarized systems.

Other lens types are available to meet specific weapon and user requirements, and they are provided by Meggitt Target Systems Corporation, and reader should refer to them. The operational infrastructure build around configuration illustrated in Fig. A.2 is the most efficient passive radar reflector available.

The Luneburg lens is used for two types of applications:

1. Reflector or passive radar reflector
2. Antenna

The electromagnetic and mechanical properties meet the needs of military and civilian applications.

- Military: radar detection of target
- Civilian:
  - Port and airport beaconing
  - Maritime and river beaconing
  - Aerial navigation assistance
  - Signaling of vehicles, ships, buoys, and obstacles
  - Microwave communication

One of the advantages of Luneburg reflector phenomena is that it increases the radar cross section (RCS) of any system which has little or none at all such as stealth airplane. Figure A.3 is a presentation of bistatic Luneburg, while Fig. A.4 shows a microwave path in Luneburg lens.

- The Luneburg reflector gives a homogeneous response inside a wide angle. It is an ideal passive responder, perfect for highlighting and eventually monitoring the radar target to which it is attached, with a high level of security.
- The Luneburg lens is the most efficient passive radar reflector available.
- The Luneburg reflector requires no power supply nor maintenance.

One military application of Luneburg lens is its implementation with stealth aircraft as a fighter jet of stealth type, presented in Fig. A.5.

As we stated above, Luneburg lens has a spherical configuration as its infrastructure and the reflectivity of a spherical lens reflector and the scattering of an

Fig. A.3 Bistatic Luneburg image

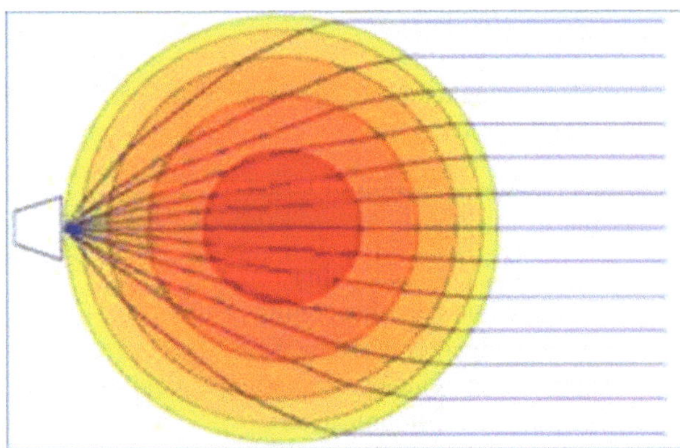

Fig. A.4 Microwave path in Luneburg lens

electromagnetic plane wave by a spherical lens reflector can be treated as a classical boundary value problem for Maxwell's equations.

No restrictions are imposed on the electrical size of reflectors and the angular size of the metallic spherical cap. The competitiveness of the spherical lens reflector against the Luneburg lens reflector can be demonstrated [1]. It has been found that spherical lens reflectors with relative dielectric constant in the range $3.4 \leq \varepsilon_r \leq 3.7$ possess better spectral performance than three- or five-layer Luneburg lens reflectors (LLR) in a wide frequency range.

The spherical lens (SL) is a homogeneous dielectric sphere which, for all dielectric constants in the range $1 \leq \varepsilon_r \leq 4$, focuses paraxial rays to a point z_{GO} outside the sphere. The distance from the center of the lens to z_{GO} is f, the paraxial focal length, which may be determined by geometrical optics, and is given in normalized form by Equation A.1.

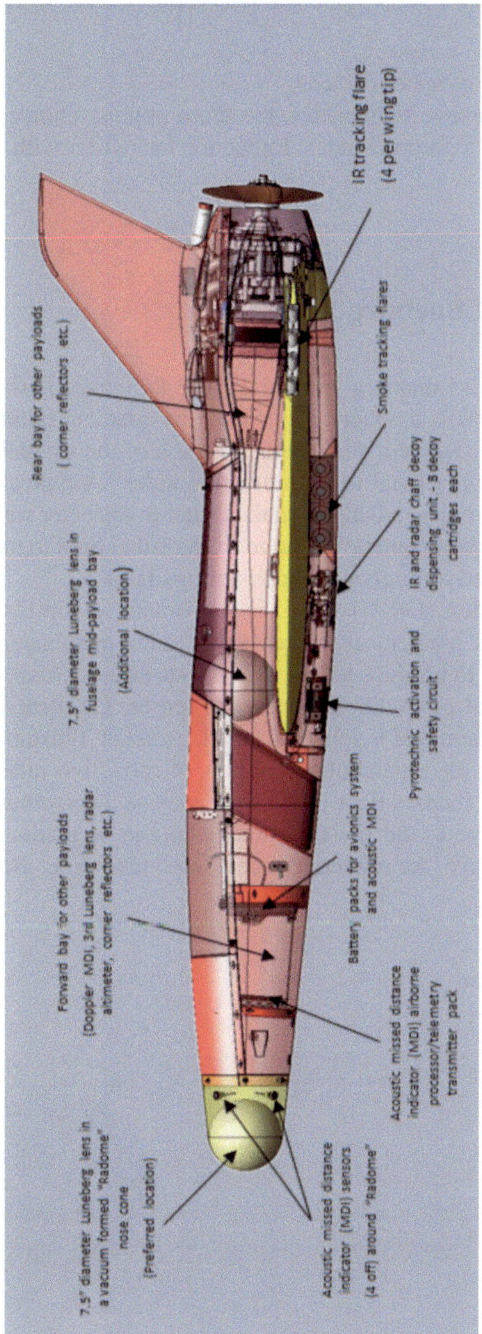

Fig. A.5 Infographic of Luneburg lens in a Banshee

$$\frac{f}{r_1} = \frac{\sqrt{\varepsilon_r}}{2\left(\sqrt{\varepsilon_r} - 1\right)} \tag{A.1}$$

where r_1 is the radius of spherical lens.

Note that at microwave frequencies, the more popular choice for the design of efficient reflectors is a stepped-index Luneburg lens (LL) with attached metallic spherical cap.

A.2 Physics of Luneburg Len

As Wikipedia states, a Luneburg lens (originally ***Lüneburg lens***, often incorrectly spelled *Luneburg lens*) is a spherically symmetric gradient-index (GRIN) lens as illustrated in Fig. A.6, where it shows cross section of the standard Luneburg lens, with blue shading proportional to the refractive index. A typical Luneburg lens's refractive index n decreases radially from the center to the outer surface. They can be made for use with electromagnetic radiation from visible light to radio waves as they all have been described throughout chapters of this book.

Note: Gradient-index (GRIN) optics is the branch of optics covering optical effects produced by a gradient of the refractive index of a material. Such gradual variation can be used to produce lenses with flat surfaces or lenses that do not have the aberrations typical of traditional spherical lenses. Gradient-index lenses may have a refraction gradient that is spherical, axial, or radial. Illustrated in Fig. A.7 is a gradient-index lens with a parabolic variation of refractive index (n) with radial distance (x). The lens focuses light in the same way as a conventional lens.

The lens of the eye is the most obvious example of gradient-index optics in nature. In the human eye, the refractive index of the lens varies from approximately

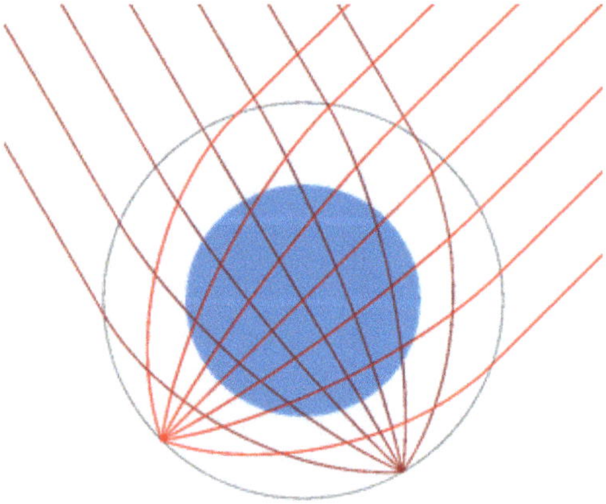

Fig. A.6 The standard Luneburg lens cross section. (Source: www.wikipedia.com)

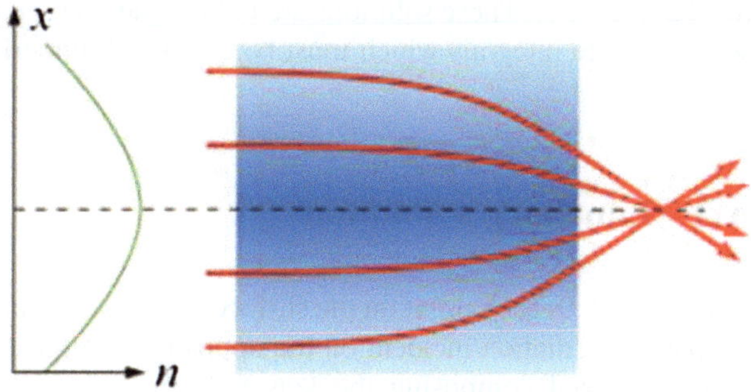

Fig. A.7 A gradient-index lens with a parabolic variation of refractive index. (Source: www.wikipedia.com)

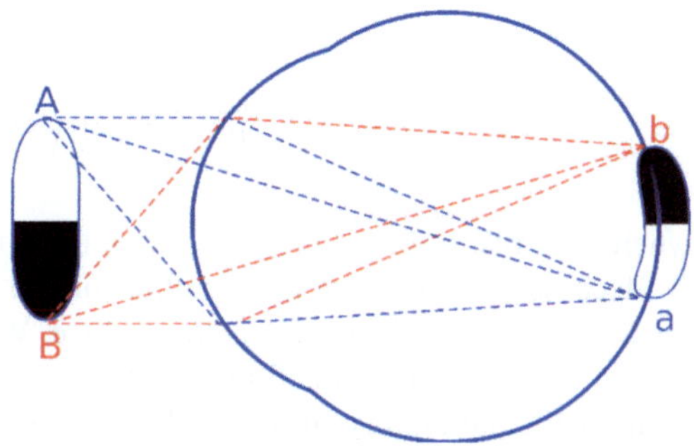

Fig. A.8 Eye focusing illustration. (Source: www.wikipedia.com)

1.406 in the central layers down to 1.386 in less dense layers of the lens. This allows the eye to image with good resolution and low aberration at both short and long distances.

To continue on the subject of Luneburg lens, per Wikipedia, we can say that for certain index profiles, the lens will form perfect geometrical images of two given concentric spheres onto each other. There are an infinite number of refractive index profiles that can produce this effect. The simplest such solution was proposed by Rudolf Luneburg in 1944. Luneburg's solution for the refractive index creates two conjugate *foci* (Fig. A.8) outside of the lens [2].

Figure A.8 is an indication of eye focusing which ideally collects all light rays from a point on an object into a corresponding point on the retina. *Foci* is an ellipse that has two focus points and in the eye is the plural of "focus," which is one focus, two *foci*. It is also an indication of the degree of clarity with which an eye or optical instrument produces an image. See focal point, a central point or region, such as the point at which an earthquake starts. The *foci* always lie on the major (longest) axis, spaced equally each side of the center. If the major axis and minor axis are of the same length, the figure is a circle, and both *foci* are at the center.

The solution takes a simple and explicit form if one focal point lies at infinity and the other on the opposite surface of the lens. J. Brown and A. S. Gutman subsequently proposed solutions which generate one internal focal point and one external focal point [3, 4]. These solutions are not unique; the set of solutions are defined by a set of definite integrals which must be evaluated numerically [5].

A.3 Luneburg's Solution

Using Fig. A.9, each point on the surface of an ideal Luneburg lens is the focal point for parallel radiation incident on the opposite side. Ideally, the dielectric constant ε_r of the material composing the lens falls from 2 at its center to 1 at its surface (or equivalently, the refractive index n falls from $\sqrt{2}$ to 1), according to Equation A.2 as:

$$n = \sqrt{\varepsilon_r} = \sqrt{2 - \left(\frac{r}{R}\right)^2} \tag{A.2}$$

where R is the radius of the lens and n again is refractive index.

Because the refractive index at the surface is the same as that of the surrounding medium, no reflection occurs at the surface. Within the lens, the paths of the rays are arcs of ellipses as illustrated in Fig. A.10.

In mathematics, an ellipse is a plane curve surrounding two focal points, such that for all points on the curve, the sum of the two distances to the focal points is a constant. As such, it generalizes a circle, which is the special type of ellipse in which the two focal points are the same.

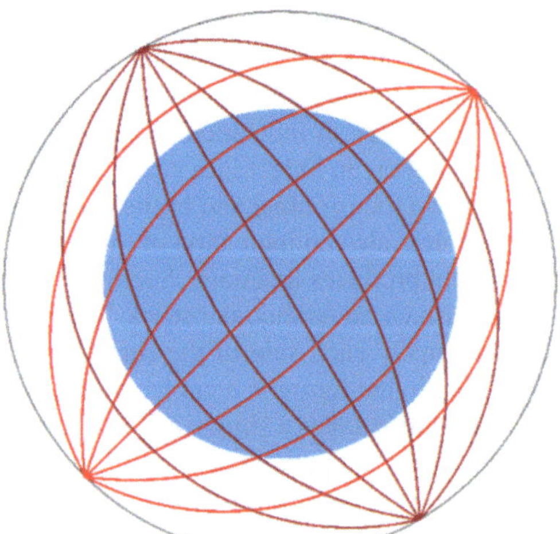

Fig. A.9 Cross section of Maxwell's fish-eye lens. (Source: www.wikipedia.com)

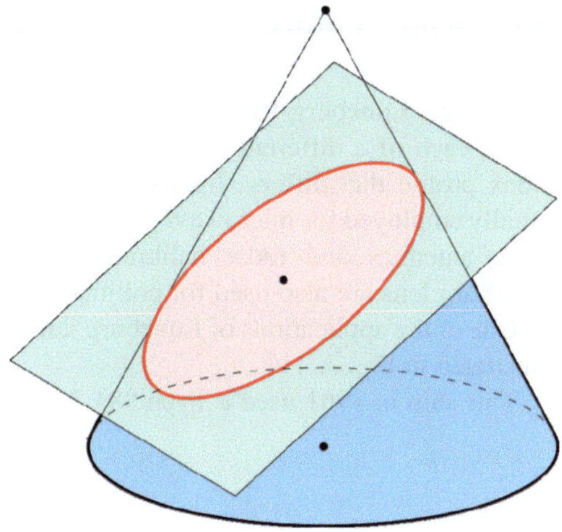

Fig. A.10 Ellipse shape. (Source: www.wikipedia.com)

A.4 Maxwell's Fish-Eye Lens

Maxwell's fish-eye lens is also an example of the generalized Luneburg lens. The fish-eye, which was first fully described by Maxwell in 1854 [6] (and therefore pre-dates Luneburg's solution), has a refractive index n varying according to Equation A.3 as:

$$n = \sqrt{\varepsilon_r} = \frac{n_0}{1 + \left(\frac{r}{R}\right)^2} \tag{A.3}$$

It focuses each point on the spherical surface of radius R to the opposite point on the same surface. Within the lens, the paths of the rays are arcs of circles.

A.5 Production and Attribution

The properties of this lens are described in one of a number of set problems or puzzles in the 1853 *Cambridge and Dublin Mathematical Journal* [7]. The challenge is to find the refractive index as a function of radius, given that a ray describes a circular path, and further to prove the focusing properties of the lens. The solution is given in the 1854 edition of the same journal [6]. The problems and solutions were originally published anonymously, but the solution of this problem (and one other) were included in Niven's *The Scientific Papers of James Clerk Maxwell* [8], which was published 11 years after Maxwell's death.

A.6 Applications

In practice, Luneburg lenses are normally layered structures of discrete concentric shells, each of a different refractive index. These shells form a stepped refractive index profile that differs slightly from Luneburg's solution. This kind of lens is usually employed for microwave frequencies, especially to construct efficient microwave antennas and radar calibration standards. Cylindrical analogues of the Luneburg lens are also used for collimating light from laser diodes.

One of the applications of Luneburg lens can be found on HMS Victor warship as illustrated in Fig. A.11.

This ship in 1961 used a Type 984 3D radar.

A.6.1 Radar Reflector

A radar reflector as pictured in Fig. A.12 in the form of retroreflector can be made from a Luneburg lens by metallizing parts of its surface. Radiation from a distant radar transmitter is focused onto the underside of the metallization on the opposite side of the lens; here it is reflected and focused back onto the radar station. A difficulty with this scheme is that metallized regions block the entry or exit of radiation on that part of the lens, but the non-metallized regions result in a blind spot on the opposite side.

Note: A retroreflector (sometimes called a retroflector or cataphote) is a device or surface that reflects radiation (light, usually) back to its source with a minimum of scattering. In a retroreflector the wavefront (see Chap. 1) of the radiation is reflected straight back to the wave's source. This works at a wide range of angle of incidence, unlike a planar mirror, which does this only if the mirror is exactly perpendicular to

Fig. A.11 Her Majesty Ship (HMS) Victors. (Source: www.wikipedia.com)

Fig. A.12 A gold first-surface corner cube retroreflector. (Source: www.wikipedia.com)

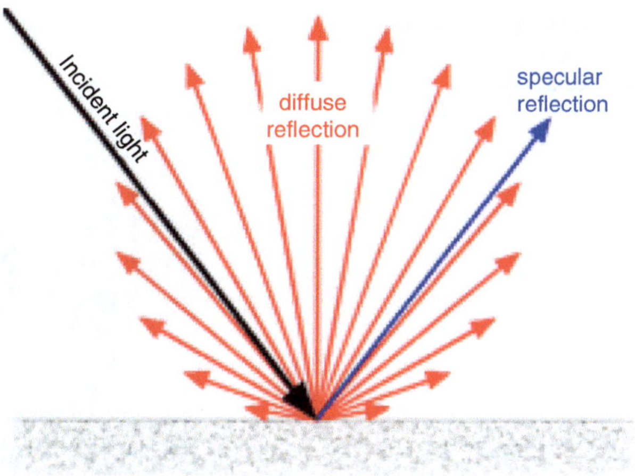

Fig. A.13 Diffused reflection. (Source: www.wikipedia.com)

the wavefront, having a zero angle of incidence. Being directed, the retroflector's reflection is brighter than that of a diffuse reflector as depicted in Fig. A.13.

Corner reflectors and cat eye reflectors are the most used kinds.

A.6.2 Microwave Antenna

A Luneburg lens can be used as the basis of a high-gain radio antenna. This antenna is comparable to a dish antenna (Fig. A.14) but uses the lens rather than a parabolic reflector as the main focusing element.

As with the dish antenna, a *feed* (see Fig. A.15) to the receiver or from the transmitter is placed at the focus, the feed typically consisting of a horn antenna (Fig. A.16).

Fig. A.14 A dish antenna configuration. (Source: www.wikipedia.com)

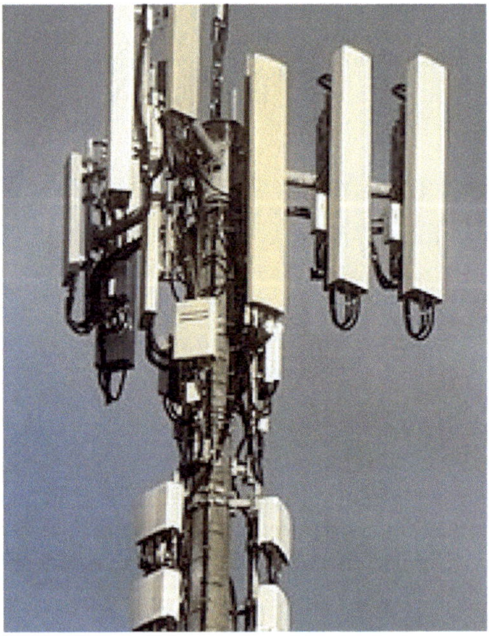

Fig. A.15 An antenna feed structure. (Source: www.wikipedia.com)

The phase center of the feed horn (Fig. A.17) must coincide with the point of focus, but since the phase center is invariably somewhat inside the mouth of the horn, it cannot be brought right up against the surface of the lens. Consequently, it is necessary to use a variety of Luneburg lens that focuses somewhat beyond its surface [9], rather than the classic lens with the focus lying on the surface.

Note that a feed horn (or feedhorn) is a small horn antenna used to convey radio waves between the transmitter and/or receiver and the parabolic reflector. In transmitting antennas, it is connected to the transmitter and converts the radio-frequency

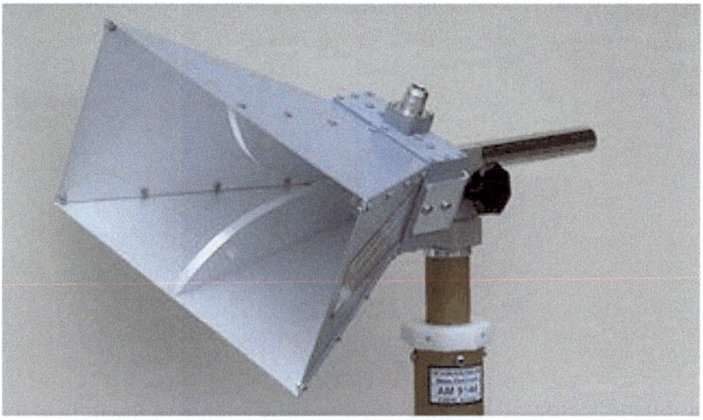

Fig. A.16 A horn antenna configuration. (Source: www.wikipedia.com)

Fig. A.17 Corrugated feed horn on a Hughes DirecWay home satellite dish. (Source: www.wikipedia.com)

alternating current from the transmitter to radio waves and feeds them to the rest of the antenna, which focuses them into a beam. In receiving antennas, incoming radio waves are gathered and focused by the antenna's reflector on the feed horn, which converts them to a tiny radio-frequency voltage which is amplified by the receiver. Feed horns are used mainly at microwave super-high frequency (SHF) (i.e., frequency range 2–30 GHz and wavelength range 1 dm to 1 cm) and higher frequencies.

A Luneburg lens antenna offers a number of advantages over a parabolic dish. Because the lens is spherically symmetric, the antenna can be steered by moving the feed around the lens, without having to bodily rotate the whole antenna. Again, because the lens is spherically symmetric, a single lens can be used with several feeds looking in widely different directions. In contrast, if multiple feeds are used

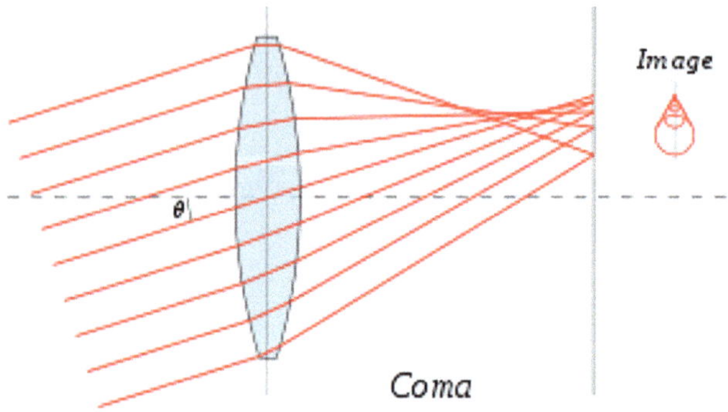

Fig. A.18 The coma or comatic aberration. (Source: www.wikipedia.com)

with a parabolic reflector, all must be within a small angle of the optical axis to avoid suffering coma (a form of de-focusing) (Fig. A.18).

Apart from offset systems, dish antennas suffer from the feed and its supporting structure partially obscuring the main element (*aperture blockage*); in common with other refracting systems, the Luneburg lens antenna avoids this problem.

A variation on the Luneburg lens antenna is the *hemispherical Luneburg lens antenna* or *Luneburg reflector antenna*. This uses just one hemisphere of a Luneburg lens, with the cut surface of the sphere resting on a reflecting metal ground plane. The arrangement halves the weight of the lens, and the ground plane provides a convenient means of support. However, the feed does partially obscure the lens when the angle of incidence on the reflector is less than about 45°.

A.7 Path of a Ray Within the Lens

For any spherically symmetric lens, each ray lies entirely in a plane passing through the center of the lens. The initial direction of the ray defines a line which together with the center point of the lens identifies a plane bisecting the lens. Being a plane of symmetry of the lens, the gradient of the refractive index has no component perpendicular to this plane to cause the ray to deviate to either one side of it or the other. In the plane, the circular symmetry of the system makes it convenient to use polar coordinates (r, θ) to describe the ray's trajectory.

Given any two points on a ray (such as the point of entry and exit from the lens), Fermat's principle asserts that the path that the ray takes between them is that which it can traverse in the least possible time. Given that the speed of light at any point in the lens is inversely proportional to the refractive index, and by Pythagoras, the time of transit between two points (r_1, θ_2) and (r_2, θ_2) is as demonstrated in Equation A.4:

$$T = \int_{(r_1,\theta_1)}^{(r_2,\theta_2)} \frac{n(r)}{c} \sqrt{(rd\theta)^2 + dr^2} = \frac{1}{c} \int_{\theta_1}^{\theta_2} n(r) \sqrt{r^2 + \left(\frac{dr}{d\theta}\right)^2} d\theta \tag{A.4}$$

where c is the speed of light in vacuum. Minimizing this T yields a second-order differential equation determining the dependence of r on θ along the path of the ray. This type of minimization problem has been extensively studied in Lagrangian mechanics, and a ready-made solution exists in the form of the Beltrami identity, which immediately supplies the first integral of this second-order equation. Substituting $\mathcal{L}(r, r') = n(r)\sqrt{r'^2 + r^2}$ (where r' represents $dr/d\theta$) into this identity gives:

$$n(r)\sqrt{r'^2 + r^2} - n(r)\frac{r'^2}{\sqrt{r'^2 - r^2}} = h \tag{A.5}$$

where h is a constant of integration. This first-order differential equation is separable, that is, it can be re-arranged so that r only appears on one side and θ only on the other [2].

$$d(\theta) = \frac{h}{r\sqrt{(n(r))^2 r^2 - h^2}} dr$$

The parameter h is a constant for any given ray but differs between rays passing at different distances from the center of the lens. For rays passing through the center, it is zero. In some special cases, such as for Maxwell's fish-eye, this first-order equation can be further integrated to give a formula for θ as a function or r. In general it provides the relative rates of change of θ and r, which may be integrated numerically to follow the path of the ray through the lens.

Appendix B: New Weapon of Tomorrow's Battlefield Driven by Hypersonic Velocity

Speed is the new stealth, and earlier this week, America's top nuclear commander described a grim scenario for US forces facing off against hypersonic weapons.

"We do not have any defense that could deny the employment of such a weapon against us," Air Force Gen. John Hyten, Commander of US Strategic Command, told the Senate Armed Services Committee on Tuesday March 20, 2018. Russian and Chinese are aggressively developing and new weapons that travel at Mach 5 or higher, which is at least five times faster than the speed of sound (hypersonic). These weapons travel in excess of 3600 miles per hour (1 mile per second), and currently, no military possesses a credible defense. Finding, tracking, and intercepting something that fast is unprecedented. Given that Russia and China have invested heavily in advanced defensive technologies that now hold most of our traditional forms of power projection at risk, this is a significant advantage—it is one that would impose major costs upon a defending nation. Recently, according to the Director of the Army's Rapid Capabilities and Critical Technologies Office (ARCCTO), the Army will field a battery of truck-borne hypersonic missiles in 2023, with a contract award in August, the service's new three-star Program Executive Officer said. The service will also field a battery of 50 kW lasers on Stryker armored vehicles by 2021, he said. A program to put a 100-plus-kilowatt laser on a heavy truck, however, is under review and may be combined with Air Force and/or Navy efforts to reach comparable power levels, Lt. Gen. Neil Thurgood told reporters in his interview. In this white paper, we are suggesting a new technology as a countermeasure against such an adversary measure and threat that is aggressively being pursued by these two nations, Russia and China, both tactically and strategically. We also briefly discuss possible physics and science of aerodynamics involved with these vehicles traveling between range of 5 Mach and higher, where we discuss current status and future direction driven by phenomena of Plasma Aerodynamics thorough possibly, Weakly Ionized Gases (WIG) program that was started by the former Soviet Republics under AJAX Vehicle and that was direct understanding of the role of plasmas in the performance of the this vehicle.

B. Zohuri, *Radar Energy Warfare and the Challenges of Stealth Technology*,
https://doi.org/10.1007/978-3-030-40619-6

Disclaimer:
All of the information contained in this paper was obtained from open sources. The opinions expressed are the authors alone.

B.1 Introduction

On December 26, 2018, Russia successfully carried out the launch of a liquid-fueled intercontinental-range ballistic missile (ICBM) carrying the Avangard hypersonic glide vehicle payload. The Avangard is a modernized Russian delivery vehicle, designed to maneuver in the upper atmosphere at speeds in excess of Mach 5. See the link below on YouTube and the artistic image depicted in Fig. B.1.

The missile that carries the Avangard hypersonic glide vehicle is the UR-100 NUUTKh, an ICBM-class missile. As *The Diplomat* reported earlier this year, Russian defense industry sources had noted that the first Russian Strategic Missile Forces regiment to operate the Avangard would oversee a test later this year [10].

"The launch was performed by an operational unit of the Strategic Missile Forces from Dombarovsky missile deployment area against a hypothetical target at the Kura range, Kamchatka Territory," the Kremlin noted in a statement.

"Flying at hypersonic speed, the glide vehicle performed vertical and horizontal maneuvers and hit the hypothetical target in time within the range's combat field," the statement added [10].

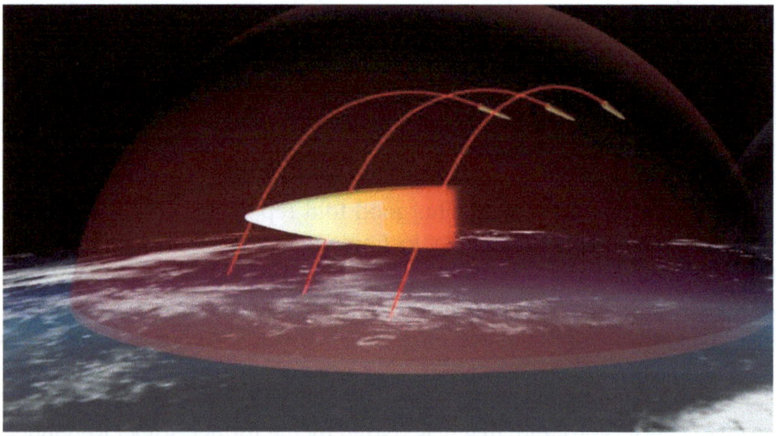

Fig. B.1 Artistic depiction of glide hypersonic warhead weapon. (Source: https://www.youtube.com/watch?v=tKa31NaYsNw)

Fig. B.2 Chinese DF-26 launch

Footage of the successful test was released immediately by Russian media, showing the missile's hot launch from a silo. The Avangard glider itself was not seen in any released footage. The missile's payload separates after the booster's powered flight into the exoatmosphere. After the resulting ballistic trajectory, the hypersonic glide vehicle descends and maneuvers in the upper atmosphere on its way to the target.

Putin emphasized the Avangard's primary purpose. "The new Avangard system is impervious to current and future air defense and missile defense systems of a potential enemy," he noted [10].

Hypersonic missiles developed by China and Russia are being used to justify the reprise of space-based missile defense systems by the Pentagon.

The United States is particularly concerned about super-fast guided missiles under development in China that could put US ships and bases at risk in Asia. China's DF-26 ballistic missile drill sent a clear message to the United States. See Fig. B.2.

In the case of the Russian hypersonic weapon Avangard, according to them, it maneuvers to bypass missile defenses en route to the target. President Vladimir Putin has declared that Russia has developed a range of new nuclear weapons that cannot be intercepted by an enemy.

Machine guns. Fighter jets. Nuclear weapons. When a new facet of military technology gains operational capability, sometimes it changes the rules of the game. Hypersonic weapons—which travel over five times the speed of sound—are difficult to detect and harder to intercept and offer that kind of potential.

B.2 The History of Hypersonic Vehicle

In 1994, Russian physicist and scientist were claiming and introduced an innovative and novel hypersonic flight vehicle concept that is known to us as AJAX or AYAKS through public domain articles and literatures. AJAX was described as a scramjet-powered vehicle driven by plasma-based technology with two purposes behind it as:

1. Combustion
2. Aerodynamics performance

Each of the above ideas has their own scientific ideology behind the purpose incorporating physics of weakly ionized gases (WIG) according to the US Air Force Research Laboratory (AFRL) and the European Office of Aerospace Research and Development (EOARD) in response to the Russian AJAX program. The WIG program was established to head on and foster United States-Eastern Bloc in order to collaborate and exchange on the recent interest on the subject of "plasma aerodynamics" science.

This field of science led to an international study and collaboration, which was stimulated by reveal and disclosure of the Soviet AJAX vehicle concept in the mid-1990s, and true belief today in Russia and China in respect to what hypersonic weapons are is the dove tail of AJAX program started by the former Soviet Union around the 1990s timeframe.

Plasma-based flow control seems very feasible, in particular for local flow control applications where power consumption is low, where new methods utilizing Magneto-Hydro-Dynamic (MHD), Flow Dynamic and Electro-Dynamic are couple to form MHD technique, which already leading to and it is employed in order to control and power extraction are already leading to innovative designs for hypersonic vehicles to address both issues of Combustion and Aerodynamic performance by increasing speed while reducing drag and friction like we can see in case of reentry vehicle.

Keep in your mind that the fundamental equations of "plasma dynamics" have been greatly simplified by MHD approximation, which uses the assumptions given and summarized below: [11]

1. Plasma is a single continuous medium of definite composition.
2. Electromagnetic forces are of the same order as gas dynamic forces.
3. The time scale of the problem is the characteristic length divided by characteristic velocity.
4. The applied electric field $\vec{E}$ is of the same order as the induced electromotive force.
5. The flow velocity is much smaller than the speed of light.

6. Maxwell's equations are unaffected by gas dynamic motion or, in other words, the magnetic field $\vec{B}$ induced by the fluid motion is small compared to the applied magnetic field.
7. Inviscid flow is assumed, but a friction term can be easily added to the momentum equation.
8. No heat loss is assumed, but a heat loss term can be easily added to the energy equation.
9. The equation of state of the gas is assumed to be the perfect gas law, but other equations of state can be implemented.
10. Body forces due to gravity are neglected.
11. Steady state is assumed.

All these above points can be simplified in generalized 3D magnetohydrodynamic (MHD) equations written as follows: [12]

Momentum Equation:

$$\rho\left(\vec{\mathrm{v}} \cdot \vec{\nabla}\right)\vec{\mathrm{v}} = \vec{J} \times \vec{B} - \nabla p \tag{B.1}$$

Mass Equation:

$$\vec{\nabla} \cdot \left(\rho\vec{\mathrm{v}}\right) = 0 \tag{B.2}$$

Energy Equation:

$$\rho\vec{\mathrm{v}} \cdot \vec{\nabla}\left(\frac{\left|\vec{\mathrm{v}}\right|^2}{2} + U\right) = -\vec{\nabla} \cdot \left(\vec{\mathrm{v}}p\right) + \vec{J} \cdot \vec{E} \tag{B.3}$$

Current Equation:

$$\vec{\nabla} \cdot \vec{J} = 0 \tag{B.4}$$

Ohm's Law:

$$\vec{J} = \sigma\left(\vec{E} + \vec{\mathrm{v}} \times \vec{B}\right) - \frac{\omega\tau}{\left|\vec{B}\right|}\vec{J} \times \vec{B} \tag{B.5}$$

In the above sets of conservation equations, the following nomenclature applies:

$\vec{B}$: vector magnetic field
$\vec{J}$: vector current density
ρ: gas density
$\vec{v}$: vector velocity of flow
U: internal energy of gas
p: flow pressure
$\vec{E}$: vector of applied electric field
$\omega\tau$: Hall parameter
σ: electric conductivity

The Russian AJAX hypersonic vehicle was the first to propose magnetohydrodynamic (MHD) energy bypass of a scramjet as a means of extending the scramjet's performance to higher Mach numbers [13].

Further analyses of the MHD concept lead to the conclusion that energy bypass of a scramjet can result in subsonic ramjet propulsion can supersede and being able to maintain speed in the range of Mach numbers between 10 and 16 [14].

A simplified thermodynamic cycle analysis of scramjet energy bypass demonstrated that the concept merits further investigation [15].

Based on these results, an examination of the feasibility of MHD energy bypass with turbojets was proposed [16].

As with the scramjet, the enthalpy into the combustor is to be reduced allowing more efficient addition of energy in the combustor without exceeding temperature limitations on the turbine materials. Preliminary 1D analysis of the energy extraction process shows that significant enthalpy extraction is possible, but this extraction also results in significant total pressure losses [17].

Today, *plasma-enhanced combustion* is behind 85% of primary energy conversion processes which are based on combustion, and this proportion is expected to remain stable in the foreseeable future. However, AJAX project triggered interest in the use of plasma discharges as a means to provide in-place, on-demand enhancement of fuel-air reactivity and thus launched the new field of plasma-assisted combustion.

To continue our discussion with flight control and reduced surface heating can also be achieved through the use of magnetic field interaction with the bow shock as illustrated in Fig. B.3 in respect to the other shock as illustrated in Fig. B.4, and concept is currently under Air Force Research Laboratory (AFRL) supervision research program for implementation of some kind of flight test conceptual study.

And conceptual images of the other three main types of shock waves, namely, are:

1. Normal shock wave
2. Oblique shock wave
3. Bow shock wave

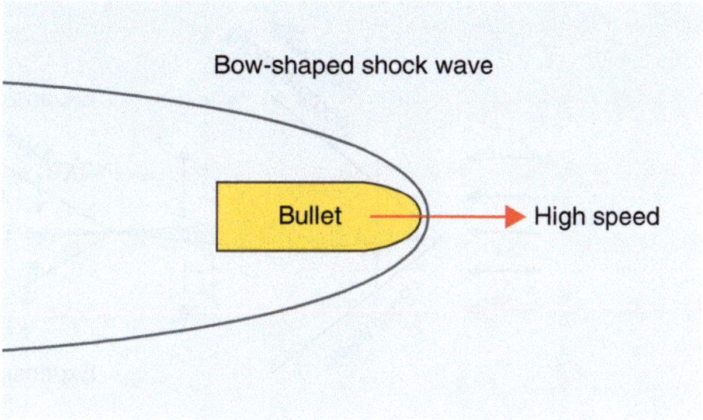

Fig. B.3 Bow-shaped shock wave

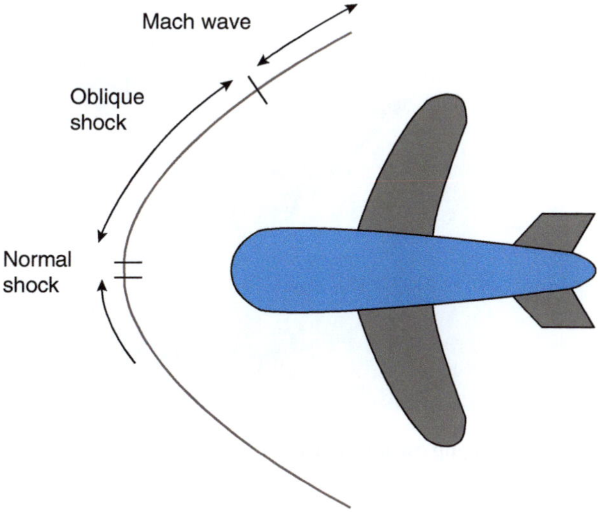

Fig. B.4 All three-shock wave presentation

Note that, while *normal shock* is a normal shock wave that occurs completely parallel to the surface as shown in Fig. B.4, the oblique shock wave is usually at angle, and the bow shock wave is completely parabolic. The shape of the nose is designed in order to generally create a bow shock (usually circular nose).

The *bow shock* wave forms when the aircraft is flying at a speed faster than the speed of sound (i.e., Mach number = 1). A bow wave is a shock wave in front of a body, such as an airfoil, or is apparently attached to the forward tip of the body as it is depicted in Fig. B.5.

The concept of reducing surface heating and consequently Drag/Friction driven by with assist from magnetic field interacting with bow shock as it was explained in preceding paragraph under continues industrial investigation (i.e. implementation of Wingtip Canard), as Illustrated in Fig. B.6.

Wingtip devices such as "canard" are intended to improve the efficiency of fixed-wing aircraft by reducing drag.

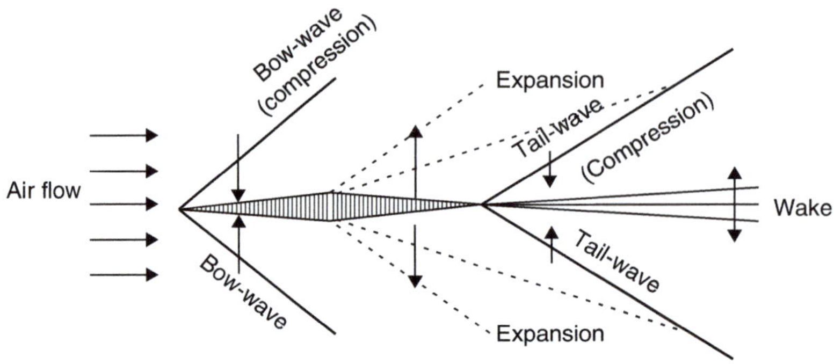

Fig. B.5 Bow shock wave illustration

Fig. B.6 Wingtip canard surface

Industry seems to be receptive to the adoption of plasma-based technology, but has reservations about technical risk, performance, reliability, and integration. There is a need to identify applications where plasma devices are significantly better than completing technologies and to demonstrate prototypes in an operational environment.

In conclusion, in order to satisfy the two mentioned purposes behind the physics and science of weakly ionized gases (WIG), there are some breakthrough areas suggested by scientist in order to take advantages of plasma-assisted combustion to increase speed of flying vehicle beyond Mach 5, a range or higher that hypersonic weapons like to travel either in glide or cruise mode (see Sect. B.5 of this Appendix), with additional assist from effectiveness of surface plasma actuator, plasma-enhanced aerodynamics, and measurement technology that are expected.

Furthermore, effectiveness of surface plasma actuator is anticipated with new electrode configurations, optimized dielectrics, and optimized high-voltage driving waveforms. In addition, breakthroughs are expected to occur with multiple-electrode configurations for thrust generation and shock wave focusing, and further breakthrough may also arise from new surface materials and their characteristics that allow changes in fundamental structure or the temporal evolution of the discharge.

More surface-based concepts can make use of plasma arrays that are capable of generating shock waves that propagate away from the surface and coalesce to generate vorticity or drive acoustic waves for control of near-surface flows, while plasma-generated far-ultraviolet radiation may also be taken under consideration for rapid near-surface additional energy driven by direct absorption and molecular dissociation of chemical element in the atmosphere such as oxygen.

Breakthrough technology in the measurement area will be enabled via the development of new devices that are capable of interacting with air and combustion environments in ways that are not feasibly practical today. However, maybe in the near future, such feasibility will be created through rapid progress in direction of developing an efficient and very short-pulsed, yet precision along with controllable, high-power, high-repetition rate lasers. These lasers will open up new methods for real-time data acquisition as well as off-body energy addition, efficient volume ionization method for MHD applications, and volumetric, selective radial production for combustion reaction and ignition control. With progress in laser technology and goal toward its miniaturization of new generation of laser at higher efficiency, operation in flight will be a practical fact.

Additionally, inspired by the National Aero-Space Program, which started in 1995 in the United Sates, research on hypersonic flight has been actively pursued in various nations. The hypersonic research has been aimed mainly at the development of a reasonable hypersonic vehicle with a supersonic combustion ramjet or SCRAMJET.

One of the most challenging concepts of this type of vehicle is the single-stage -to-orbit vehicle or SSTO vehicle which would take off horizontally and fly up to orbital speeds in the atmosphere. A trans-atmospheric air-breathing vehicle, which needs to provide sufficient air for the engine operation, must be accelerated at relatively lower altitudes compared with the conventional rocket boosters. On the other hand, in order to avoid excessive dynamic pressure and aerodynamic heating for the structures and materials, higher altitudes are desirable.

Also bear in mind that the important aspect of object flying at hypersonic speed is the creation of plasma sheath around the vehicle during flight time period. From science of plasma physics point of view, when a vehicle flying in the atmosphere at high velocities becomes surrounded by regions of ionized gas that affect the propagation of electromagnetic waves to and from the vehicle, the kinetic energy in a hypersonic free stream is converted to the internal energy of the gas across the strong bow shock wave, creating very high temperatures in the shock layer near the nose. If the temperature is high enough, ionization is present, and a large number of free electrons are produced throughout the shock layer.

Downstream of the nose region, a boundary layer grows along the surface of the vehicle. Since the Mach number at the outer edge of the boundary layer is still high, the intense frictional dissipation within the hypersonic boundary layer creates high temperatures and causes chemical reactions.

The ions and electrons produced in the high temperature air around the vehicle create a plasma sheath, which interacts with electromagnetic waves propagating to and from the vehicle. If the attenuation of the electromagnetic waves due to the plasma sheath is excessively high, then a communication blackout occurs.

However, because of these constraints, the trajectory of a hypersonic vehicle will be constrained into a very narrow region. An important question which should be taken into account for the trajectory is the interference between the electromagnetic waves used for communication and the plasma sheath around the vehicle. This is an important event that needs to be dealt with when it comes to detecting an incoming hypersonic weapon at far field or distance using a high-power microwave (HPW) either as a weapon of countermeasure against such measure or as we said yet to detect it, beside obstacle within geodesic distance the electromagnetic wave of this device in transverse electromagnetic (TEM) mode to go back and forth from source to target and back to source.

However, to overcome this matter, we suggest application a new family of waves known as scalar longitudinal wave (SLW), which is discussed by Zohuri [18, 19], and it is briefly explained in Sect. B.8 of this Appendix.

B.3 Weakly Ionized Plasmas via MHD and Electrohydrodynamics Driving Hypersonic Flows

The following argument applies to weakly ionized plasma via magnetohydrodynamics controlling hypersonic flows.

Theoretical analysis and fundamental aspects of high-speed flow control using electric and magnetic field applying MHD are well understood and established by the scientists [17].

There is a growing interest in using weakly ionized gases (plasmas) and electric and magnetic fields in high-speed aerodynamics. Wave and viscous drag reduction, thrust vectoring, reduction of heat fluxes, sonic boom mitigation, boundary-layer and turbulent transition control, flow turning, and compression, onboard power generation, and scramjet inlet control are among plasma and MHD technologies that can potentially enhance performance and significantly change the design of supersonic and hypersonic vehicles [17].

Meanwhile, despite many studies devoted to these new technologies, a number of fundamental issues have not been adequately addressed. Any plasma created in gas flow and interacting with electric and magnetic fields would result in gas heating. This heating can certainly have an effect on the flow and, in some cases, can be used advantageously. However, a more challenging issue is whether significant nonthermal effects of plasma interaction with electric and magnetic fields can be used for high-speed flow control.

In conventional MHD of highly conducting fluid, electric and magnetic effects give rise to ponderomotive force terms $\nabla(\varepsilon_0 E^2/2)$ and $\nabla(B^2/2\mu_0)$, which can be interpreted as gradients of electric and magnetic field pressures. These ponderomotive forces are successfully utilized for plasma containment in fusion devices and also play an important role in astrophysics. One might hope that these forces can also be used for control of high-speed flow of ionized air. However, the great importance of ponderomotive forces in fusion and astrophysical plasmas is due to the fact that those plasmas are fully, or almost fully, ionized and, therefore, are highly conductive. In contrast, high-speed air encountered in aerodynamics is not naturally ionized, even in boundary layers and behind shocks if the flight Mach number is below about 12, due to the low static temperature. Therefore, ionization has to be created artificially, using various electric discharges or high-energy particle beams [20, 21]. In most conditions, the artificially created plasmas are weakly ionized, with ionization fraction ranging from 10^{-8} to 10^{-5}. Because of the low ionization fraction and electrical conductivity, interaction of the plasma with electromagnetic fields and transfer of momentum and energy to or from the bulk neutral gas can be quite inefficient. Further information can be found in reference by Sergey Macheret et al. [17].

In conclusion, the principal difficulty in high-speed flow with hypersonic regime flow control using electric and magnetic fields is that the relatively cold gas has to be ionized in electric discharge or by electron beams, which requires large power inputs and results in low ionization fraction and electrical conductivity. The low ionization fraction means that, although electrons and ions can interact with electromagnetic fields, transfer of momentum and energy to or from the bulk neutral gas can be small compared with momentum and energy carried by the high-speed flow.

B.4 What Is a Hypersonic Weapon?

A hypersonic weapon is a missile that travels at Mach 5 or higher, which is at least five times faster than the speed of sound. This means that a hypersonic weapon can travel about 1 mile per second. For reference, commercial airliners fly sub-sonically, just below Mach 1, whereas modern fighter jets can travel supersonically at Mach 2 or Mach 3.

B.5 What Types of Hypersonic Weapons Are in Development?

There are two types of weapons emerging:

1. Hypersonic cruise missiles
2. Hypersonic glide vehicles

See Fig. B.7.

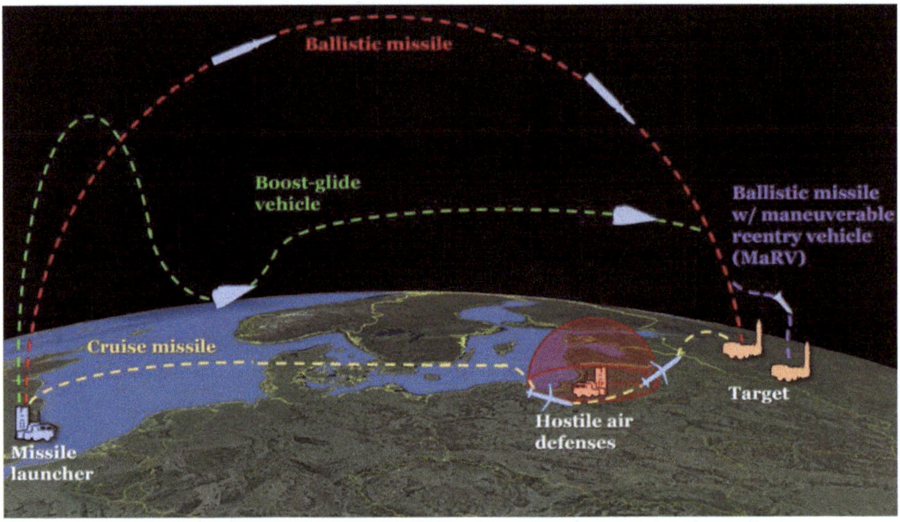

Fig. B.7 Notional flight paths of hypersonic boost-glide missile, ballistic missile, and cruise missile. (Courtesy of CSBA Graphic)

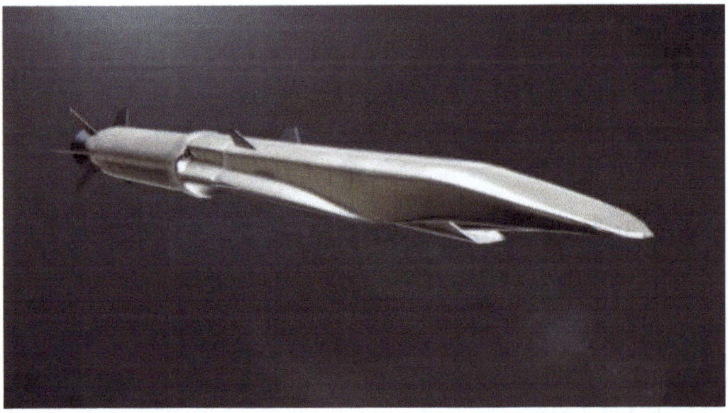

Fig. B.8 Cruise-type hypersonic weapon

Hypersonic cruise missiles are powered all the way to their targets using an advanced propulsion system called a SCRAMJET. These are very, very fast. You may have as little as 5–20 min from the time it's launched until the time it strikes, for anticipated standoff ranges. See Fig. B.8.

Hypersonic cruise missiles can fly at altitudes up to 100,000 ft., whereas hypersonic glide vehicles can fly above 100,000 ft.

Hypersonic glide vehicles are placed on top of rockets, launched, and then glide in the upper atmosphere.

Both are like a plane with no engine. They use aerodynamic forces to maintain stability, to fly along, and to maneuver as well. Furthermore, because they are maneuverable, they can keep their targets a secret up until the last few seconds of their flights.

B.6 What Are Some Technical Requirements Needed for Hypersonic Weapons?

Once they reach Mach 5 in flight mode within the atmosphere, they cannot use traditional jet engines to make such vehicles go faster.

They need a completely different design to unclutter the flow path and sustain combustion of the supersonic airflow inside the engine.

The answer becomes a supersonic combustion ramjet (SCRAMJET), which can operate between Mach 5 and Mach 15.

In order to maintain sustained hypersonic flight, a vehicle must also endure the extreme temperatures of flying at such speeds. The air impinging on the frontal surfaces of the vehicles becomes a plasma.

One can think of it as flying into a blowtorch. It is very similar circumstance to the return of reentry capsules coming back from orbiting the moon, traveling with or without astronauts in them. The faster a vehicle flies, the pressure and temperature rise exponentially. Thus, they need to have materials that can withstand high temperature over a long period of time during the journey of such a weapon to the designated target.

In conclusion, as we have discussed, for hypersonic vehicle to be able to travel at a speed above 5 Mach and higher, these types of vehicles need to travel to an environment very close to plasma condition either ahead of object or be surrounded by plasma such as plasma sheath that is created either by means of weakly ionized gas schema or shock wave ahead of the object or even some plasma actuator. Thus, disturbing such an environment around the vehicle, one can interrupt the pattern of flight trajectory via longitudinal scalar wave (LSW) or some other means, since transverse electromagnetic (TEM) wave is not able to penetrate the plasma sheath.

B.7 Which Countries Are Developing Hypersonic Weapons?

"The U.S., Russia and China are ahead of other nations in developing hypersonic weapons," Richard Speier, adjunct staff with Rand Corporation, told CNBC [22].

Speier, who worked to initiate the Pentagon's Office of Counterproliferation Policy, added that France, India, and Australia are also developing military uses of hypersonic technology.

"Japan and various European countries such as France recently are working on civilian uses of the technology, such as space launch vehicles, or civilian airliners, but civilian uses can be adapted for military purposes," Speier noted.

It is very commonly asserted that there is an arms race in hypersonic technology and that the United States is losing. There is certainly an arms race, but it is not obvious or clear and we are not convinced the United States is losing. Experts often argue the United States is behind in this technology because Russia and China appear to be testing more frequently. This is true, but in many ways, the United States is running a different race from Russia and China.

Russia and China appear to be focused primarily on the delivery of nuclear warheads, and in this case, accuracy can be relaxed. The United States is interested in the delivery of non-nuclear warheads, and here, accuracy is absolutely critical for the weapon to be militarily effective. The United States wants to be landing weapons within a few meters of the target with enhanced circular error probable (CEP). US goals are much more demanding than Russian and Chinese goals.

The United States also has a very long history of testing in this area, which gives the United States an advantage in its current efforts. For instance, the most successful US boost-glide weapon research and development (R&D) program, the Advanced Hypersonic Weapon (AHW), has seen a glider tested over about 4000 km. China, by contrast, appears to have been testing boost-glide weapons at a range of less than 2000 km.

So, to summarize, there are two considerations here: the US history in this area and the inherently more demanding technology that the United States is pursuing. When you take those factors into account, it is easy to conclude the United States is not behind in this competition.

However, the development of hypersonic weapons in the United States has been largely motivated by technology, not by strategy.

In other words, technologists have decided to try and develop hypersonic weapons because it seems like they should be useful for something, not because there is a clearly defined mission need for them to fulfill. The first-order task for the Department of Defense (DOD) is, therefore, to decide what missions it has that warrant hypersonic weapons. Then we can have a conversation about what the most cost-effective way of achieving those goals are. Is it indeed hypersonic weapons, or is there a better alternative? The real priority task here is for the DOD to develop a strategy for the acquisition of hypersonic weapons, and that isn't possible until it is decided what these weapons might be used for.

B.7.1 What Is the Significance of Russia's Latest Missile Efforts?

In his speech at the beginning of March, President Vladimir Putin presented an extraordinary list of new weapons that he claims Russia is developing or has deployed. This list included a number of hypersonic capabilities [10].

The most significant is a boost-glide weapon called Avangard. This maneuvering weapon, according to Putin, has been designed to defeat US missile defenses. Since Putin's speech, Russia has indicated that the Avangard glider is going to be deployed on at least two different kinds of ballistic missiles and will carry nuclear warheads. It's possible that this weapon could, in the future, be used for the delivery of non-nuclear warheads, if its accuracy can be refined. But in the short term, its only purpose appears to be the delivery of nuclear warheads.

Second, Putin announced a novel boost-glide weapon, called Kinzhal, which means "dagger" in Russian. This weapon is launched from an aircraft and has a shorter range than Avangard. The media reporting we've seen coming out of Russia suggests that this weapon is also nuclear armed. Perhaps counterintuitively, however, the development of nuclear-armed boost-glide vehicles by Russia should be less worrying to us than the development of non-nuclear boost-glide vehicles. Russia already has the capability to deliver nuclear weapons to US and allied targets—and, frankly, we cannot totally deny it that capability.

Russian nuclear-armed boost-glide vehicles do not, therefore, change the status quo. If, however, Russia developed boost-glide weapons with non-nuclear warheads, it would present a new and potentially very significant security threat to the United States and its allies. Such weapons would allow Russia to threaten, with non-nuclear warheads, targets in Europe and eventually the continental United States that, previously, it could only have destroyed with nuclear weapons.

B.7.2 Is China Also Testing and Using Similar Hypersonic Missiles?

China, like Russia, is developing boost-glide weapons and hypersonic cruise missiles—but let us focus on the boost-glide part of China's program. China is developing a glider that's named WU-14 by the Pentagon and, it's been reported, DF-ZF by China (Fig. B.9) [23]. This glider has been tested repeatedly—at least seven times—over a range of up to 2000 km, which makes its range substantially shorter than the US Advanced Hypersonic Weapon. It's not clear whether it will be armed with a nuclear warhead or a non-nuclear warhead or could accommodate either. On balance, it is likely that in the first instance, it will be armed with nuclear warheads (though the evidence is far from conclusive). Perhaps, over time, China will subsequently develop a non-nuclear-armed glider.

Fig. B.9 DF-ZF carried by DF-17. (Source: www.wikipedia.com)

Very much like Russia, China already has the ability to attack US and allied targets with nuclear weapons. So Chinese nuclear-armed boost-glide weapons would merely serve to reinforce the status quo. By contrast, if China develops non-nuclear boost-glide weapons, especially if those could hit the continental United States, it would present the United States with a new and very real technical and military challenge.

The DF-ZF is a Chinese hypersonic glide vehicle (HGV), previously denoted by the Pentagon as WU-14 and currently officially operational on October 1, 2019, in the 70th anniversary of the People's Republic of China. The DF-ZF is designed to be mounted on a DF-17, a type of ballistic missile specifically designed to carry HGVs [23].

B.7.3 How Could the United States Defend Against Hypersonic Weapons

Currently we do not have any effective means of detection and defense mechanisms or for that matter countermeasures against hypersonic weapons because of the way they fly, i.e., they are maneuverable and fly at an altitude that our current defense systems are not designed to operate. For example, an ICBM missile has a predictable trajectory. One can obtain enough information via a remote sensing platform such as the Defense Support Program (DSP) that target acquisition radars can engage in order to intercept a ballistic object.

A ballistic missile is like a fly ball in baseball; the outfielder knows exactly where to catch it because its path is determined by momentum and gravity.

Since hypersonic weapons are maneuverable and, therefore, unpredictable, they are difficult to defend against. However, there are potential ways to address hypersonic weapons, but they will be very expensive.

As an example, the Missile Defense Agency is proposing to develop a space-based sensor system that would be able to track hypersonic glide vehicles globally (this would be one of the first steps in defending against these new missiles) or to have powerful radar detection that can dwell on a target during its long trajectory from apogee to target. These types of radar need to detect such weapons way beyond the range of traditional radar systems that are in operation by the militaries around the world today. Their operating range is around a couple hundred miles within the line of sight from the radar source to the target of interest.

However, the traditional detection radar beyond few hundred miles has its own deficiency beyond the line of sight due to the nature of geodesic path that they have to propagate their electromagnetic wave.

However, a new version of a detection radar that is driven by a high-power microwave (HPM) source can reach out further, but it has its own drawbacks.

The problem of microwave breakdown near antennas at high altitudes must be considered in order to find out what the limitation on transmission conditions is [24].

Electrical breakdown is often associated with the failure of solid or liquid insulating materials used inside high-voltage transformers or capacitors in the electricity distribution grid, usually resulting in a short circuit or a blown fuse. Electrical breakdown can also occur across the insulators that suspend overhead power lines, within underground power cables, or lines arcing to nearby branches of trees.

Airborne radar systems may initiate electrical discharges in front of the antennas at high altitudes because, at ultra-high frequencies, the electric field required to break down air at low pressures is, in general, much less than that required at atmospheric pressure. The processes which determine ultra-high-frequency (UHF) breakdown have been discovered and verified during the past decade. These have been applied to determining optimum transmission conditions for high flying radar as an example.

Breakdown takes place when an electric field is applied to a gas, and the free electrons move in the direction of the field, constituting a current. There are always a small number of electrons present in any collection of gas because of ionization by cosmic rays or some other phenomenon, such as the photoelectric effect. If the electric field is gradually increased from zero, the gas first appears to obey Ohm's law until the field becomes large enough to impart sufficient energy to some of the electrons to produce secondary electrons by collision. If the electric field is sufficiently great so that many secondary electrons are produced, there will come a point at which the gas will become highly conducting. For a very minute change in voltage or field near this value, the electron concentration and current will change by many orders of magnitude, and the gas will start to glow as it can be seen in Fig. B.9 [18].

Electrical breakdown occurs within a gas when the dielectric strength of the gas is exceeded. Regions of intense voltage gradients can cause nearby gas to partially ionize and begin conducting. This is done deliberately in low-pressure discharges such as in fluorescent lights. The voltage that leads to electrical breakdown of a gas is approximated by Paschen's law [24, 25].

Note: Paschen's law is an equation that gives the breakdown voltage, that is, the voltage necessary to start a discharge or electric arc, between two electrodes in a gas as a function of pressure and gap length. It is named after Friedrich Paschen who discovered it empirically in 1889 [25].

It is often asserted that it is impossible to defend against hypersonic weapons because they go too fast. That's empirically not true. The United States has already developed fairly effective "point defenses"—like Patriot air defense missile and Terminal High Altitude Area Defense (THAAD)—that can defend small areas against ballistic missiles, which are actually moving faster than hypersonic weapons. (We don't normally class ballistic missiles as a type of hypersonic weapon because they have no ability to maneuver.) So, speed, in and of itself, is not an insuperable barrier for missile defenses. Those point-defense systems, and particularly THAAD, could very plausibly be adapted to deal with hypersonic missiles. The disadvantage of those systems is that they can only defend small areas. To defend the whole of the continental United States, you would need an unaffordable number of THAAD batteries. The United States has deployed one missile defense system, the Ground-Based Midcourse System, that is designed to try to defend the whole of the United States against ballistic missiles [20]. For a variety of technical reasons, however, using these "area defenses" to deal with hypersonic weapons is more or less impossible.

Therefore, it is a nuanced picture when it comes to defenses—you can probably defend small areas fairly effectively against gliders, but it is likely to be much more challenging to defend large areas.

In contrast, a new technology and technical approach looking at a scalar wave driving an energy wave in longitudinal mode has been proposed by the first author here Zohuri [18]. This approach seems very reasonable and makes sense at least in theory. A small amount of funding needs to be secured to take the theoretical concept into the experimental stage and make a prototype of apparatus that is driven based on the technology known as a scalar longitudinal wave [19]. Until such experiments occur, the actual capabilities are only speculative.

B.8 What Is the Scalar Wave?

As it has been understood, the scalar longitudinal wave (SLW) or simply scalar wave (SW) does not have the characteristic of electromagnetic (EM) wave and does not behave like EM. However, it is understood from classical electrodynamics (CED) that electromagnetic wave has both electric $\vec{E}$ and magnetic $\vec{B}$ fields and power flow in EM wave is driven by means of the Poynting vector, as it is presented by Equation B.1 in the following form as:

$$\vec{S} = \vec{E} \times \vec{B} \ \text{W/m}^2 \tag{B.6}$$

Analyzing Equation B.1 indicates that the energy per second crossing a unit area whose normal is oriented in the direction of $\vec{S}$ per definition of two vectors' cross-product is flowing in electromagnetic (EM) form.

On the other hand, a scalar wave (SW) has no time-varying $\vec{B}$ field. And in some cases, it also has no $\vec{E}$ field. Thus, it has no energy propagated in the EM wave mode as transversal electromagnetic (TEM) form and shape. Furthermore, with some authors in the field of electromagnetic physics, based on non-classical effects of electrodynamics or quantum electrodynamics (QED), they often speak of electromagnetic waves not being based on oscillations of electric and magnetic fields.

"For example, it is claimed that there is an effect of such waves on biological systems and the human body. Even medical devices are sold which are assumed to work on the principle of transmitting any kind of information via 'waves' which have a positive effect on human health. In all cases, the explanation of these effects is speculative, and even the transmission mechanism remains unclear because there is no sound theory on such waves, often subsumed under the notion 'scalar waves'" [18].

Additionally, it must be recognized, however, that any vector could be added that could integrate to zero over a closed surface and the Poynting theorem still applies. Thus, there is some ambiguity in even stating the form of Equation B.1, which is the total EM energy flow.

In order to establish the longitudinal potential waves, we develop our theory of electromagnetic waves with vanishing field vectors, which is taking place in a "vacuum state." This state also plays a role in the microscopic interaction with matter, so we can restrict our consideration to classical electrodynamics in order to understand better. With $\vec{E}$ and $\vec{B}$ parameters designating the classical electric and magnetic field vectors, then in vacuum they can be written as:

$$\vec{E} = 0 \tag{B.7}$$

and

$$\vec{B} = 0 \tag{B.8}$$

Then, the only possibility to find electromagnetic effects is by the potentials. These are defined as vector (magnetic) potential $\vec{A}$ and scalar (electric) potential ϕ constituting the "force" fields $\vec{E}$ and $\vec{B}$, and both can be derived as follows.

As we know, Maxell's electromagnetic wave propagation in linear media (i.e., matter), namely, Equations B.9 through B.12 (i.e., their empirical basis with Equation B.9 plus three other Equations B.10 through B.12) in the case of transverse waves, for which the field pointers oscillate perpendicular to the direction of propagation with which we are already familiar with, namely:

$$\nabla \times \vec{H} = \vec{J} + \frac{\partial \vec{D}}{\partial t} \tag{B.9}$$

$$\nabla \times \vec{E} = -\frac{\partial \vec{B}}{\partial t} \tag{B.10}$$

$$\nabla \cdot \vec{D} = \rho \tag{B.11}$$

$$\nabla \cdot \vec{B} = 0 \tag{B.12}$$

In a homogeneous and isotropic with current density $\vec{J} = 0$ as well as $\vec{B} = \mu\vec{H}$ where $\vec{H}$ is the magnetic field intensity and $\vec{D} = \varepsilon\vec{E}$ where $\vec{D}$ is referring to the electric displacement. In these relations, the following definitions are also applied as:

$\varepsilon =$ electric permittivity of the medium
$\mu =$ magnetic permeability of the medium

With speed of light being defined as c, then it can be written as $\mu\varepsilon = 1/c^2$.

Given the above conditions and definitions, Equation B.4 can be written as a new form as:

$$\frac{1}{\mu}\nabla \times \vec{B} = \varepsilon\frac{\partial \vec{E}}{\partial t} \tag{B.13}$$

Since the magnetic induction has zero divergence, it may always be represented as the curl of a vector potential from electromagnetic (EM) point of view and can be written as:

$$\vec{B} = \vec{\nabla} \times \vec{A} \tag{B.14}$$

Using Equation B.14 for expression of $\vec{B}$ in Equation B.10 of Maxwell's equation, we obtain the following result as:

$$\begin{aligned} \nabla \times \vec{E} &= -\frac{\partial}{\partial t}\left(\nabla \times \vec{A}\right) \\ \nabla \times \vec{E} + \frac{\partial}{\partial t}\nabla \times \vec{A} &= 0 \end{aligned} \tag{B.15}$$

Assuming sufficient continuity of the fields to interchange the spatial and temporal differentiations, this can be written as:

$$\nabla \times \left[\vec{E} + \frac{\partial \vec{A}}{\partial t}\right] = 0 \tag{B.16}$$

The vector $\vec{E} + \partial\vec{A}/\partial t$ thus has zero curl and can be written as the gradient of a scalar as:

$$\vec{E} = -\nabla\phi - \frac{\partial \vec{A}}{\partial t} \tag{B.17}$$

In this case ϕ is the scalar (electric) potential and $\vec{A}$ is the (magnetic) vector potential in Equations B.14 and B.17.

These potentials satisfy wave equations which are very similar to those satisfied fields. The wave equation for $\vec{A}$ is derived by substituting the expressions given in Equations B.9 and B.12 for $\vec{B}$ and $\vec{E}$ into Equation B.13, with result:

$$\begin{aligned} &\frac{1}{\mu}\nabla \times \left(\nabla \times \vec{A}\right) = \varepsilon\frac{\partial}{\partial t}\left\{-\nabla\phi - \frac{\partial \vec{A}}{\partial t}\right\} \\ &\frac{1}{\mu}\nabla \times \nabla \times \vec{A} + \varepsilon\frac{\partial}{\partial t}\left\{-\nabla\phi - \frac{\partial \vec{A}}{\partial t}\right\} = 0 \end{aligned} \tag{B.18}$$

By using vector identity $\nabla \cdot \ \nabla \ - \nabla^2$ for $\nabla \times \ \nabla \times$ and multiplying it by μ in the second form of Equation B.18 and using $\mu\varepsilon = 1/c^2$, we obtain the following result as:

$$-\nabla^2\vec{A} + \frac{1}{c^2}\frac{\partial^2 A}{\partial t^2} + \nabla\nabla \cdot \vec{A} + \frac{1}{c^2}\nabla\frac{\partial\phi}{\partial t} = 0 \tag{B.19}$$

Equation B.19 is taking place under vacuum conditions or homogeneous media where current density $\vec{J} = 0$.

Until now only the curl of vector potential $\vec{A}$ has been specified; the choice of the divergence of $\vec{A}$ is still arbitrary. It is clear from Equation B.19 that imposing the so-called Lorenz gauge condition, where:

$$\nabla \cdot \vec{A} + \frac{1}{c^2}\frac{\partial\phi}{\partial t} = 0 \tag{B.20}$$

results in a considerable simplification. If this condition is satisfied, then $\vec{A}$ satisfies the wave equation as:

$$\underbrace{\nabla^2\vec{A} - \frac{1}{c^2}\frac{\partial^2\vec{A}}{\partial t^2} = 0}_{\text{(Vector Potential Wave)}} \quad \text{(Vector potential wave)} \tag{B.21}$$

Furthermore, using Equation B.17 in Equation B.11 for vacuum or homogeneous media $\rho = 0$ with $\vec{D} = \varepsilon\vec{E}$, we obtain:

$$-\varepsilon\left[\nabla \cdot \nabla\phi + \nabla \cdot \frac{\partial\vec{A}}{\partial t}\right] = 0 \tag{B.22}$$

By interchanging the order of the divergence and the time derivative operating on $\vec{A}$ and using the Lorenz condition Equation B.20, everything leads to:

$$\nabla^2\phi - \frac{1}{c^2}\frac{\partial^2\phi}{\partial t^2} = 0 \quad \text{(Scalar potential wave)} \tag{B.23}$$

Thus, by imposing the Lorenz condition, both the scalar and vector potentials are forced to satisfy inhomogeneous wave equations of similar forms. However, the problem of finding the general solution of the inhomogeneous scalar wave equation is analogous to finding the general solution of Poisson's equation.

However, a solution appears to exist for the special case of $\vec{E} = 0$, $\vec{B} = 0$, and $\nabla \times \vec{A} = 0$, for a new wave satisfying:

$$\begin{cases} \vec{A} = \nabla S \\ \phi = -\dfrac{1}{c^2}\dfrac{\partial S}{\partial t} \end{cases} \tag{B.24}$$

S then satisfies:

$$\nabla^2 S - \frac{1}{c^2}\frac{\partial^2 S}{\partial t^2} = 0 \tag{B.25}$$

Mathematically S is a "potential" with a wave equation, one that suggests propagation of this wave even through $\vec{E} = \vec{B} = 0$ and the Poynting theorem indicates no electromagnetic (EM) power flow.

From Equation B.6 and condition of establishing it in above, there is the suggestion of a solution to Maxwell's equations involving a scalar wave with potential S that can propagate without Poynting vector EM power flow. But the question then arises as to where the energy is drawn from to sustain such a flow of energy.

A vector that integrates to zero over a closed surface might be added in the theory, as suggested in the beginning of Sect. 9 of the book by Zohuri [18]. Another is the possibility of drawing energy from the vacuum, assuming net energy could be drawn from "free space." As suggested by quantum mechanics (QM), it allows random energy in free space, but conventional or classical electromagnetics (CEM) theory has not allowed this to date.

Random energy in free space that is built of forces fields that sum to zero is a possible approach. If so, these might be a source of energy to drive the *S* wave drawn from "free space."

Note that with condition of $\vec{E} = \vec{B} = 0$, both Equations B.14 and B.17 will reduce to:

$$\nabla \phi = -\frac{\partial \vec{A}}{\partial t} \tag{B.26}$$

and

$$\nabla \times \vec{A} = 0 \tag{B.27}$$

From Equation B.26, one follows immediately that the vector potential $\vec{A}$ is vortex-free, representing a laminar flow. The gradient of the scalar potential is coupled to the time derivative of the vector potential, so both are not independent of one another [18].

The scalar wave (SW) could be accompanied by a vector potential $\vec{A}$. A scalar wave is a non-linear, non-Hertzian, standing wave capable of supporting significant effects including carrying information and inducing higher levels of cellular energy.

Scalar waves can be created by wrapping electrical wires around a figure eight in the shape of a Möbius coil. When an electric current flows through the wires in opposite directions, the opposing electromagnetic fields from the two wires cancel each other and create a scalar wave.

The scalar waves cannot be detected directly because they do not impart energy and momentum to matter. On the other hand, they impart phase shifts to matter, and they may be detected through interference means. Because of their elusive nature, they may also be called scalar vacuum waves. The underlying scalar field is already known to physicists in the context of quantum field theory and is known as the scalar gauge field. It should be noted at this time that other researchers have reported the observation of fields which behave qualitatively similar to the predicted scalar fields. The extension of the forceless field concept to the nucleonic field should yield higher-order fields with even more interesting properties than the scalar fields. This matter is under investigation.

The scalar longitudinal wave (SLW) does not damp out with $1/r^2$ dependence, where r is distance between the source of the SLW and its target. The dispersion of such a wave far from the source is correlated to $1/r$ as is shown in Equation B.28 with strength source S_0.

$$S = \frac{S_0}{r} e^{j(\omega t - kr)} \tag{B.28}$$

The power per unit area of the wavefront is proportional to the square of the strength S of the wave. However, the total power must be independent of distance from the source. The area of the spherical surface over which the wave is spread at a distance r from the source is r^2 in the case of the *near field*. Hence, S^2r^2 is constant, but far from the source (i.e., far field), the strength S of such a wave is described by Equation B.22 which is an exact solution of the linear scalar wave equation.

$$\begin{cases} S = \frac{S_0}{r}\left\{e^{-jk[r-(a/2)\cos\theta]} - e^{-jk[r+(a/2)\cos\theta]}\right\} \\ S = \frac{2jS_0}{r}\left(e^{-jkr}\right)\sin\ [(ka/2)\cos\theta] \end{cases} \tag{B.29}$$

Equation B.29 is derived based on Fig. B.10 with some mathematical manipulation as well, assuming Equation B.1 holds exactly and that the source to the left is equal and opposite to the source to the right i.e., two sources of scalar waves dueling on target, as depicted in Fig. B.11.

The strengths of the two waves are not exactly equal and opposite at the point P for two reasons (Fig. B.12):

1. The right-hand source is nearer to P, and hence the wave from this source has a slightly greater strength than the wave from the source to the left.
2. Because the two sources are at slightly different distances, the waves from them are not exactly 180° out of phase at P.

In this Appendix we do not show details of analysis that get to the result in Equation B.2; however, the signals from the two sources may arrive at point P in phase, so that the strengths add, or 180° out of phase, so that the strengths cancel.

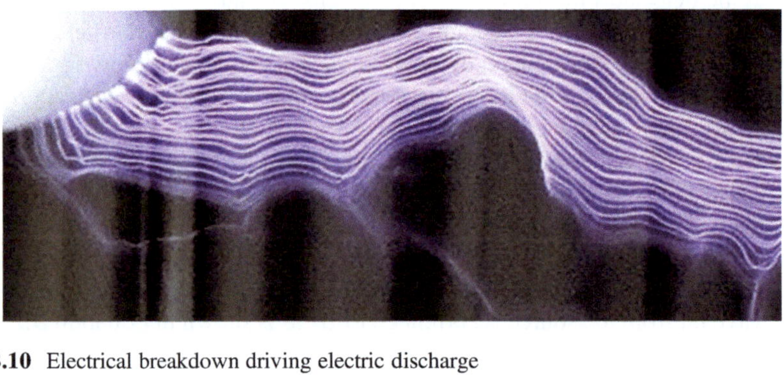

Fig. B.10 Electrical breakdown driving electric discharge

Fig. B.11 Two different sources of wave

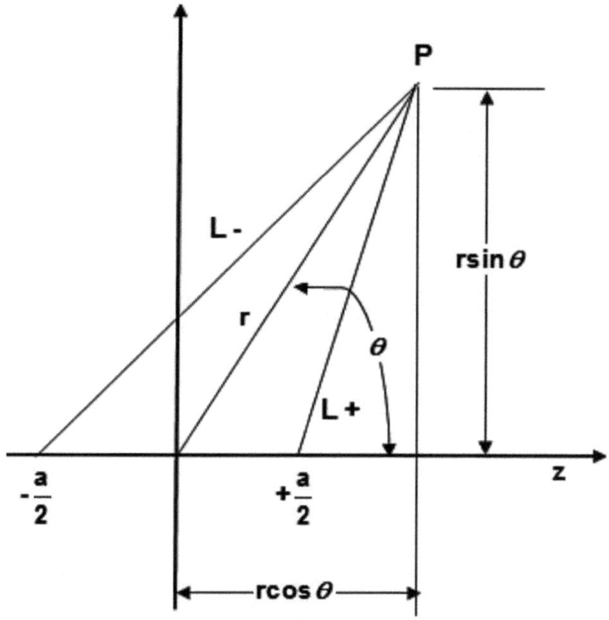

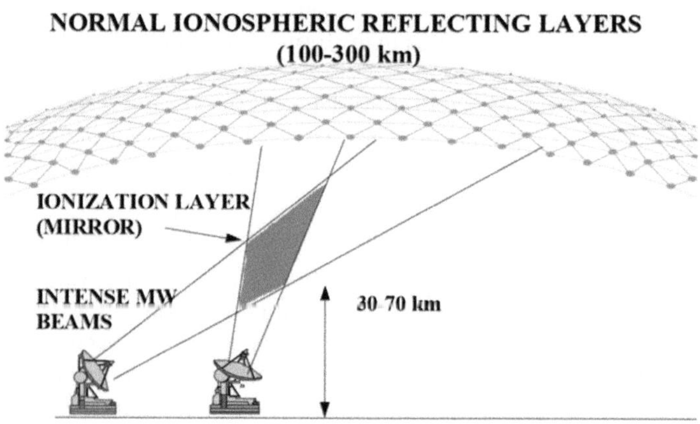

Fig. B.12 Crossbeam approach generation artificial ionospheric mirror

B.9 Transmitters and Receivers for Scalar Longitudinal Wave (SLW)

Horst Eckardt [10] in his paper suggests the following procedure for transmitting and receiving scalar longitudinal wave, where a sender for longitudinal potential waves has to be a device which avoids producing $\vec{E}$ and $\vec{B}$ fields but sends out oscillating

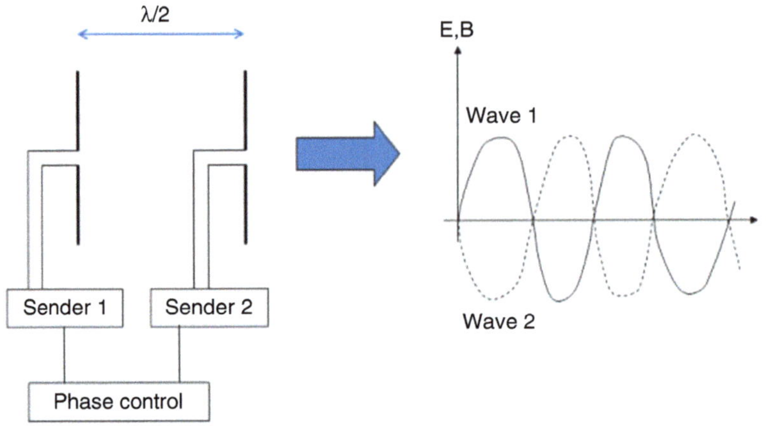

Fig. B.13 Suggestion for a transmitter of longitudinal potential waves

potentials. He discusses two propositions how this can be achieved technically. In the first case, we use two ordinary transmitting antennas (with directional characteristic) with a distance of half a wavelength (or an odd number of half waves). This means that ordinary electromagnetic waves cancel out, assuming that the near field is not disturbing significantly. Since the radiated energy cannot disappear, it must propagate in space and is transmitted in the form of potential waves. This is depicted in Fig. B.13.

A more common example is a bifilar flat coil, for example, from the patent of Tesla [21]; see Fig. B.14, second drawing. The currents in opposite directions effect annihilation of the magnetic field component, while an electric part may remain due to the static field of the wires.

Construction of a receiver is not so straightforward. In principle no magnetic field can be retrieved directly from $\vec{A}$ due to Equation B.27. The only way is to obtain an electrical signal by separating both contributing parts in Equation B.20 so that the equality of Equation B.17 is outweighted and an effective electric field remains which can be detected by conventional devices. A very simple method would be to place two plates of a capacitor in distance of half a wavelength (or odd multiples of it). Then the voltage in space should have an effect on the charge carriers in the plates, leading to the same effect as if a voltage had been applied between the plates. The real voltage in the plates or the compensating current can be measured (Fig. B.15).

The "tension of space" operates directly on the charge carriers, while no electric field is induced. The $\partial\vec{A}/\partial t$ part is not contributing because the direction of the plates is perpendicular to it, i.e., no significant current can.

Another possibility of a receiver is to use a screened box (Faraday cage). If the mechanism described for the capacitor plates is valid, the electrical voltage part of the wave creates charge effects which are compensated immediately due to the high conductivity of the material. As is well-known, the interior of a Faraday cage is free

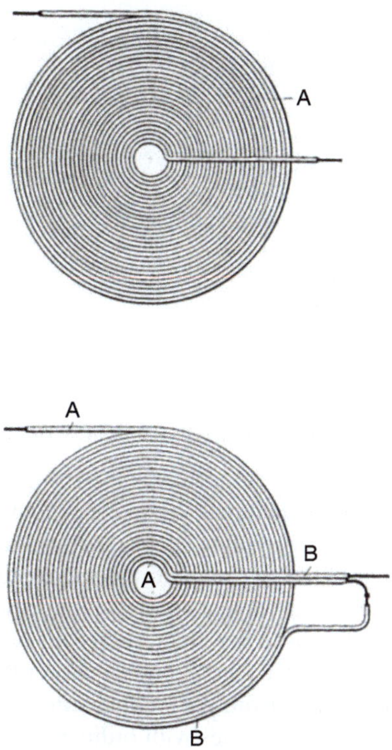

Fig. B.14 Tesla coils according to the patent of his [21]

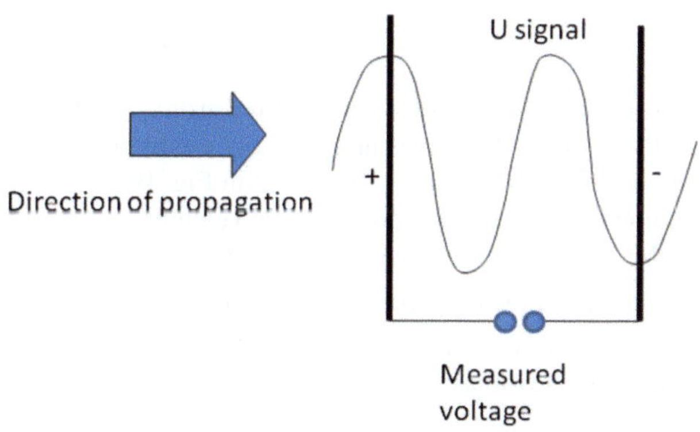

Fig. B.15 Suggestion for a receiver of longitudinal potential waves (capacitor) [10]

of electric fields. The potential is constant because it is constant on the box surface. Therefore, only the magnetic part of the wave propagates in the interior where it can be detected by a conventional receiver; see Fig. B.16.

Another method of detection is using vector potential effects in crystalline solids. As is well-known from solid-state physics, the vector potential produces excitations within the quantum mechanical electronic structure, provided the frequency is near to the optical range. Crystal batteries work in this way. They can be engineered

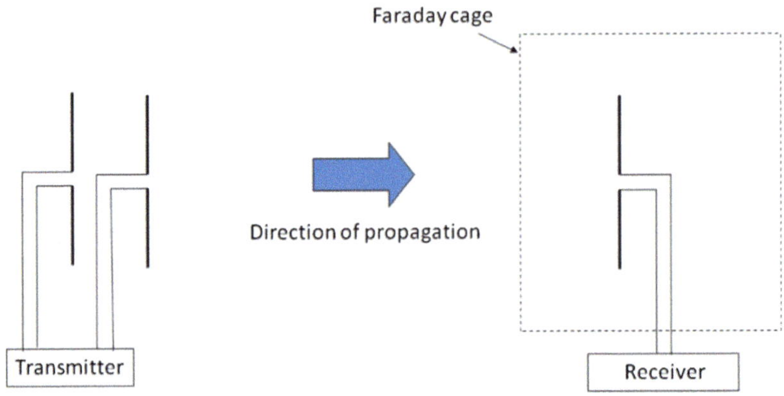

Fig. B.16 Suggestion for a receiver of longitudinal potential waves (Faraday cage) [10]

through chemical vapor deposition of carbon. In the process you get strong light-weight crystalline shapes that can handle lots of heat and stress (by high currents). For detecting longitudinal waves, the excitation of the electronic system has to be measured, for example, by photoemission or other energetic processes in the crystal.

All these are suggestions for experiments with longitudinal waves. Additional experiments can be performed for testing the relation between wave vector k which is defined from the wavelength $\lambda = 2\pi/k$ and frequency ω to check if this type of waves propagates with ordinary velocity of light c. See Horst Eckardt [10].

$$c = \frac{\omega}{k} \tag{B.30}$$

As pointed out by Eckardt and Lindstrom [26], the speed of propagation depends on the form of the waves and possibly could be a non-linear step function as well. The experimental setup is illustrated in Fig. B.11, and it can directly being used for finding the $\omega(k)$ relation due to the wavelength and frequency that could be measured at the same time, and analysis can be seen from the solution to Equation B.16, whereas a simple example we assume is a sine-like behavior of vector potential $\vec{A}$ in the direction of x-axis with direction $\vec{k}$ wave vector, space coordinate vector $\vec{x}$, and time frequency ω as:

$$\vec{A} = A_0 \sin\left(\vec{k} \cdot \vec{x} - \omega t\right) \tag{B.31}$$

Substituting this solution into Equation B.20, we obtain:

$$\frac{\partial \phi}{\partial t} = \nabla \phi = A_0 \omega \cos\left(\vec{k} \cdot \vec{x} - \omega t\right) \tag{B.32}$$

This condition has to be met for any potential ϕ. We make the following approach as:

$$\phi = \phi_0 \sin\left(\vec{k} \cdot \vec{x} - \omega t\right) \tag{B.33}$$

in order to find

$$\frac{\partial \phi}{\partial t} = \nabla \phi = k\phi_0 \cos\left(\vec{k} \cdot \vec{x} - \omega t\right) \tag{B.34}$$

Comparing Equation B.34 with Equation B.32, we can see that the constant A_0 can be defined as:

$$A_0 = k\frac{\phi_0}{\omega} \tag{B.35}$$

It is obvious that the waves of $\vec{A}$ and ϕ have the same phases, and naturally with these results, we can now consider the energy density of such a combined wave, and generally speaking, it can be given as:

$$\mathcal{E} = \frac{1}{2}\varepsilon_0 \vec{E}^2 + \frac{1}{2\mu_0}\vec{B}^2 \tag{B.36}$$

From Equations B.20 and B.21, we can observe that the magnetic field disappears identically, but the electric field is a vanishing sum of two terms which are different from zero [18].

These two terms evoke an energy density ε of space where the wave propagates. This cannot be obtained out of the force fields (these are zero) but must be computed from the constituting potentials. As discussed in [22] we have to write:

$$\mathcal{E} = \frac{1}{2}\varepsilon_0 \left(\dot{\vec{A}}^2 + (\nabla \phi)^2\right) \tag{B.37}$$

With help from Equations B.31 and B.32, it follows that:

$$\varepsilon = \varepsilon_0 k^2 \phi_0^2 \cos^2\left(\vec{k} \cdot \vec{x} - \omega t\right) \tag{B.38}$$

This is an oscillating function, meaning that the energy density varies over space and time in phase with the propagation of the wave. All quantities are depicted in Fig. B.11. Energy density is maximal where the potentials cross the zero axis. There is a phase shift of 90° between both.

Further, analysis of Fig. B.16 indicates that there is an analogy between longitudinal potential waves and acoustic waves.

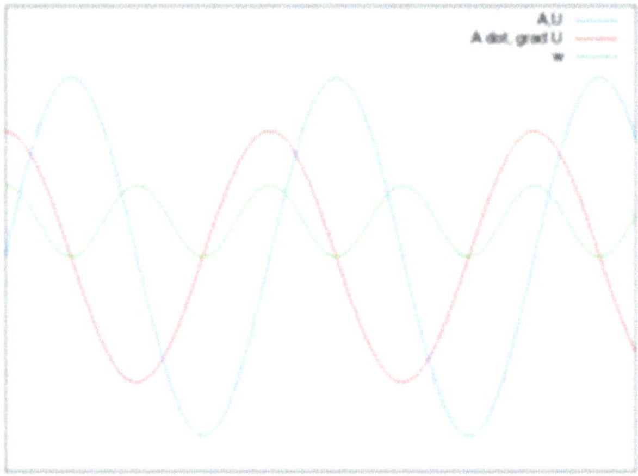

Fig. B.17 Phases of potentials $\vec{A}$ and ϕ and energy density ε

It is well-known that acoustic waves in air or solids are mainly longitudinal too. The elongation of molecules is in the direction of wave propagation as shown in Fig. B.17. This is a variation in velocity. Therefore, the magnetic vector potential can be compared with a velocity field. The differences of elongation evoke a local pressure difference. Where the molecules are pressed together, the pressure is enhanced, and vice versa.

From conservation of momentum, the force $\vec{F}$ in a compressible fluid is:

$$\vec{F} = \frac{\partial u}{\partial t} + \frac{\Delta p}{\rho} \tag{B.39}$$

where u is the velocity field, p the pressure, and ρ the density of the medium.

This is in full analogy to Equation B.12. In particular we see that in the electromagnetic case, spacetime must be "compressible"; otherwise, there were no gradient of the scalar potential.

As a consequence, space itself must be compressible, leading us to the principles of general relativity. See Fig. B.18.

B.10 Scalar Waves as Weapons

Since the scalar wave carries no energy in its beam, it can only manipulate the energy available in any scenario. The object of targeting a scalar wave at a hypersonic vehicle is to manipulate the plasma impinging on the surface of the vehicle. The scalar wave may amplify or attenuate the forces on the surface of the vehicle in such a way that the structural integrity of the vehicle is compromised. Or it may perturb the plasma in such a way that the vehicle is driven off target or malfunctions. Because the scalar wave travels at the speed of light and oscillates at megahertz

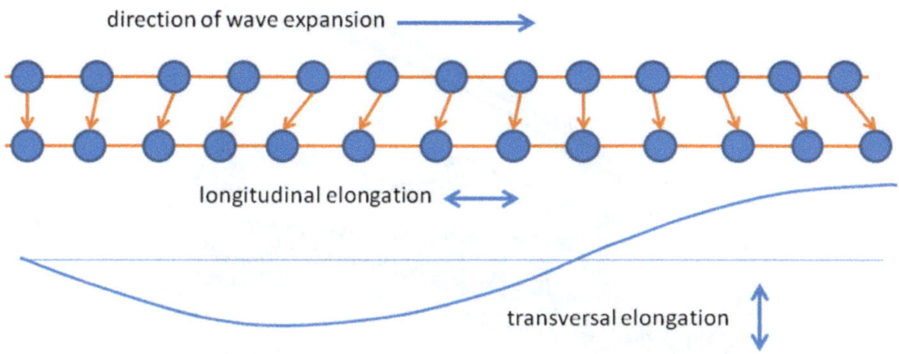

Fig. B.18 Schematic representation of longitudinal and transversal waves

frequencies, the hypervelocity technology is a significant advantage. However, the plasma generated by the hypervelocity vehicle is required to give the scalar wave something to manipulate.

B.11 Scalar Waves Superweapon Conspiracy Theory

According to Tom Bearden [27], the scalar interferometer is a powerful superweapon that the Soviet Union used for years to modify weather in the rest of the world. It taps the quantum vacuum energy, using a method discovered by T. Henry Moray in the 1920s. However, some conspiracy theorists believe Bearden is an agent of disinformation on this topic; thus, we leave this matter to the reader to make their own conclusions and be able to follow up their own finding, and this paper does not claim any of these matters are false or true.

However, in the 1930s, Tesla announced other bizarre and terrible weapons: a death ray, a weapon to destroy hundreds or even thousands of aircraft at hundreds of miles range, and his ultimate weapon to end all war—the Tesla shield—which nothing could penetrate. However, by this time no one any longer paid any real attention to the forgotten great genius. Tesla died in 1943 without ever revealing the secret of these great weapons and inventions. Tesla called this superweapon a scalar potential howitzer or death ray as artistically depicted in Fig. B.19. Such a weapon apparently was demonstrated by Soviets at their Sary Shagan missile range during the peak of the Strategic Defense Initiative (SDI) time period. It was mentioned during Strategic Arms Limitation Talk (SALT) treaty negotiations.

According to Bearden, in 1981 the Soviet Union had discovered and weaponized the Tesla scalar wave effects. Brezhnev undoubtedly was referring to it in 1975 when the Soviet side at the SALT talks suddenly suggested limiting the development of new weapons "more frightening than the mind of man had imagined." One of these weapons is the Tesla howitzer recently completed at Sary Shagan ballistic missile range near the Sino-Soviet border in Southern Russia, according to a high-level US

Fig. B.19 Scalar potential interferometer. (Source: Multimode Tesla Weapon)

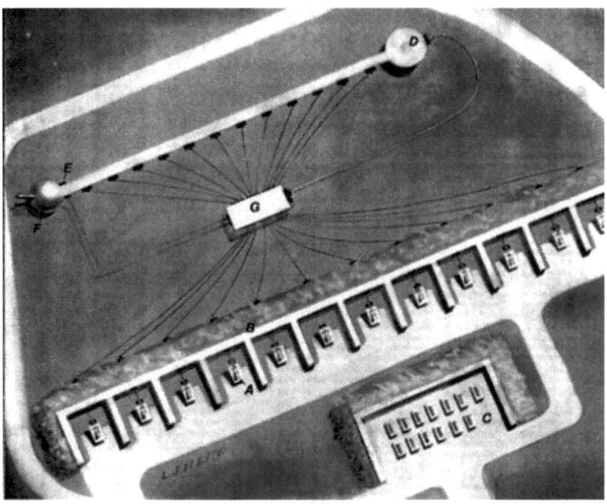

Fig. B.20 Aviation Week & Space Technology July 28, 1980, page 48. (Source: The photo is taken by US High-Resolution Reconnaissance Satellite KH-11)

official, and presently considered to be either a high-energy laser or a particle beam weapon. (See Aviation Week & Space Technology, July 28, 1980, p. 48 for an artistic conception.) A retouched photograph of the Sary Shagan installation extracted from Aviation Week is shown as Fig. B.20.

Bearden claims that the Sary Shagan howitzer actually is a huge Tesla scalar interferometer with four modes of operation. One continuous mode is the Tesla shield, which places a thin, impenetrable hemispherical shell of energy over a large defended area. The three-dimensional shell is created by interfering two Fourier expansion, three-dimensional scalar hemispherical patterns in space so they pair-couple into a dome-like shell of intense, ordinary electromagnetic energy. The air molecules and atoms in the shell are totally ionized and thus highly excited, giving off intense, glowing light. Anything physical which hits the shell receives an enormous discharge of electrical energy and is instantly vaporized—it goes pfft! like a bug hitting one of the electrical bug killers now so much in vogue. See Fig. B.15.

Bearden goes on further to say that, if several of these hemispherical shells are concentrically stacked, even the gamma radiation and EMP from a high-altitude nuclear explosion above the stack cannot penetrate all the shells due to repetitive absorption and reradiation and scattering in the layered plasmas.

Bearden speculates about many other effects of scalar waves [28].

Appendix C: Digital Signal Processing for Radar Applications

As we described in Chap. 1 of this book, radar technology is used heavily in military applications. Ground-based radar is used for long-range threat detection and air traffic control. Ship-based radar provides surface-to-surface and surface-to-air observation. Airborne radar is utilized for threat detection, surveillance, mapping, and altitude determination. Finally, missile radars are used for tracking and guidance. Digital signal processing is a methodology or technique that is enhancing radar detection by far out, and in this Appendix, we discuss the fundamentals of radar measurement and signal processing as an introductory to our reader.

C.1 Introduction

As we discussed in Chap. 1, there are many commercial aviation applications of radar such as air traffic control (ATC) long-range surveillance, terminal air traffic monitoring, surface movement tracking, and weather surveillance. Additionally, short-range radar is increasingly being used in automotive applications for collision avoidance, driver assistance, and autonomous driving. Specialized radars can also be used to provide imaging through fog, clutter environment, walls, and even underground.

Modern radars produce complicated pulses that present significant measurement challenges. Improvements to range, resolution, and immunity to interference have motivated numerous coding schemes, frequency- and phase-modulated pulses, frequency-chirped pulses, and narrow pulses with high overall bandwidth.

Here we briefly describe the fundamentals of radar measurement and signal analysis, by describing the two fundamental types that are operating and transmitting rate under circumstances as:

1. Continuous-wave radar
2. Pulse radar

B. Zohuri, *Radar Energy Warfare and the Challenges of Stealth Technology*,
https://doi.org/10.1007/978-3-030-40619-6

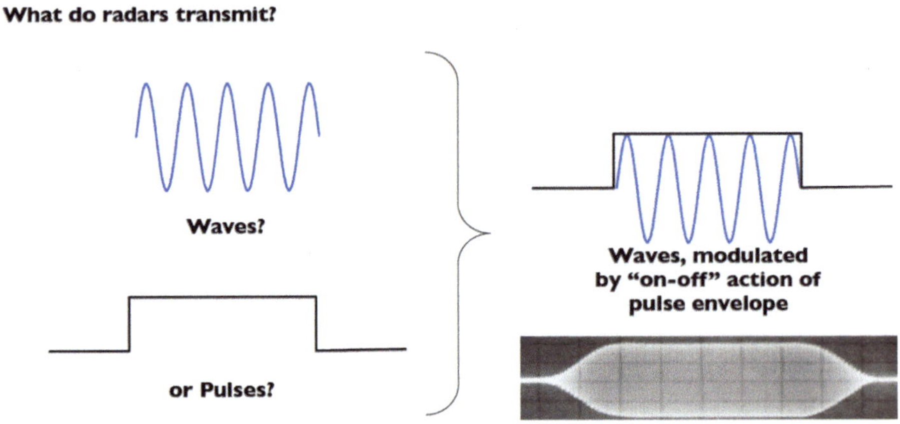

Fig. C.1 Waves versus pulses

Each of these two types radar is briefly introduced as the following sub-sections below. However, one thing we should understand is the *waves* versus *pulses* and what does radar transmit. And these two modes of transmission are depicted in Fig. C.1.

Moreover, when it comes to pulse radar type, the question is that "How many cycles are in a typical pulse?", and typical long pulse length normally is 480 μs.

Furthermore, modern radars are very diverse and have different applications such as:

- Military radars
- Imaging radars
- Radar gun
- Automotive radars
- Civil aviation radars
- Weather radars
- Ground-penetrating radars

All of the above radar categories are described in Chap. 1.

C.2 Continuous-Wave Radar

Radar systems can use continuous-wave (CW) signals or, more commonly, low-duty-cycle pulsed signals. CW radar applications can be simple unmodulated Doppler speed sensing systems such as those used by police and sports-related radars or may use modulation to sense range as well as speed. Modulated CW applications have many specialized and military applications such as maritime/naval applications, missile homing, and radar altimeters. The detection range of CW radar systems is relatively short, due to the constraints of continuous radio-frequency (RF) power.

There is no minimum range, however, which makes CW radar particularly useful for close-in applications.

C.3 Pulse Radar

Although there are several continuous transmission types of radar, primarily Doppler, the great majority of radars are pulsed. There are two general categories of pulse radar, moving target indicator (MTI) and pulsed Doppler. MTI radar is a long-range, low pulse repetition frequency (PRF) radar used to detect and track small (~2 m^2) moving targets at long distances (up to ~30 km) by eliminating ground clutter (aka chaff). MTI is useful when velocity is not a big concern (i.e., "just tell me if something is moving"). Pulsed Doppler radar, in contrast, utilizes a high PRF to avoid "blind speeds" and has a shorter "unambiguous" range (~15 km), has high resolution, and provides detailed velocity data. It is used for airborne missile approach tracking, air traffic control, and medical applications (e.g., blood flow monitoring).

RF pulse characteristics such as those illustrated below reveal a great deal about a radar's capability. Electronic Warfare (EW) and Electronic INTelligence (ELINT) experts specialize in the study of these pulsed signals. Pulse characteristics provide valuable information about the type of radar producing a signal and what its source might be—sailboat, battleship, passenger plane, bomber, missile, etc.

As Fig. C.2 indicates, pulse radar typically uses very-low-duty-cycle RF pulses (<10%). Range and resolution are determined by the pulse repetition frequency (PRF), pulse width (PW), and transmit power.

A wide PW generally provides better range, but poor resolution. Conversely, a narrow PW has less range, but better resolution. This relationship constitutes one of the fundamental trade-offs in radar engineering. Pulse compression with a modulated carrier is often used to enhance resolution while maintaining a narrow PW allowing for higher power and longer range.

The pulse repetition interval (PRI) is the time the pulse cycle takes before repeating. It is equal to the reciprocal of the pulse radar frequency (PRF) or pulse

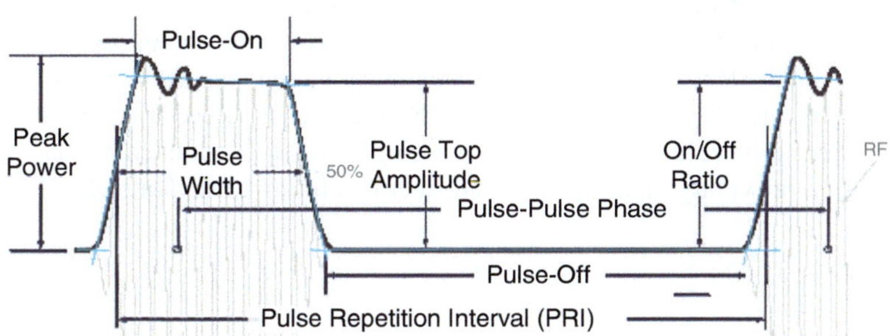

Fig. C.2 Typical pulse radar

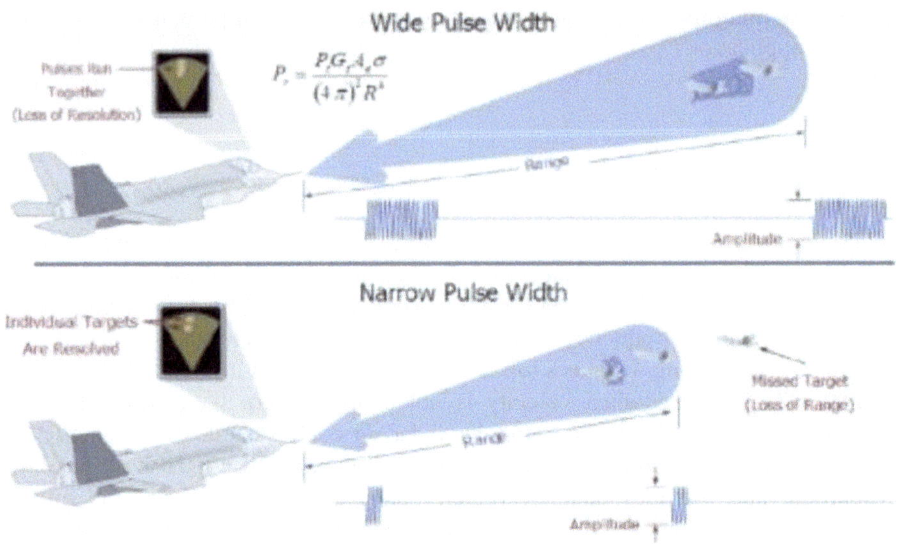

Fig. C.3 Pulse compression signal

repetition rate (PRR), the number of transmitted pulses per second. PRI is important because it determines the maximum unambiguous range or distance of the radar. In fact, pulse-off time may actually be a better indication of the radar system's maximum design range.

Traditional radar systems employ a transmit/receive (T/R) switch to allow the transmitter and receiver to share a single antenna. The transmitter and receiver take turns using the antenna. The transmitter sends out pulses and during the off-time, the receiver listens for the return echo. The pulse-off time is the period the receiver can listen for the reflected echo. The longer the off-time, the further away the target can be without the return delay putting the received pulse after the next transmitted pulse. This would incorrectly make the target appear to be reflected from a nearby object. To avoid this ambiguity, most radars simply use a pulse-off time that is long enough to make echo returns from very distant objects so weak in power, they are unlikely to be erroneously detected in the subsequent pulse's off-time.

Figure C.3 illustrates the need for pulse compression to obtain good range and resolution. Wider PWs have higher average power, which increases range capability.

However, wide PWs may cause echoes from closely spaced targets to overlap or run together in the receiver, appearing as a single target. Modulated pulses mitigate these issues, providing higher power and finer resolution to separate closely spaced targets.

C.3.1 Pulse Radar

Another consideration for the maximum range of a radar is the transmitted power. Peak power is a measure of the maximum instantaneous power level in the pulse. Power droop, pulse top amplitude, and overshoot are also of interest.

Pulse top amplitude (power) and pulse width (PW) are important for calculating the total energy in a given pulse (power × time). Knowing the duty cycle and the power of a given pulse, the average RF power transmitted can be calculated (pulse power × duty cycle).

A scenario of pulse radar is depicted in Fig. C.4, where all the above parameters are defined:

$$\text{Duty Cycle} = \frac{\text{Pulse Length}}{\text{Pulse Repetition Interval}} \tag{C.1}$$

$$\text{Average Power} = \text{Peak Power} * \text{Duty Cycle} \tag{C.2}$$

The pulse repetition frequency (PRF) is the number of pulses of a repeating signal in a specific time unit, normally measured in pulses per second. The term is used within a number of technical disciplines, notably radar.

In radar, a radio signal of a particular carrier frequency is turned on and off; the term "frequency" refers to the carrier, while the PRF refers to the number of switches. Both are measured in terms of cycle per second, or hertz (Hz). The PRF is normally much lower than the frequency. For instance, a typical World War II radar like the Type 7 GCI radar (Fig. C.5) had a basic carrier frequency of 209 MHz (209 million cycles per second) and a PRF of 300 or 500 pulses per second. A related measure is the pulse width, the amount of time the transmitter is turned on during each pulse.

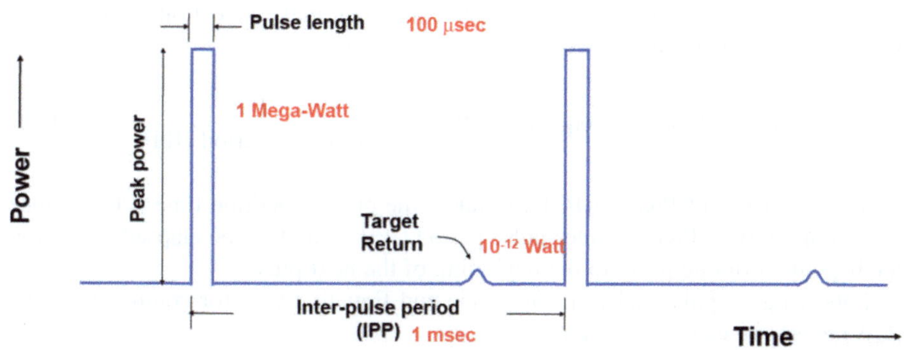

Fig. C.4 Pulse radar characteristics

Fig. C.5 Type 7 GCI air defense metric search radar

The Type 7 was a metric radar operating in the 1.5 m wave band used for ground-controlled interception (GCI). The usual operating frequency was 209 MHz, though later equipments operated on 193 and 200 MHz.

This was a parallel development of the Chain Home Low (CHL) equipment by the addition of a height-finding capability and a plan position indicator (PPI) display.

The PRF is one of the defining characteristics of a radar system, which normally consists of a powerful transmitter and sensitive receiver connected to the same antenna. After producing a brief pulse of radio signal, the transmitter is turned off in order for the receiver units to hear the reflections of that signal off distant targets. Since the radio signal has to travel out to the target and back again, the required inter-pulse quiet period is a function of the radar's desired range. Longer periods are required for longer-range signals, requiring lower PRFs. Conversely, higher PRFs produce shorter maximum ranges, but broadcast more pulses, and thus radio energy, in a given time. This creates stronger reflections that make detection easier. Radar systems must balance these two competing requirements.

$$\text{Pulse Repition Frequency (PRF)} = \frac{1}{\text{Inter Pulse Period (IPP)}} \tag{C.3}$$

The reciprocal of PRF (or PRR) is called the pulse repetition time (PRT), pulse repetition interval (PRI), or inter-pulse period (IPP), which is the elapsed time from the beginning of one pulse to the beginning of the next pulse.

Following on pulse radar characteristic and Equation C.1, for continuous-wave (CW) radar, we can express that:

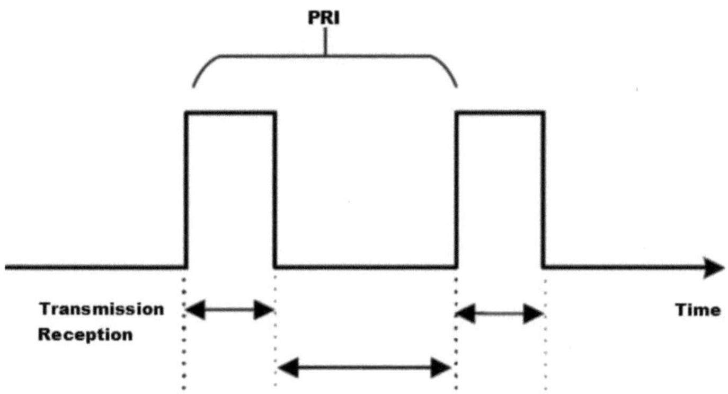

Fig. C.6 Pulse repetition interval

$$\text{Continuous Radar (CW) Radar} = \text{Duty Cycle} = 100\%(\text{Always on})$$

In summary, pulse repetition interval (PRI) can be described as Fig. C.6 and what is listed below:

- Pulse repetition interval (PRI) is defined as the time interval between consequent pulses.
- Pulse repetition frequency (PRF) is given as Equation C.3 or per Equation C.4 below:

$$\text{PRF} = 1/\text{PRI} \tag{C.4}$$

- Duty cycle is defined as the time proportion of PRI in which the transmission takes place as Equations C.1 or C.5 here:

$$\text{Duty Cycle} = \text{T}/\text{PRI} \tag{C.5}$$

- If the same antenna is used for transition and reception, the duty cycle gives a measure of how long the radar is "blind."

C.4 Radar Equation

The radar equation defines many of the engineering trade-offs encountered by radar designers, and in Chap. 1 of this book, we provided the variation of it for various radar types. However, the simple radar equation is provided here as Equation C.6.

$$P_r = \frac{P_t G_t A_r \sigma}{(4\pi)^2 R_t^2 R_r^2} \tag{C.6}$$

where:

P_t = transmitted pulse power
G_t = transmitting antenna gain
A_r = area of the receiving antenna
σ = target cross section (aka reflectivity)
R_t = range from the transmitter antenna to the target
R_r = range of the target to the receiving antenna

Thus, Equation C.6 simply relates the expected receive power (P_r) to the transmitted pulse power (P_t), based on transmitting antenna gain (G_t), area of the receiving antenna (A_r), target cross section (aka reflectivity) σ, range from the transmitter antenna to the target (R_t), and range of the target to the receiving antenna (R_r).

Unlike many communications systems, radar systems suffer from very large signal path losses. The round-trip distance is twice that of a typical communications link, and there are losses associated with the radar cross section and reflectivity of targets. As you can see from the radar equation, the range term is raised to the fourth (i.e., $R_t = R_r$) power in the denominator, underscoring the tremendous signal power losses radar signals experience.

Using the radar equation, the received signal level can be calculated to determine if sufficient power exists to detect a reflected radar pulse. Combining multiple pulses to accumulate greater signal power and average out the noise is also helpful for increasing the detection range.

The radar Equation C.6 can have a different format for stationary radar such as weather radar to measure clutter and why there is ground clutter on the radar.

You may have noticed green on the radar, even when it is completely dry outside. This is what we call anomalous propagation or better known as ground clutter.

This occurs when the radar beam goes out, but it is refracted or deflected. Radars usually scan high above the clouds, and the signal that is returned is normally representative of that, but sometimes the radar beam is only seeing ground level.

Radars are made to observe rain, hail, and snow. Sometimes there are other things that get detected such as birds, bugs, objects close to the ground, and dust.

Ground clutter is usually from objects close to the ground since the radar beam starts close to the ground the further out the radar beam goes to higher elevations as you move away from the radar site. See Fig. C.7.

In this case we can assume that the range from the transmitter antenna to the target (R_t) and range of the target to the receiving antenna (R_r) are equal to each other, namely, $R_t = R_r = R$, as it can be seen in Fig. C.7. Thus, under this circumstance, Equation C.6 can be written as a new form as presented in Equation C.7 for *radar range equation* for *receive power* P_{receive}.

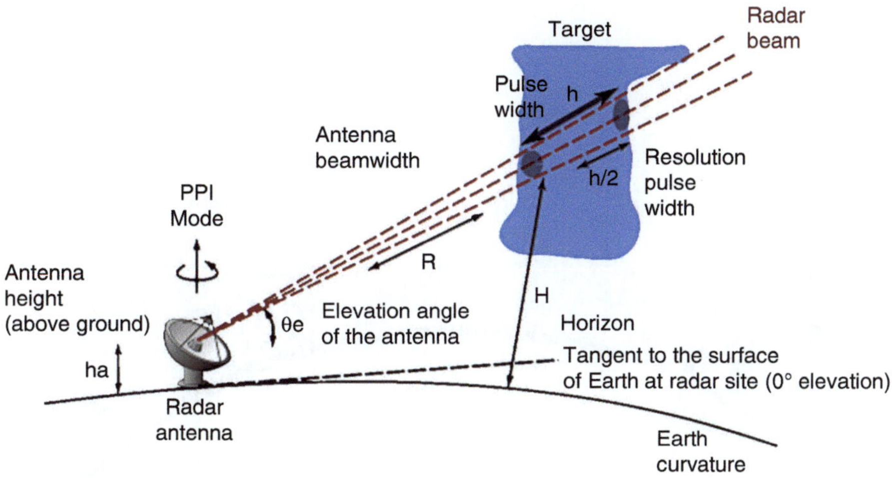

Fig. C.7 Stationary radar site and antenna beam

$$P_{\text{receive}} = \frac{P_t G_t A_r \sigma F^4 \left(t_{\text{pulse}}/T\right)}{\left[(4\pi)^2 R^4\right]} \tag{C.7}$$

where again:

P_t = transmitted power
G_t = transmitting antenna gain
A_r = receiving antenna aperture area
σ = radar cross section, which is a function of target geometric cross section, reflectivity of surface, and directivity of reflections
F = pattern propagation factor unity in vacuum, accounts for multipath, shadowing, and other factors
t_{pulse} = duration of receive pulse
T = duration of transmit interval
R = range between radar and target

To close our discussion on radar equation as well as as far as weather radar is concerned, we can state that as the radar beam is being deflected, it can sometimes hold the beam close to the surface and the beam can still travel for long distances. As illustrated in Fig. C.8, this is referred to as *super-refraction*.

Other times, the deflection is so strong that it sends the radar beam back down to the Earth's surface. As illustrated in Fig. C.9, this is known as *ducting*.

There are three types of clutter, surface, volume, and point clutter. Surface clutter is returns from the ground and sea.

Volume clutter examples are rain, snow, and hail. Finally, point clutters are when birds or tall buildings obstruct the radar beam.

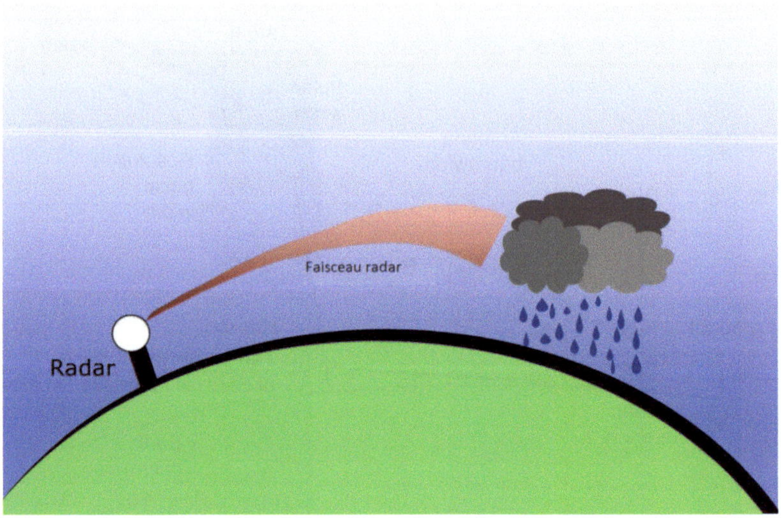

Fig. C.8 The super-refraction illustration

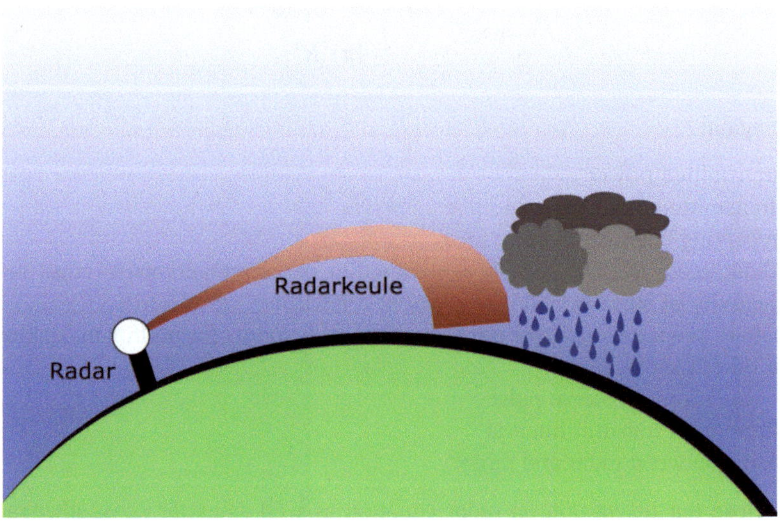

Fig. C.9 The ducting illustration

Ground clutter can be caused by an inversion. An inversion is a layer of warm air up above cool air.

However, ground clutter is most common in the morning hours.

So, the next time you see green on radar, even when the skies are clear, just know it might be from various objects in the sky.

In conclusion of this section, we may say that ground clutter is a common reference that will be made when examining weather radar. What typically needs to be observed on weather radar are hydrometeors such as rain, hail, and snow. Other substances that do not necessarily need to be detected on weather radar are also detected such as bugs, particulates, birds, objects at ground level, aircraft, and dust.

Ground clutter shows up best near the radar site since the radar beam trajectory takes it from close to the surface at the radar site to higher elevations moving away from the radar site. Thus, a ring of ground clutter objects can show up near the radar site. Most ground cluttering matter will be relatively close to ground level. Software programs can remove much of the ground clutter, but ground clutter can still be an issue, especially when the density profile of the atmosphere helps keep the radar beam near the ground.

One way the ground clutter issue is dealt with is by the use of multiple radars. This can help determine the characteristics of the ground clutter and if hydrometeors are also present in that area.

C.5 Pulse Width

Pulse width is an important property of radar signals. The wider a pulse, the greater the energy contained in the pulse for a given amplitude. The greater the transmitted pulse power, the greater the reception range capability of the radar.

Greater pulse width also increases the average transmitted power. This makes the radar transmitter work harder. The difference in decibels between the pulse power and average power level is easily calculated using ten times the log of the pulse width divided by the pulse repetition interval.

Range is therefore limited by the pulse characteristics and propagation losses. The PRI and duty cycle set the maximum allowed time for a return echo, while the power or energy transmitted must overcome the background noise to be detected by the receiver.

Pulse width also affects a radar's minimum resolution. Echoes from long pulses can overlap in time, making it impossible to determine the nature of the target or targets. A long pulse return may be caused by a single large target, possibly an airliner, or multiple smaller targets closely spaced, possibly a tight formation of fighter aircraft. Without sufficient resolution, it is impossible to determine the number of objects that actually make up the echo return. Narrow pulse widths mitigate the overlapping of echoes and improve resolution at the expense of transmit power.

As such, pulse width affects two very important radar system capabilities—resolution and detection range. These two qualities trade off against each other. Wider pulse longer-range radars offer less resolution, whereas narrow pulse shorter-range radars have finer resolution.

Narrow pulses also require greater bandwidth to correctly transmit and receive. This makes the pulse's spectral nature also of interest, which must be considered in the overall system design as you can see in Fig. C.10 as illustrated below.

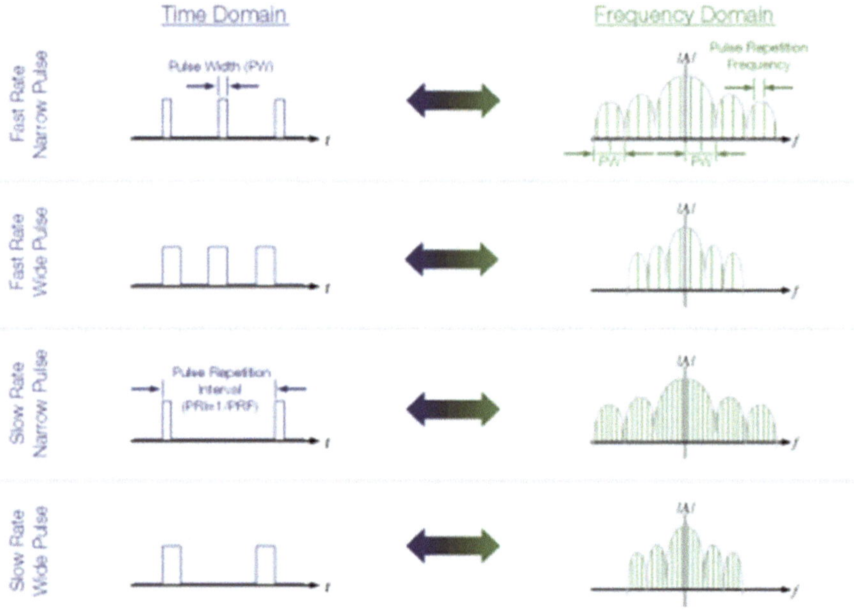

Fig. C.10 Overall pulse width system illustration

While unmodulated pulse radars are relatively simple to implement, they have drawbacks including relatively poor range resolution. To make the more efficient use of transmit power and optimize range resolution, radars modulate pulses using the following modulation techniques:

- *Linear Frequency Modular Chirps*—The simplest and most common pulse modulation scheme is the linear frequency modular (LFM) chirp. Sweeping the carrier frequency throughout a pulse results in every part of a pulse being distinct and discernable. This enables pulse compression techniques in the receiver to improve range resolution and transmit power efficiency.
- *Phase Modulation*—Phase modulation can also be used to differentiate segments of a pulse and is often implemented as a version of binary phase-shift keying (BPSK). There are specific phase coding schemes, such as Barker codes, to ensure orthogonality of the coding and range resolution.
- *Frequency Hopping*—This approach involves several frequency hops within a pulse. When each frequency has a corresponding filter with the appropriate delay in the receiver, all segments can be compressed together in the receiver. If the frequency hopping sequence remains the same for all pulses, then the receiver compression can even be implemented with a simple surface acoustic wave (SAW) filter. The variable frequency pattern used by hopping pulses can reduce susceptibility to spoofing and jamming and help with interference problems.
- *Digital Modulation*—Digital signal processing enables more complex pulse modulations. For example, more effective anti-spoofing can be accomplished

Table C.1 Frequency spectrum of pulse (FSP)

Spectrum of single pulse	
Spectrum of slowly repeating pulse (low PRF) Spectrum of rapidly repeating pulse (high PRF)	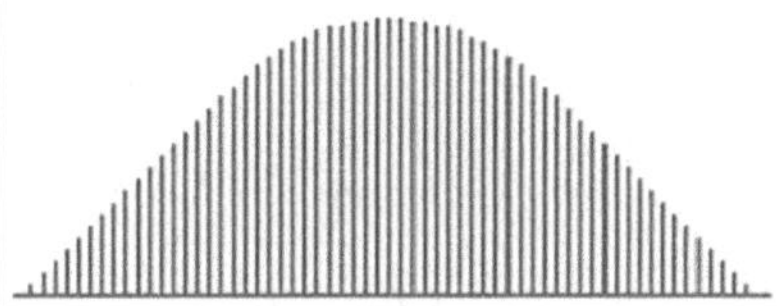
Line spacing equal to PRF	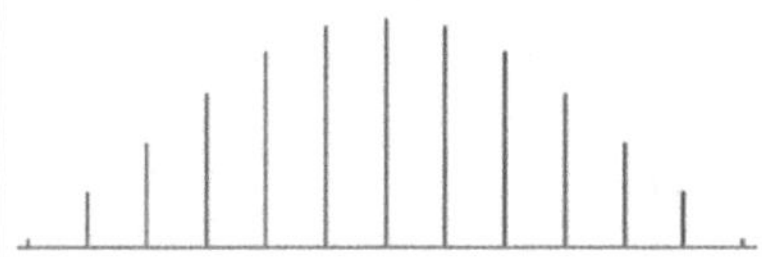

using M-ary PSK or QAM modulations that resemble noise rather than coherent frequencies to make detection more difficult. Other information can be encoded into digital modulations as well.

All the above points can be covered under frequency spectrum of pulse (FSP) Table C.1 and depicted accordingly.

Note that side-looking radars can produce a large number of pulses, thus increasing radar sensitivity. If a coherent radar is used, improved sensitivity and resolution can be obtained by using Doppler filter banks or digital fast Fourier transform (FFT) processing. If the platform motion compared with the aperture length is sufficiently large, platform-motion compensation will be required.

Ship detection can be improved by rapidly scanning the antenna so that sea clutter is decorrelated and surface-target returns are integrated or leave a pattern of returns indicating their track. In some cases, frequency agility can also be utilized to decorrelate clutter and integrate ship target returns. Scan-to-scan video cancelation can be utilized for detecting moving targets overland if their scan-to-scan motion is of the order of the radar pulse width.

Also, it is worth to mention from electronic design point of view of pulse radar, where transistor and resistor for devices, we can state the following expression.

There is a thermal time constant associated with the numerous thermally resistive layers between the transistor junction and the heat sink or cold plate to which the device is attached. This occurs because each layer (silicon, ceramic, transistor flange) not only has a thermal resistance but also exhibits a thermal capacity. Since the overall thermal time constant for a typical L-band power transistor may be on the order of hundreds of microseconds, the trade-off between peak and average power versus device size can be significant for typical radar pulse widths in the 20- to 1000-jjis range. Devices that operate for short pulse and low-duty-cycle applications, such as distance measuring equipment (DME), TACAN, and identification, friend,

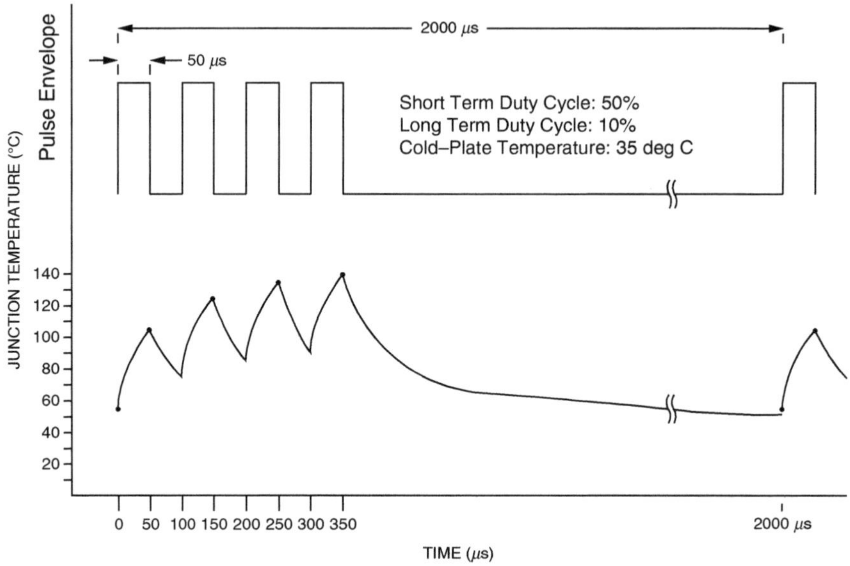

Fig. C.11 Transient thermal response of a class C-biased silicon power transistor [29]

or foe (IFF) systems, differ in design from the devices that have been designed for the longer pulse widths and moderate-duty-cycle waveforms that are more typical for surveillance radars. Very high duty cycles or CW operation dictates careful thermal design.

Figure C.11 is a transient thermal response of a class C-biased silicon power transistor for a pulsed RF input.

An illustration of the thermal-time-constant effect, as it relates to a train of RF pulses, is shown in Fig. C.11. Table C.2 illustrates some reported device applications and their general performance characteristics [29].

In summary, a radar system uses a radio-frequency electromagnetic signal reflected from a target to determine information about that target. In any radar system, the signal transmitted and received will exhibit many of the characteristics described in Fig. C.12.

The diagram in Fig. C.12 shows the characteristics of the transmitted signal in the time domain. Note that in this and in all the diagrams, the x-axis is exaggerated to make the explanation clearer.

Furthermore, the pulse width τ (or pulse duration) of the transmitted signal is the time, typically in microseconds, each pulse lasts. If the pulse is not a perfect square wave, the time is typically measured between the 50% power levels of the rising and falling edges of the pulse.

The pulse width must be long enough to ensure that the radar emits sufficient energy so that the reflected pulse is detectable by its receiver. The amount of energy that can be delivered to a distant target is the product of two things: the peak output power of the transmitter and the duration of the transmission. Therefore, pulse width

Table C.2 System applications of microwave power transistors

System	Frequency, MHz	Pulse/duty	Transistor performance		
			Peak power, W	Gain, dB	Efficiency, %
OTH	5–30	CW	130	14.0	60
NAVSPASUR	217	CW	100	9.2	72
AN/SPS-40	400–450	60 μs at 2%	450	8.0	60
PAVE PAWS	420–450	16 ms at 20%	115	8.5	65
BMEWS	420–450	16 ms at 20%	115	8.5	65
AN/TPS-59	1215–1400	2 ms at 20%	55	6.6	52
RAMP	1250–1350	100 μs at 10%	105	7.5	55
MARTELLO S723	1235–1365	150 μs at 4%	275	6.3	40
MATCALS	2700–2900	100 μs at 10%	63	6.5	40
AN?SPS-48	2900–3100	40 μs at 4%	55	5.9	32
AN?TPQ-37	3100–3500	100 μs at 25%	30	5.0	30
HADR	3100–3500	800 μs at 23%	50	5.3	35

Reprinted with permission from E. D. Ostroff et. al., "Solid-State Transmitters," Artech House, Norwood, Mass., 1985

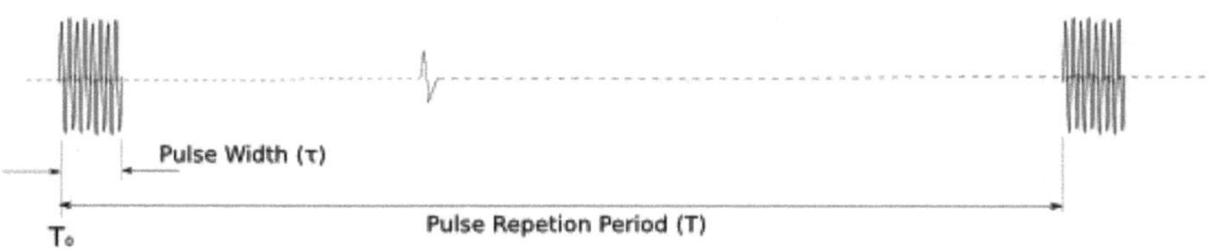

Fig. C.12 Pulse radar characteristics

constrains the maximum detection range of a target. See illustration in Fig. C.13, where pulse repetition period (PRP) is depicted.

Pulse width also constrains the range discrimination, that is, the capacity of the radar to distinguish between two targets that are close together. At *any* range, with similar azimuth and elevation angles and as viewed by a radar with an unmodulated pulse, the range resolution is approximately equal in distance to half of the pulse duration times the speed of light (approximately 300 m per microsecond).

Pulse width also determines the radar's dead zone at close ranges. While the radar transmitter is active, the receiver input is blanked to avoid the amplifiers being swamped (saturated) or (more likely) damaged. A simple calculation reveals that a radar echo will take approximately 10.8 μs to return from a target 1 statute mile away (counting from the leading edge of the transmitter pulse (T_0) (sometimes known as transmitter main bang)). For convenience, these figures may also be expressed as

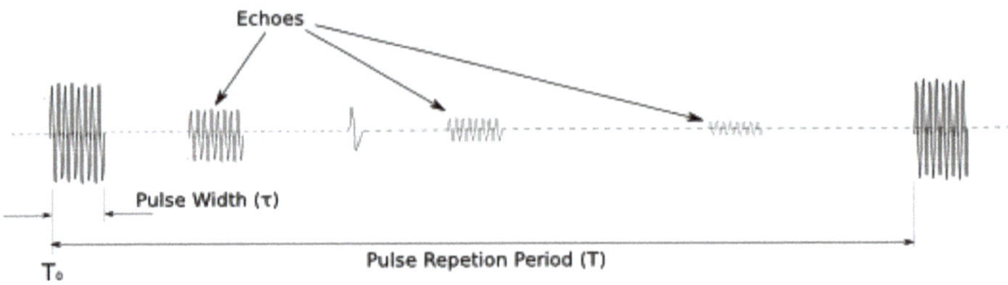

Fig. C.13 Pulse repetition period (PRP)

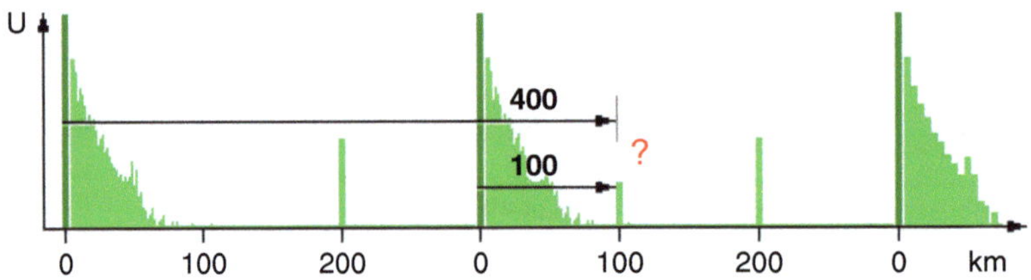

Fig. C.14 A second-sweep echo in a distance of 400 km assumes a wrong range of 100 km

1 nautical mile in 12.4 μs or 1 km in 6.7 μs. (For simplicity, all further discussion will use metric figures.) If the radar pulse width is 1 μs, then there can be no detection of targets closer than about 150 m, because the receiver is blanked.

All this means that the designer cannot simply increase the pulse width to get greater range without having an impact on other performance factors. As with everything else in a radar system, compromises have to be made to a radar system's design to provide the optimal performance for its role.

C.6 Unambiguous Range

A problem with pulse radars and range measurement is how to unambiguously determine the range to the target if the target returns a strong echo. This problem arises because of the fact that pulse radars typically transmit a sequence of pulses. The radar receiver measures the time between the leading edges of the last transmitting pulse and the echo pulse. It is possible that an echo will be received from a long-range target after the transmission of a second transmitting pulse as illustrated in Fig. C.14.

In this case, the radar will determine the wrong time interval and therefore the wrong range. The measurement process assumes that the pulse is associated with the second transmitted pulse and declares a much reduced range for the target. This is called range ambiguity and occurs where there are strong targets at a range in excess of the pulse repetition time. The pulse repetition time defines a maximum unambiguous range. To increase the value of the unambiguous range, it is necessary to increase the pulse repetition time (PRT); this means to reduce the pulse repetition frequency (PRF).

Echo signals arriving after the reception time are placed either into the:

- Transmit time where they remain unconsidered since the radar equipment is not ready to receive during this time
- Following reception time where they lead to measuring failures (ambiguous returns)

Unambiguous range can be described for:

1. Single PRF
2. Multiple PRF
3. Maximum unambiguous range

Each of the above categories is defined as follows:

1. Single PRF
   In simple systems, echoes from targets must be detected and processed before the next transmitter pulse is generated if range ambiguity is to be avoided. Range ambiguity occurs when the time taken for an echo to return from a target is greater than the pulse repetition period (T); if the interval between transmitted pulses is 1000 μs, and the return time of a pulse from a distant target is 1200 μs, the apparent distance of the target is only 200 μs. In sum, these "second echoes" (i.e., Fig. C.15) appear on the display to be targets closer than they really are.

   Consider the following example: if the radar antenna is located at around 15 m above sea level, then the distance to the horizon is pretty close (perhaps 15 km). Ground targets further than this range cannot be detected, so the PRF can be quite high; a radar with a PRF of 7.5 kHz will return ambiguous echoes from targets at about 20 km or over the horizon. If, however, the PRF was doubled to 15 kHz, then the ambiguous range is reduced to 10 km, and targets beyond this range would only appear on the display after the transmitter has emitted another pulse. A target at 12 km would appear to be 2 km away, although the strength of the echo might be much lower than that from a genuine target at 2 km.

   Considering Fig. C.16 we can present the maximum unambiguous range equation as it is written in Equation C.8, where it defines the maximum distance to locate a target.

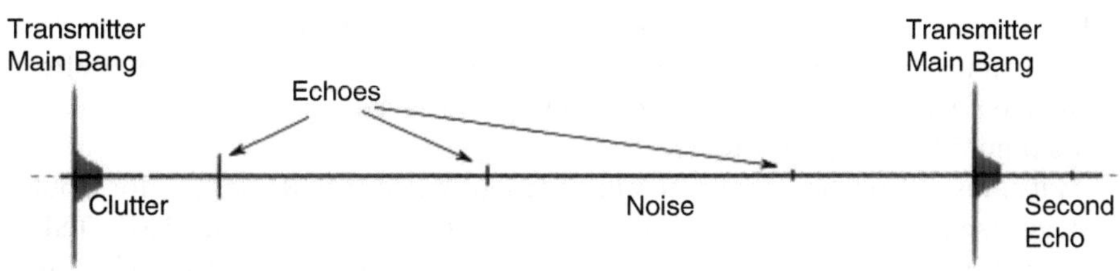

Fig. C.15 Second echoes characteristics

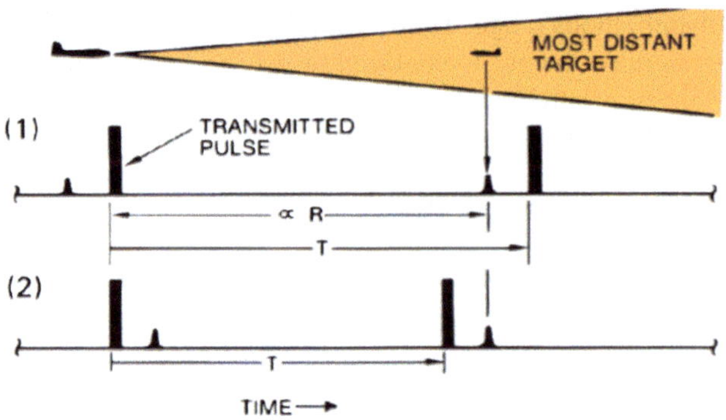

Fig. C.16 Maximum unambiguous range depiction

$$R_{\text{Max Unambiguous}} = \frac{c\text{PRI}}{2} = \frac{c}{2\text{PRF}} \tag{C.8}$$

where:

c = speed of light
IPP = inter-pulse period (s)
PRF = pulse repetition frequency

Part (1) of Fig. C.16 says that, if inter-pulse period (IPP), T, is long enough for all echoes from one pulse to be received before the next pulse is transmitted, echoes may be presumed to belong to pulse that immediately precedes them. (2) but not if T is shorter than this.

2. Multiple PRF
   Modern radars, especially air-to-air combat radars in military aircraft, may use PRFs in the tens-to-hundreds of kilohertz and stagger the interval between pulses to allow the correct range to be determined. With this form of staggered PRF, a *packet* of pulses is transmitted with a fixed interval between each pulse, and then another *packet* is transmitted with a slightly different interval. Target reflections appear at different ranges for each *packet*; these differences are accumulated, and then simple arithmetical techniques may be applied to determine true range. Such radars may use repetitive patterns of *packets* or more adaptable *packets* that respond to apparent target behaviors. Regardless, radars that employ the technique are universally coherent, with a very stable radio frequency, and the pulse *packets* may also be used to make measurements of the Doppler shift (a velocity-dependent modification of the apparent radio frequency), especially when the PRFs are in the hundreds-of-kilohertz range. Radars exploiting Doppler effects in this manner typically determine relative velocity first, from the Doppler effect, and then use other techniques to derive target distance.
3. Maximum Unambiguous Range
   At its most simplistic, maximum unambiguous range (MUR) or R_{max} for a pulse stagger sequence may be calculated using the total sequence period (TSP). TSP is defined as the total time it takes for the pulsed pattern to repeat. This can be found by the addition of all the elements in the stagger sequence. The formula is derived from the speed of light and the length of the sequence.

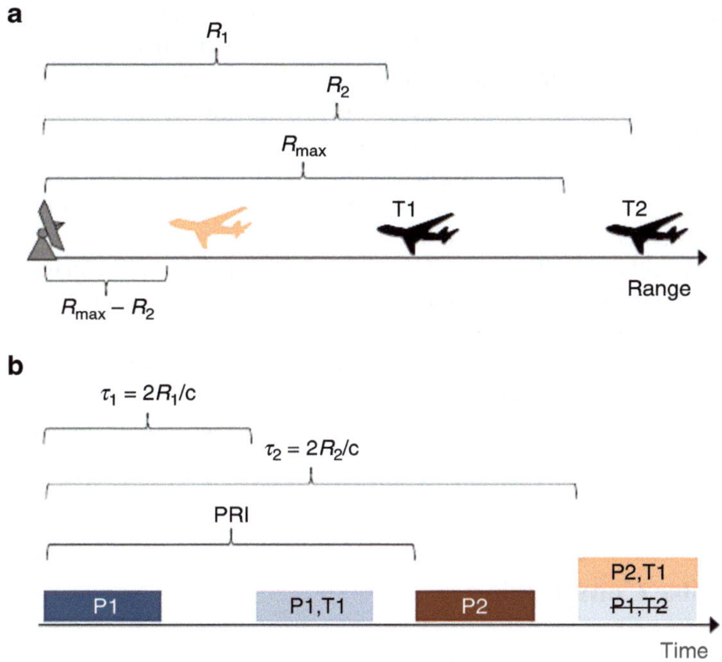

Fig. C.17 (**a**) Maximum unambiguous range of two targets. (**b**) Maximum unambiguous range two targets analysis

$$\text{MUR} = R_{\max} = [0.5 * c * \text{TSP}] \tag{C.9}$$

where c is the speed of light, usually in meters per microsecond, and TSP is the addition of all the positions of the stagger sequence, usually in microseconds. However, in a stagger sequence, some intervals may be repeated several times; when this occurs, it is more appropriate to consider TSP as the addition of all the unique intervals in the sequence.

The MUR defines the maximum distance to locate a target and radar is not able to discriminate between echoes from and older and the current transmission as illustrated in Fig. C.17a, b as well as a new form of Equation C.9 in the form of Equation C.10:

$$R_{\max} = \frac{c\text{PRI}}{2} = \frac{c}{2\text{PRF}} \tag{C.10}$$

where:

PRI = pulse repetition interval
PRF = pulse radar frequency

In Fig. C.17a, radar and two real targets (dark), one in (T1) and one out (T2) of unambiguous range, second target (T2) appears in closer range (light).

In Fig. C.17b, transmitted (dark) and received pulses (light) at the radar in time, radar confuses the echo from fist pulse to second target (P1,T2) to an echo from second pulse (P2) and a target at a closer range ($R_{max} - R_2$).

Also, it is worth remembering that there may be vast differences between the MUR and the maximum range (the range beyond which reflections will probably be too weak to be detected) and that the maximum *instrumented* range may be *much* shorter than either of these. A civil marine radar, for instance, may have user-selectable maximum *instrumented* display ranges of 72 or 96 or rarely 120 nautical miles, in accordance with international law, but maximum unambiguous ranges of over 40,000 nautical miles and maximum detection ranges of perhaps 150 nautical miles. When such huge disparities are noted, it reveals that the primary purpose of staggered PRF is to reduce "jamming," rather than to increase unambiguous range capabilities.

C.7 Pulse Compression

Basic pulse radar using time-of-flight to measure target range has limitations. For a given pulse width, the range resolution is limited to the distance over which the pulse travels. When multiple targets are at nearly equal distance from the radar, the return from the farthest target will overlap the return from the first target. In this situation, the two targets can no longer be resolved from each other with just simple pulses.

Using a short pulse width is one way to improve distance resolution. However, shorter pulses contain proportionately less energy, preventing reception at greater range due to propagation losses.

Increasing the transmit power is often impractical, such as for aircraft radar due to power constraints.

The answer to these challenges is pulse compression. If a pulse can be effectively compressed in time, then the returns will no longer overlap. Pulse compression allows low amplitude returns to be "pulled" out of the noise floor. It is achieved by modulating the pulse in the transmitter so different parts of the pulse become more discernable. The actual time compression is accomplished by the radar receiver.

The most common pulse compression technique is linear frequency modulation (LFM). An LFM pulse or chirp is one where the pulse begins at one carrier frequency and then ramps linearly, up or down to an end frequency. Compression is achieved in the receiver by passing the signal through a matched filter as shown below. This filter is designed to have a delay characteristic matched to the LFM frequency range and delays portions of the modulated signal proportional to the carrier frequency (Fig. C.18).

When a pulse entering the receiver is a target return, there will likely be multiple close reflections due to the different surfaces of the target. If the compression signal

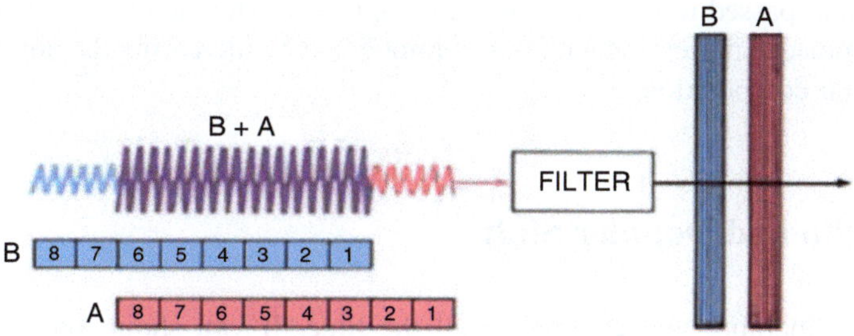

Fig. C.18 Linear frequency modulation (LFM)

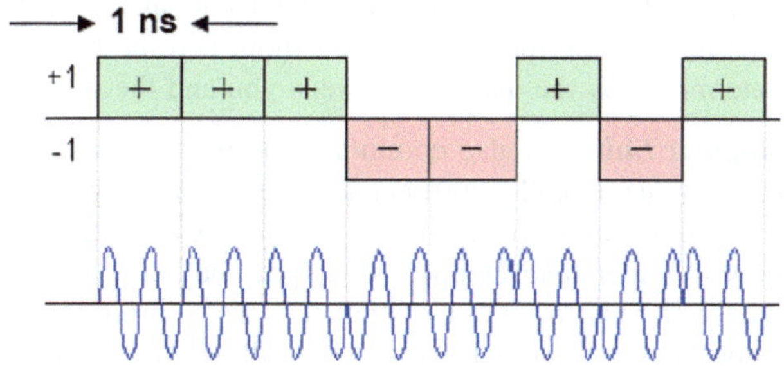

Fig. C.19 BPSK modulation

processor has sufficient resolution, it can separate each of these reflections into discrete narrow pulses.

Another common pulse compression technique employs binary phase-shift keying, or binary phase-shift keying (BPSK) modulation, using a Barker-coded sequence as shown below (Fig. C.19). Barker codes are unique binary patterns that autocorrelate against themselves at only one point in time. Barker codes can vary in length from 2 to 13 bits, providing a corresponding compression ratio of 2–13. The compression is achieved in the receiver by sensing the autocorrelation of the Barker sequence within the detected pulse.

Now that you have a basic understanding of radar signals, it is time to move on to measuring them. In the next installment, we'll look at the lifecycle of radar measurement tasks, which can vary significantly depending on the job to be done and type of radar you are characterizing [30].

In summary, pulse compression is another signal processing function that is predominantly being performed digitally in radar systems. However, at this writing, many systems still exist with analogue-delay-line pulse compressors. In these systems, analogue pulse compression is performed at an intermediate frequency (IF), followed by the analogue-to-digital converter (ADC) in the processing chain. Because pulse compression increases the signal-to-noise ratio (SNR) of the signal, performing it before sampling increases the dynamic range requirement of the ADC. In a digital pulse compression system, the ADC precedes the pulse compressor and only has to accommodate the precompression dynamic range of the signal, which can be a significantly lower requirement. The digitized signal is converted to baseband and passed to the digital pulse compressor. The increased dynamic range due to the pulse compression gain is accommodated by increasing the number of bits in the digital computation.

C.8 Echo and Doppler Shift

Echo: Is something you experience all the time. If you shout into a well or a canyon, the echo comes back a moment later. The echo occurs because some of the sound waves in your shout reflect off of a surface (either the water at the bottom of the well or the canyon wall on the far side) and travel back to your ears. The length of time between the moment you shout and the moment that you hear the echo is determined by the distance between you and the surface that creates the echo.

Doppler Shift: Is also common. You probably experience it daily (often without realizing it). Doppler shift occurs when sound is generated by, or reflected off of, a moving object. Doppler shift in the extreme creates sonic booms (see below). Here's how to understand Doppler shift (you may also want to try this experiment in an empty parking lot). Let us say there is a car coming toward you at 60 miles per hour (mph) and its horn is blaring. You will hear the horn playing one "note" as the car approaches, but when the car passes you the sound of the horn will suddenly shift to a lower note. It's the same horn making the same sound the whole time. The change you hear is caused by Doppler shift.

Calculating Depth with Echo: When you shout into a well, the sound of your shout travels down the well and is reflected (echoes) off the surface of the water at the bottom of the well. If you measure the time it takes for the echo to return and if you know the speed of sound, you can calculate the depth of the well fairly accurately. See Fig. C.20.

Here's what happens. The speed of sound through the air in the parking lot is fixed. For simplicity of calculation, let us say its 600 mph (the exact speed is determined by the air's pressure, temperature, and humidity). Imagine that the car is standing still, it is exactly 1 mile away from you, and it toots its horn for exactly 1 min. The sound waves from the horn will propagate from the car toward you at a rate of 600 mph. What you will hear is a 6-s delay (while the sound travels 1 mile at 600 mph) followed by exactly 1 min worth of sound.

Doppler detection and intuitive approach is illustrated in Fig. C.21 as well.

Sonic Boom: While we are here on the topic of sound and motion, we can also understand sonic booms. Say the car was moving toward you at exactly the speed of sound—700 mph or so. The car is blowing its horn. The sound waves generated by the horn cannot go any faster than the speed of sound, so both the car and the horn are coming at you at 700 mph, so all of the sound coming from the car "stacks up." You hear nothing, but you can see the car approaching. At exactly the same moment the car arrives, so does all of its sound and it is *loud*! That is a sonic boom. See illustration in Fig. C.22.

Fig. C.20 Well and shouting image

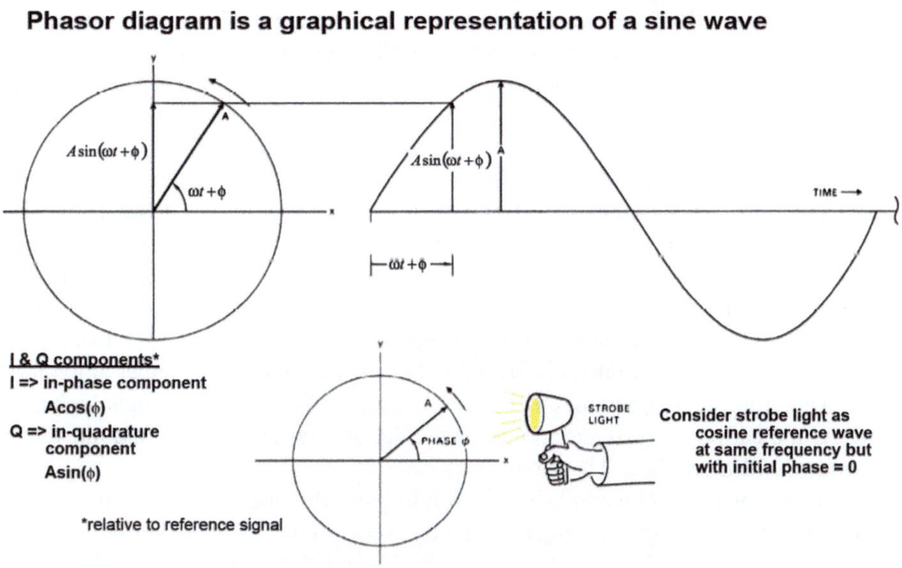

Fig. C.21 Doppler detection principle

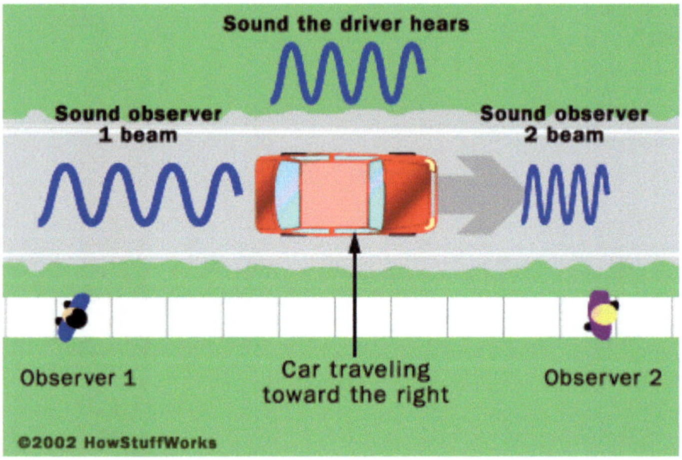

Fig. C.22 Sound of car traveling

The same phenomenon happens when a boat travels through water faster than waves travel through the water (waves in a lake move at a speed of perhaps 5 mph—all waves travel through their medium at a fixed speed). The waves that the boat generates "stack up" and form the V-shaped bow wave (wake) that you see behind the boat. The bow wave is really a sonic boom of sorts. It is the stacked-up combination of all of the waves the boat has generated. The wake forms a V shape, and the angle of the V is controlled by the speed of the boat.

Considering Fig. C.22 and Doppler shift, the person behind the car hears a lower tone than the driver because the car is moving away. The person in front of the car hears a higher tone than the driver because the car is approaching.

Now let us say that the car is moving toward you at 60 mph. It starts from a mile away and toots its horn for exactly 1 min. You will still hear the 6-s delay. However, the sound will only play for 54 s. That's because the car will be right next to you after 1 min, and the sound at the end of the minute gets to you instantaneously. The car (from the driver's perspective) is still blaring its horn for 1 min. Because the car is moving, however, the minute's worth of sound gets packed into 54 s from your perspective. The same numbers of sound waves are packed into a smaller amount of time. Therefore, their frequency is increased, and the horn's tone sounds higher to you. As the car passes you and moves away, the process is reversed, and the sound expands to fill more time. Therefore, the tone is lower.

You can combine echo and Doppler shift in the following way. Say you send out a loud sound toward a car moving toward you. Some of the sound waves will bounce off the car (an echo). Because the car is moving toward you, however, the sound waves will be compressed. Therefore, the sound of the echo will have a higher pitch than the original sound you sent. If you measure the pitch of the echo, you can determine how fast the car is going.

Appendix D: Monostatic, Bistatic, and Multistatic Radars

As we described in Chap. 1 of this book, radar technology is used heavily in military applications. All types of radar were discussed as well including monostatic, bistatic, and multistatic briefly. In this Appendix we expand upon these types of radar with more details.

D.1 Introduction

Most radar has its transmitting and receiving antennas in essentially the same location, as illustrated in Fig. D.1a. This is referred to as monostatic radar. Radars often use the same antennas for transmitting and receiving, and so is monostatic by definition. Other radars have their transmitting and receiving antennas close together, compared with the target range. These generally have the same characteristics as monostatic radar and are included in that class. Advantages of monostatic radar are the common use of radar hardware at a single site, illumination of the same region of space by the transmitting and receiving antennas, and simplified radar coordination.

With bistatic radar, the transmitting and receiving antennas are separated, as illustrated and shown in Fig. D.1b. This may be done to avoid interference between the transmitted and received signals; to allow multiple receivers to operate with a single transmitter; to permit light, non-radiating receivers to operate with the heavy transmitters located elsewhere; to take advantage of the large bistatic RCS characteristics of targets (see Sect. 3.4); or to exploit bistatic geometry. An example of the latter is in intrusion-detection radar, where the region protected lies between the transmitting and receiving antennas. With bistatic radar, it is usually necessary to coordinate operation of the transmitting and receiving sites, to provide multiple receive beams to cover the transmitted beam region, and to take the bistatic geometry into account in the signal processing. See Appendix C for more details on the subject of signal processing.

B. Zohuri, *Radar Energy Warfare and the Challenges of Stealth Technology*,
https://doi.org/10.1007/978-3-030-40619-6

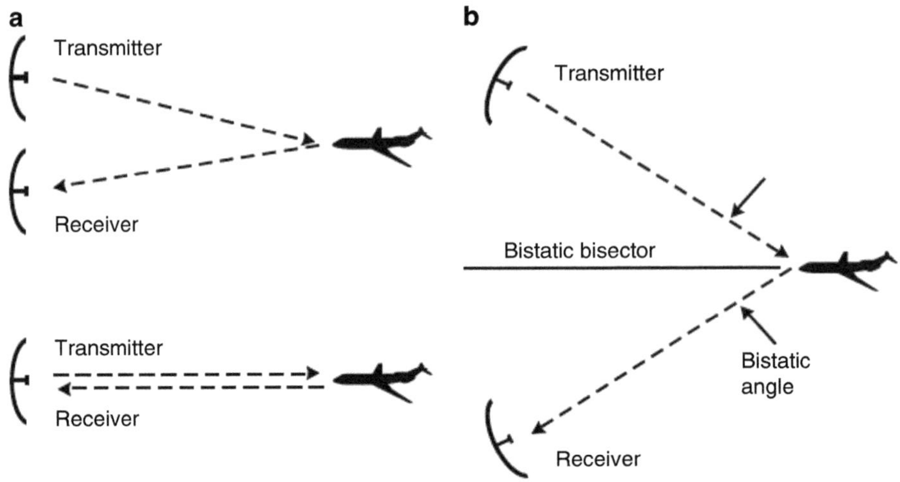

Fig. D.1 Illustrations of (**a**) monostatic and (**b**) bistatic radar geometry. (Source: www.wikipedia.com)

Fig. D.2 Operating principle of monostatic and multistatic. (Source: www.wikipedia.com)

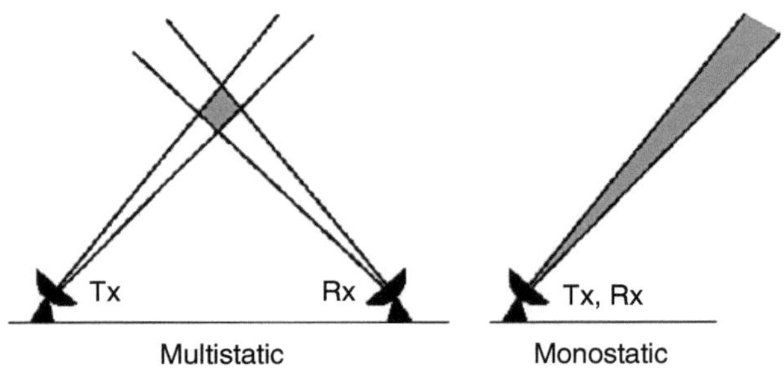

It has already been shown how faceting the dispersion of radar reflections away from the conventional transmitter/receiver unit can help drastically reduce an air asset's RCS. What needs to be emphasized is that the electromagnetic energy is still there; it is simply redirected to directions other than the radar unit and thus considered useless for conventional systems [31].

Now, what if that scattered energy is picked up by receiver units in various other directions? Provided that the received signal is accurately correlated with the original emission from the radar transmitter, the successive bearings from which it is received can be compared, and a pretty accurate estimate of the reflection point can be deduced. Aircraft that make heavy use of faceting, such as the SR-71 or the F-117, can thus be detected with a fair probability of success. Such radar systems are called bistatic (in the case of a single transmitter and a single receiver) or multistatic if the number of Tx or Rx units is greater—typically one transmitter coupled to multiple receiver sets as illustrated in terms of monostatic radar in Fig. D.2.

Putting this theory into practice requires several steps. To begin with, each successive radar pulse must be uniquely identifiable in order to correctly perform the spatial correlation between the outbound signal and the inbound returns.

This is an already-existing practice in modern pulse Doppler radar systems and is thus a modest technical challenge. More difficult is the combination of all the signal correlations into a meaningful positional estimate.

A radar return may arrive at the transmitter from a variety of directions other than the "true" reflection, as a result of both multipath or mirror effects and other factors such as anomalous atmospheric propagation, signal distortion due to interference, etc.

Sorting out the true-bearing returns from the fakes is a difficult task even for straightforward conventional radar sets, and it becomes even more complex in the case of multistatic receivers. A simple tracking algorithm may try to follow the consistent returns and wash-out spikes that seem inconsistent with the target's expected motion; this is the simplest of examples, and the software associated with such functions can be expected to be mind-numbingly complex.

As conventional monostatic radars divide their area of air surveillance into segments successively scanned by their main beam (lobe), so do multistatic systems as demonstrated in Fig. D.3.

The difference here is that the intersections of the segments can intersect between nodes of the system, thus forming surveillance "cells." These cells are then monitored in rapid succession for any reflection of a signal consistent with the one originally emitted by the Tx element of the system.

An early example of this type is the French RIAS experimental radar, set up since the mid-1970s in the French island of Levant, to explore to potential benefits of the principle. This uses a single transmitter element in the metric band with a number of

Fig. D.3 Operating principle of multistatic radar systems. (Source. www.wikipedia.com)

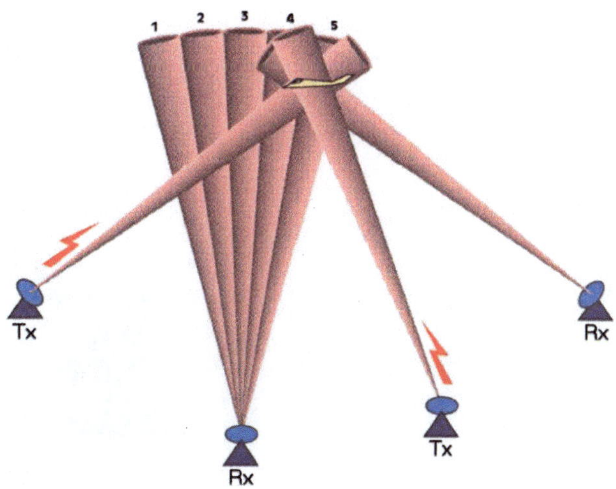

receivers (a series of dipoles spaced 15 m apart and forming two co-axial rings, the outside ring being 400 m in diameter) to provide three-dimensional target data. At least three receivers are needed to provide a 2D fix, and another one is needed for a height estimate. The main problem at the time of the system's inception was the limited computing power then available. In the early tests, with an IBM Cyber-360 mainframe handling the monitoring of the multiple cells, it took nearly a week to fully process the input of just 2 min of surveillance. From the mid-1980s, however, the replacement of this system with a Cray-II supercomputer enabled the signal processing to be performed in near real time.

With computing power being so abundant and cheap nowadays, this restriction can be assumed not to present an issue anymore.

D.2 Monostatic Radar

When both transmitting and receiving antennas are placed close to each other in one radar station at the same location for the detection of an object, it is known as monostatic radar. There is only one radar, containing the same Tx and Rx antenna located on the ground for the detection of an aircraft. The typical geometry of monostatic radar is shown in Fig. D.4. Transmitting and receiving gain in the case of monostatic radar would be the same as G_t and G_r; thus, they must be equal to each other as they are shown as Equation D.1:

$$G = G_r = G_t \tag{D.1}$$

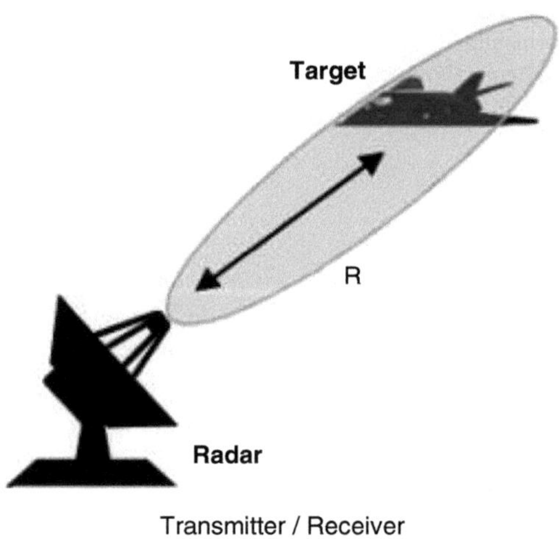

Fig. D.4 A typical monostatic radar principle. (Source: www.wikipedia.com)

Then, Equation D.1 can be written as Equation D.2:

$$P_r = \frac{P_t G^2 \sigma \lambda^2}{(4\pi)^3 R^4} \tag{D.2}$$

The maximum range of a radar, R_{max}, is the distance beyond which the target cannot be detected. It occurs when the received signal power P_r just equals the minimum detectable signal S_{min}. Rearranging terms and substituting $S_{min} = P_r$ in Equation D.2 give:

$$R_{max} = \left[\frac{P_t G^2 \sigma \lambda^2}{(4\pi)^3 S_{min}}\right]^{1/4} \tag{D.3}$$

In the above equation, the following parameter definitions do apply:

P_t = transmitter power
λ = transmitted wavelength
σ = radar cross section, or scattering coefficient, of the target
G = gain of the antenna

Although this form of the radar equation excludes many important factors and usually predicts high values for maximum range, it depicts the relationship between the maximum radar range and the target's radar cross section (RCS).

Monostatic radar is a radar in which the transmitter and receiver are collocated. This is the conventional configuration for a radar, but the term is used to distinguish it from a bistatic radar or multistatic radar.

D.3 Bistatic Radar

When both transmitting and receiving antennas are placed apart from each other at some considerable distance, it is called bistatic radar. A system in which there are one transmitter and multiple separated receivers is known as multistatic radar. The geometry of bistatic radar is shown in Fig. D.5. If α the bistatic angle is small, then bistatic radar cross section (RCS) is similar to that of monostatic RCS.

The following equation is the right equation in the case of bistatic radar as:

$$P_r = \frac{P_t G_t G_r \lambda^2 \sigma_b}{(4\pi)^3 D_t^2 D_r^2} \tag{D.4}$$

where σ_b is the bistatic radar cross section (in m^2), D_t the distance between the target and transmitter, and D_r the distance between the target and receiver.

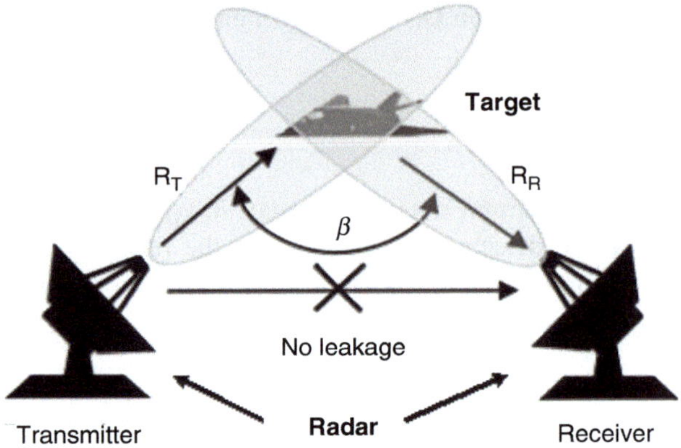

Fig. D.5 A typical bistatic radar principle. (Source: www.wikipedia.com)

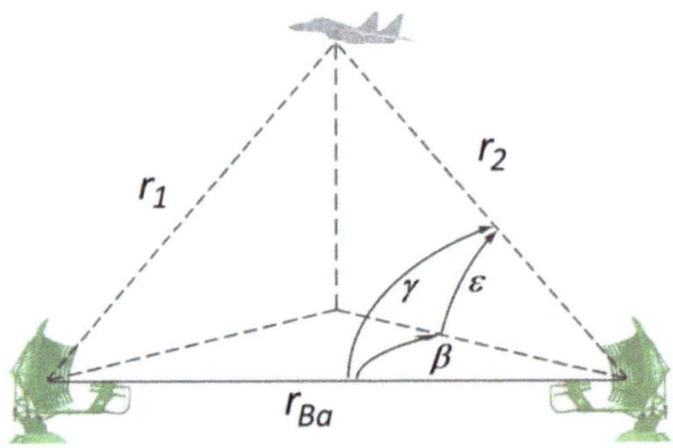

Fig. D.6 A bistatic radar block diagram. (Source: www.wikipedia.com)

Bistatic radar is a radar system comprising a transmitter and receiver that are separated by a distance comparable to the expected target distance. Conversely, a radar in which the transmitter and receiver are collocated is called a monostatic radar. A system containing multiple spatially diverse monostatic radar or bistatic radar components with a shared area of coverage is called multistatic radar. Many long-range air-to-air and surface-to-air missile systems use semi-active radar homing, which is a form of bistatic radar. See Fig. D.6.

Furthermore, as part of specific classes of bistatic radar, some radar systems may have separate transmitting and receiving antennas, but if the angle subtended between transmitter, target, and receiver (the bistatic angle) is close to zero, then they would still be regarded as monostatic or *pseudo-monostatic*. For example, some very-long-range HF radar systems may have a transmitter and receiver which are separated by a few tens of kilometers for electrical isolation, but as the expected target range is of the order 1000–3500 km, they are not considered to be truly bistatic and are referred to as pseudo-monostatic.

The principal advantages of bistatic and multistatic radar include:

- Lower procurement and maintenance costs (if using a third-party transmitter).
- Operation without a frequency clearance (if using a third-party transmitter).

- Covert operation of the receiver.
- Increased resilience to electronic countermeasures, as waveform being used and receiver location are potentially unknown.
- Possible enhanced radar cross section of the target due to geometrical effects.
- Separate receiver is very light and mobile, while transmitter can be very heavy and powerful (surface-to-air missile).

The principal disadvantages of bistatic and multistatic radar include:

- System complexity
- Costs of providing communication between sites
- Lack of control over transmitter (if exploiting a third-party transmitter)
- Harder to deploy
- Reduced low-level coverage due to the need for line of sight from several locations

There are more other types of specific classes of bistatic radar that were described either in Chap. 1 of this book or here as well. We did talk about pseudo-monostatic radars already above, and now we describe their classification here, and they are:

1. Forward scatter radars
   In some configurations, bistatic radars may be designed to operate in a fence-like configuration, detecting targets which pass between the transmitter and receiver, with the bistatic angle near 180°. This is a special case of bistatic radar, known as a forward scatter radar, after the mechanism by which the transmitted energy is scattered by the target. See Fig. D.7.
   In forward scatter, the scattering can be modeled using Babinet's principle and is a potential countermeasure to stealth aircraft as the radar cross section (RCS) is determined solely by the silhouette of the aircraft seen by the transmitter and is unaffected by stealth coatings or shaping.
   The RCS in this mode is calculated as $\sigma = 4\pi A^2/\lambda^2$, where A is the silhouette area and λ is the radar wavelength. However, target may vary from place to place location, and tracking is very challenging in forward scatter radars, as the information content in measurements of range, bearing, and Doppler becomes very low (all these parameters tend to zero, regardless of the location of the target in the fence).
2. Multistatic radar
   A multistatic radar system is one in which there are at least three components—for example, one receiver and two transmitters, or two receivers and one transmitter, or multiple receivers and multiple transmitters. It is a generalization of the bistatic radar system, with one or more receivers processing returns from one or more geographically separated transmitter. See Sect. D.4.
3. Passive radar
   A bistatic or multistatic radar that exploits non-radar transmitters of opportunity is termed a passive coherent location system or passive covert radar. See Fig. D.8.

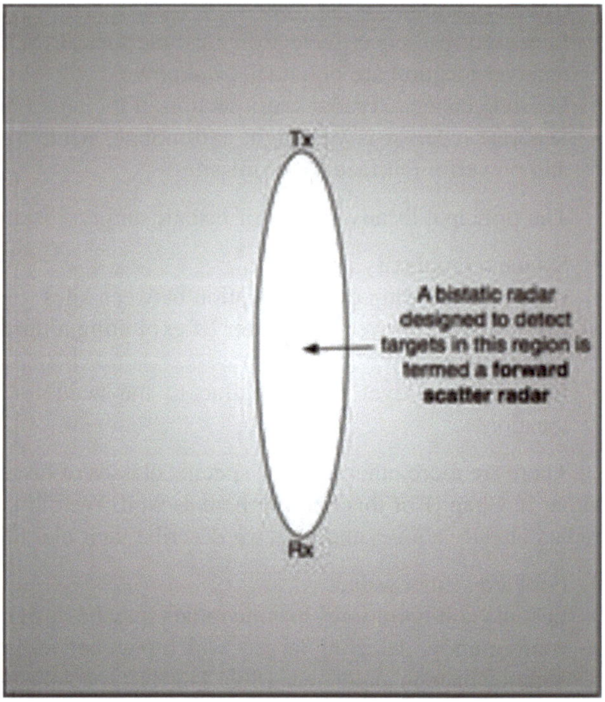

Fig. D.7 Illustration of forward scatter geometry. (Source: www.wikipedia.con)

Fig. D.8 Bistatic radar passive receiver system from NCSIST of Taiwan. (Source: www.wikipedia.con)

Any radar which does not send active electromagnetic pulse is known as passive radar. Passive coherent location also known as PCL is a special type of passive radar, which exploits the transmitters of opportunity especially the commercial signals in the environment.

D.4 Multistatic Radar

A multistatic radar system is one in which there are at least three components—for example, one receiver and two transmitters, or two receivers and one transmitter, or multiple receivers and multiple transmitters. It is a generalization of the bistatic radar system, with one or more receivers processing returns from one or more geographically separated transmitter. See Fig. D.9.

A multistatic radar system contains multiple spatially diverse monostatic radar or bistatic radar components with a shared area of coverage. An important distinction of systems based on these individual radar geometries is the added requirement for some level of data fusion to take place between component parts. The spatial diversity afforded by multistatic systems allows different aspects of a target to be viewed simultaneously. The potential for information gain can give rise to a number of advantages over conventional systems.

Multistatic radar is often referred to as "multisite" or "netted" radar and is comparable to the idea of macro-diversity in communications. A further subset of multistatic radar with roots in communications is that of multiple-input multiple-output (MIMO) radar. See Fig. D.10.

Note that multiple-input multiple-output (MIMO) radar is an advanced type of phased array radar employing digital receivers and waveform generators distributed across the aperture. MIMO radar signals propagate in a fashion similar to multistatic radar.

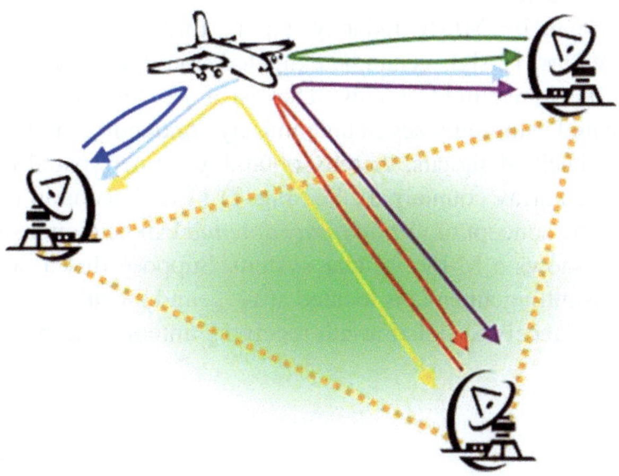

Fig. D.9 A multistatic radar block diagram. (Source: www.wikipedia.com)

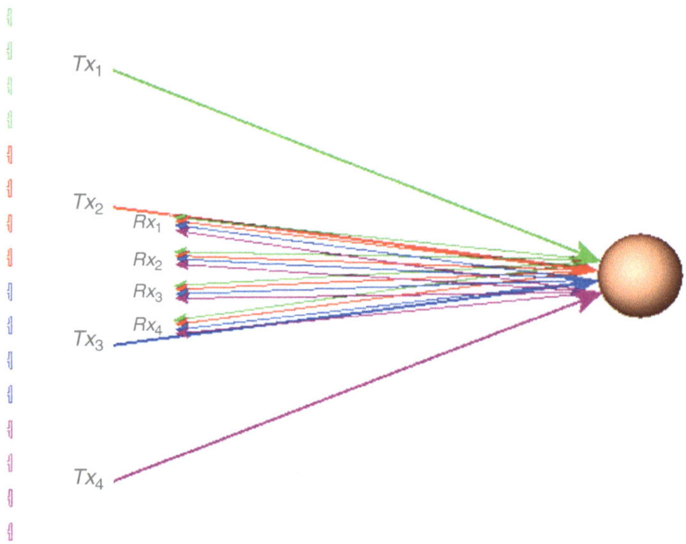

Fig. D.10 A MIMO system. (Source: www.wikipedia.com)

As Fig. D.10 illustrates, in a MIMO system, the transmitting signals from the single transmitters are different. As a result, the echo signals can be re-assigned to the source. This gives an enlarged virtual receive aperture.

However, instead of distributing the radar elements throughout the surveillance area, antennas are closely located to obtain better spatial resolution, Doppler resolution, and dynamic range [32]. MIMO radar may also be used to obtain low probability of intercept (LPI) radar properties [33].

In a traditional phased array system, additional antennas and related hardware are needed to improve spatial resolution. MIMO radar systems transmit mutually orthogonal signals from multiple transmitting antennas, and these waveforms can be extracted from each of the receiving antennas by a set of matched filters. For example, if a MIMO radar system has three transmitting antennas and four receiving antennas, 12 signals can be extracted from the receiver because of the orthogonality of the transmitted signals. That is, a 12-element virtual antenna array (VAA) is created using only seven antennas by conducting digital signal processing (DSP) on the received signals, thereby obtaining a finer spatial resolution compared with its phased array counterpart. See Fig. D.11 as a scenario of virtual array analysis.

The concept of *virtual array antenna* can be demonstrated as pictured in Fig. D.9 that shows a M-by-N radar system. Suppose that a target is located at u, the mth transmitting antenna is located at $x_{T,\ m}$, and the nth receiving antenna is located at $x_{R,\ n}$. The received signal at nth receiving antenna can be expressed as:

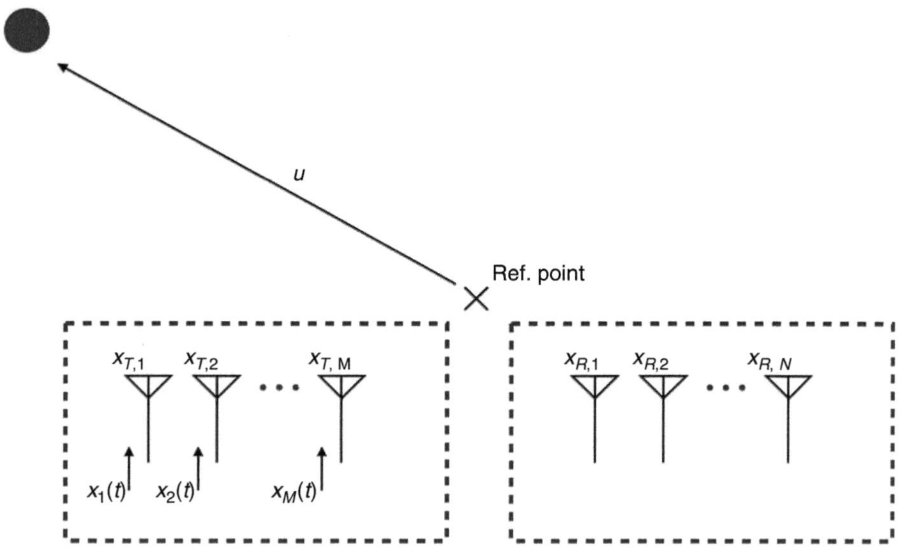

Fig. D.11 Scenario of virtual array analysis. (Source: www.wikipedia.com)

$$y_n(t) = \sum_{m=1}^{M} x_m(t) e^{j\frac{2\pi}{\lambda} u^{\mathrm{T}} (x_{\mathrm{T},m} + x_{\mathrm{R},n})} \tag{D.5}$$

As we can see, if $\{x_m(t), m = 1\sim M\}$ is an orthogonal set-type matrix, we can extract M signal from nth receiving antenna, each of which contains the information of an individual transmitting path ($u^{\mathrm{T}}(x_{\mathrm{T},\ m} + x_{\mathrm{R},\ m})$).

In order to make a comparison between phased array radars and MIMO radars, the relationship between transmitting/receiving antenna arrays and virtual arrays is discussed in [33, 34]. If the placements of the transmitting and receiving antenna array are expressed as two vectors h_{T} and h_{R}, respectively, the placement vector of the virtual array is equal to the convolution of h_{T} and h_{R} as Equation D.6 and presentation in Fig. D.12:

$$h_{\mathrm{V}} = h_{\mathrm{T}} * h_{\mathrm{R}} \tag{D.6}$$

Picture in Fig. D.12 shows the examples of antenna geometry to form a virtual array. In the first example, two uniformly distributed antenna arrays form a five-element virtual array despite having six antennas in total. In the second example, a nine-element virtual array is obtained by increasing the distance between the transmitting antennas, implying that a better spatial resolution can be achieved.

To estimate the direction of arrival of the targets according to the N∗M signals, methods like MUSIC (algorithm) and maximum likelihood estimation are commonly used with good results [35, 36].

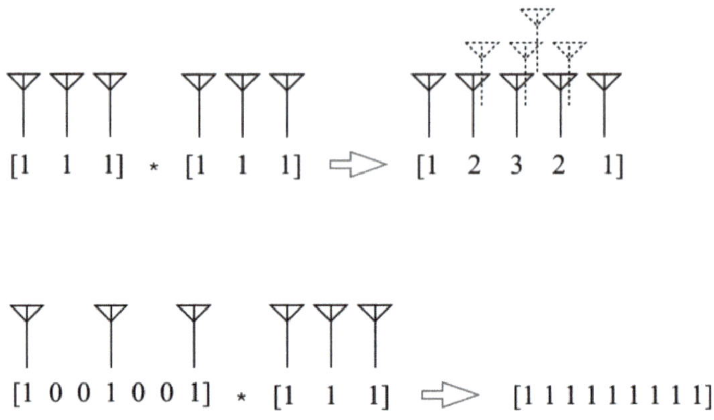

Fig. D.12 Scenario of antenna geometry to and from virtual array. (Source: www.wikipedia.com)

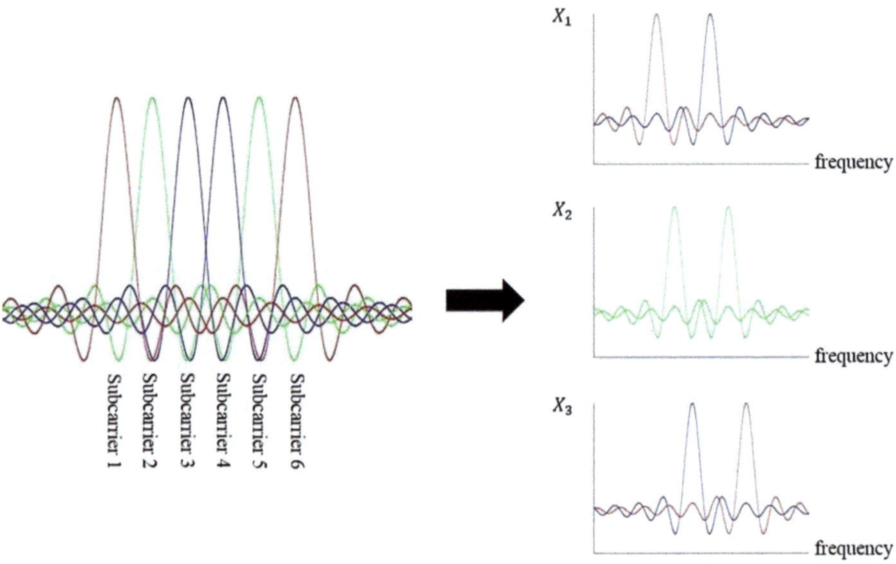

Fig. D.13 Regular subcarrier assignment to generate orthogonal signals. (Source: www.wikipedia.com)

For the aspect of *orthogonal signal*, we can state that there are a variety of orthogonal signal sets used in the field of MIMO radar. One of the proposed signal sets is the *spectrally interleaved multi-carrier signal*, which is a modified version of orthogonal frequency-division multiplexing signal [37]. In this approach, the total amount of available subcarriers is distributed among different transmitting antennas in an interleaved way. See Fig. D.13.

Another proposed signal set is orthogonal chirp signal, which can be expressed as:

$$x_m(t) = \exp\left[2\pi\left(f_{m,0}t + \frac{1}{2}kt^2\right)\right] \tag{D.7}$$

By choosing different initial frequencies $f_{m,\,0}$, these chirp waveforms can be made orthogonal [38].

Recent advances in digital signal processing (DSP) and the constant development of computational capabilities suggest that it may be feasible for next-generation radar systems to incorporate multiple-input multiple-output (MIMO) technology. With new approach to artificial intelligence (AI) technology and its augmentation in support of the computation for digital signal processing, the MIMO are becoming more and more attractive in radars in particular the multistatic types and their place in electronic warfare (EW) [39].

The superiority of a MIMO radar against other radar schemes lies in its waveform diversity, which in essence means that a MIMO radar can simultaneously emit several diverse, possibly linearly independent waveforms via multiple antennas, in contrast to existing radar systems that transmit scaled versions of the same, predefined waveform [40]. In particular, there are two principal types of MIMO radar, those that incorporate colocated antennas [41] and systems equipped with widely separated antennas (bistatic, multistatic) [42]. MIMO radar technology provides direct applicability of adaptive beamforming [43], waveform design and power allocation, higher angular resolution, ability to acquire the target's geometrical characteristics through the radar cross section (RCS), and multiple target detection [40].

However, in order to combat multiple source interference in a radar field, while achieving high detection performance using minimum power consumption, the system should adopt an optimal resource allocation strategy. A centralized approach to resource allocation is possible using convex optimization techniques, for example. Nevertheless, centralized control may not be desirable or will have implementation difficulties in a multistatic radar network, and thus, it is preferred to consider autonomous decentralized resource allocation schemes. A natural and efficient tool to achieve this is game theory, which provides a framework for analyzing coordination and conflict between rational but selfish players [39].

Using Fig. D.14, Anastasios Deligiannis et al. [39] are showing how this game gets played through the game-theoretic analysis; we have characterized the behavior of the radars in a cluster.

Specifically, the theoretical results based on Karush-Kuhn-Tucker conditions showed that in a cluster, the number of radars that exactly achieve the desired SINR is equal to the number of radars that are actively transmitting.

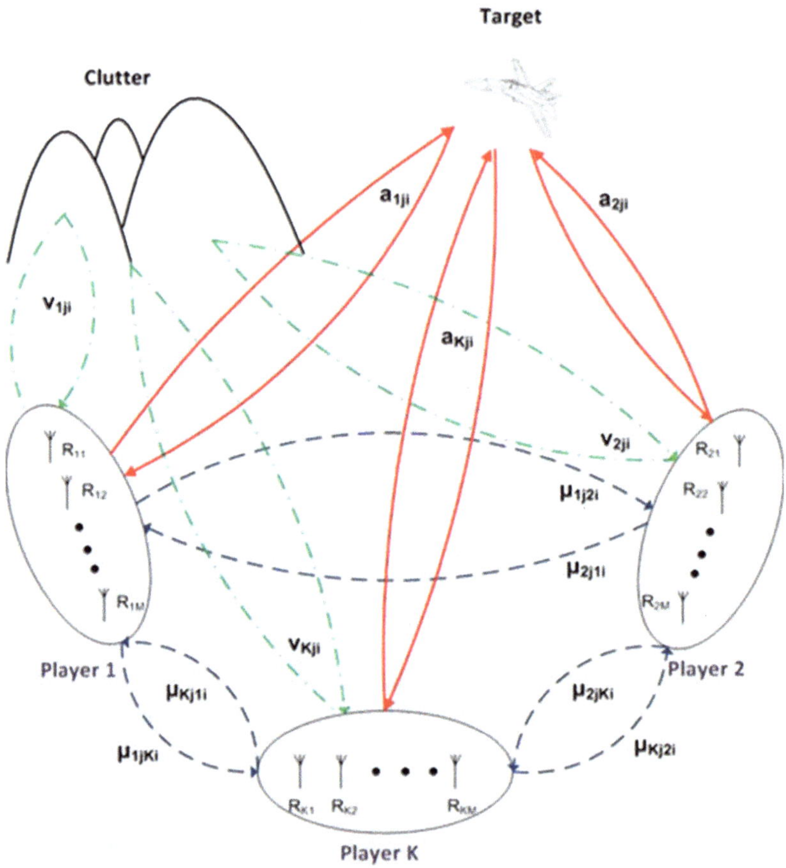

Fig. D.14 A distributed MIMO radar network with K clusters and their corresponding channel gains [39]

D.4.1 Multistatic Radar Characteristics

Since multistatic radar may contain both monostatic and bistatic components, the advantages and disadvantages of each radar arrangement will also apply to multistatic systems. A system with *N* transmitters and *M* receivers will contain *NM* of these component pairs, each of which may involve a differing bistatic angle and target radar cross section. The following characteristics are unique to the multistatic arrangement, where multiple transmitter-receiver pairs are present:

1. Detection
   Increased coverage in multistatic radar may be obtained via the spreading of the radar geometry throughout the surveillance area—such that targets might be more likely to be physically closer to transmitter-receiver pairs and thus attain a higher signal-to-noise ratio (SNR).

Spatial diversity may also be beneficial when combining information from multiple transmitter-receiver pairs which have a shared coverage. By weighting and integrating individual returns (such as through likelihood ratio-based detectors), detection can be optimized to place more emphasis on stronger returns obtained from certain monostatic or bistatic radar cross section values or from favorable propagation paths, when making a decision as to whether a target is present. This is analogous to the use of antenna diversity in an attempt to improve links in wireless communications.

This is useful where multipath or shadowing effects might otherwise lead to the potential for poor detection performance if only a single radar is used. One notable area of interest is in sea clutter and how diversity in reflectivity and Doppler shift might prove beneficial for detection in a maritime environment.

Many stealth vehicles are designed to reflect radar energy away from expected radar sources in order to present as small a return to a monostatic system as possible. This leads to more energy being radiated in directions that are only available to multistatic receivers.

2. Resolution

   Resolution may benefit from spatial diversity, due to the availability of multiple spatially diverse down-range profiles. Conventional radar typically has a much poorer cross-range resolution compared to down-range resolution; thus, there is potential for gains through the intersection of constant bistatic range ellipses. See Fig. D.15.

   This involves a process of associating individual target detections to form a joint detection. Due to the un-cooperative nature of the targets, there is potential, if several targets are present, for ambiguities or "ghost targets" to be formed. These can be reduced through an increase in information (e.g., use of Doppler information, increase in down-range resolution, or addition of further spatially diverse radars to the multistatic system).

3. Classification

   Target features such as variation in the radar cross section or jet engine modulation may be observed by transmitter-receiver pairs within a multistatic system. The gain in information through observation of different aspects of a target may improve classification of the target. Most existing air defense systems utilize a series of networked monostatic radars, without making use of bistatic pairs within the system.

4. Robustness

   Increased survivability and "graceful degradation" may result from the spatially distributed nature of multistatic radar. A fault in either transmitter or receiver for a monostatic or bistatic system will lead to a complete loss of radar functionality. From a tactical point of view, a single large transmitter will be easier to locate and destroy compared to several distributed transmitters. Likewise, it may be increasingly difficult to successfully focus jamming on multiple receivers compared to a single site.

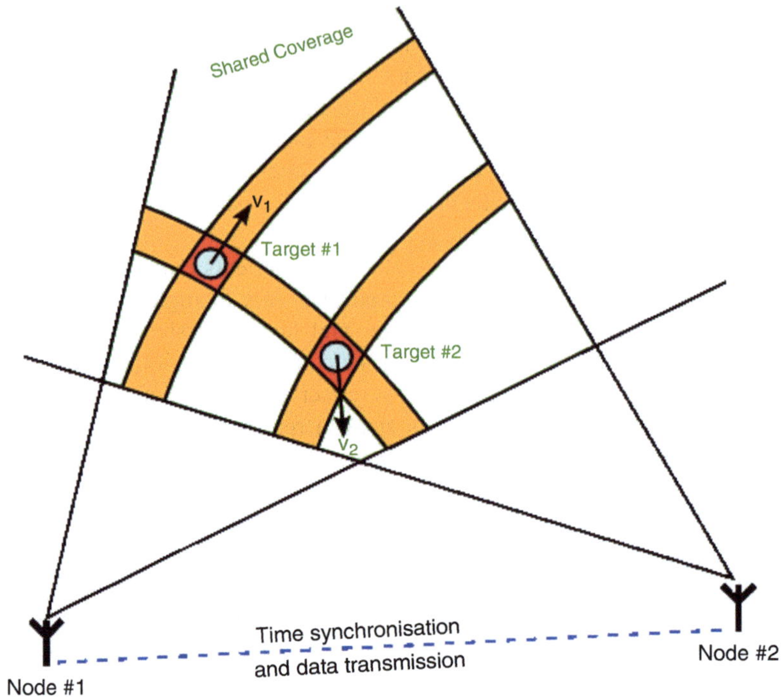

Fig. D.15 Resolving multiple targets using multistatic radar. (Source: www.wikipedia.com)

5. Spatiotemporal synchronization
   To deduce the range or velocity of a target relative to a multistatic system, knowledge of the spatial location of transmitters and receivers is required. A shared time and frequency standard also must be maintained if the receiver has no direct line of sight of the transmitter. As in bistatic radar, without this knowledge, there would be inaccuracy in the information reported by the radar. For systems exploiting data fusion before detection, there is a need for accurate time and/or phase synchronization of the different receivers. For plot-level fusion, time tagging using a standard GPS clock (or similar) is more than sufficient.
6. Communications bandwidth
   The increase in information from the multiple monostatic or bistatic pairs in the multistatic system must be combined for benefits to be realized. This fusion process may range from the simple case of selecting plots from the receiver closest to a target (ignoring others), increasing in complexity to effectively beamforming through radio signal fusion. Dependent on this, a wide communications bandwidth may be required to pass the relevant data to a point where it can be fused.

7. Processing requirements
   Data fusion will always mean an increase in processing compared to a single radar. However, it may be particularly computationally expensive if significant processing is involved in data fusion, such as attempts to increase resolution.

D.5 Low Probability of Intercept (LPI) Radar

In the modern battlefield, radars face increasingly serious threats from electronic attack and ARMs. An important feature of modern radar systems is the ability "to see and not to be seen." Low probability of intercept radar has a powerful detection capability while simultaneously itself being not easily detected by electronic reconnaissance equipment [48].

LPI radar is a class of radar systems that possess certain performance characteristics that make them nearly undetectable by today's intercept receivers. A low probability of intercept features prevents the radar from tripping off alarm systems or passive radar detection equipment in a target. These features include:

- Using an antenna with a narrow beam and low side-lobes that are hard to spot from off its boresight
- Only transmitting radar pulses when necessary
- Reducing the transmitted pulse power
- Spreading the radar pulses over a wide band so there will only be a very small signal on any one band
- Varying transmission parameters such as:
  - Pulse form
  - Frequency
  - Pulse repetition frequency (PRF)
- Jumping around in an unpredictable fashion, not staying in one place long enough to register
- Using an intrapulse modulation with an inconspicuously waveform (e.g., a pseudo-random bit pattern)

The function of an LPI radar is to prevent its interception by an electronic support receiver. This objective is generally achieved through the use of a radar waveform that is mismatched to those waveforms for which an electronic support receiver is tuned. Consequently, a conventional electronic support receiver can only detect an LPI radar at very short ranges.

The LPI radar transmits a low-power intrapulse-modulated waveform so that the range of the detected target can be determined with a good range resolution. This modulation may be phase- or frequency-modulated or pseudo-random and noise-like modulation. A typical LPI radar has a switchable pulse-power output of up to 1 W. A conventional pulse radar needs at least 10 kW for a similar detection range of targets.

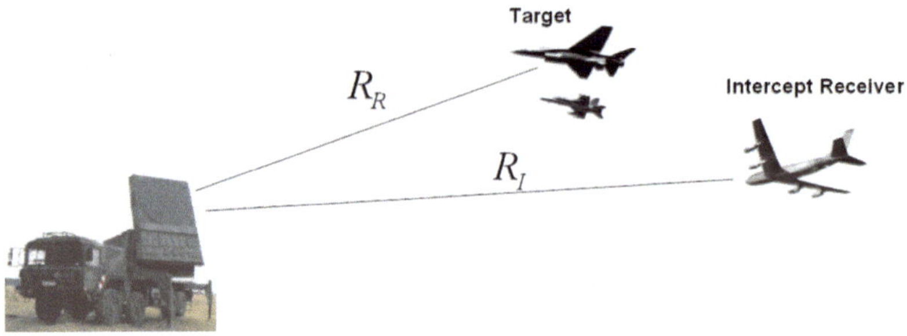

Fig. D.16 The geometry of radar, target, and intercept receiver [32]

This allows the LPI radar to achieve a processing gain, with respect to an electronic support receiver, that is equal to the time-bandwidth product of the radar waveform. This processing gain allows the LPI radar as a primary radar and fourth-root dependence of the two-way travel of electromagnetic waves to overcome the range-squared advantage of the electronic support receiver in conventional situations.

However, LPI radars are limited to short-range applications. A relatively long transmitted pulse width still applies to the transmission, which requires the duplexer to remain aligned to the transmitter throughout the pulse, and the receiver is switched off during this time. Therefore, many LPI radars have separate transmitting and receiving antennas that are co-mounted. Some recent LPI radars can use instead of the pulse radar technology a frequency-modulated continuous-wave (FM-CW) radar mode to reduce the transmitted power substantially [44].

Whether or not a radar is LPI depends on the purpose or mission of the radar, the kind of receiver that is trying to detect it, and the applicable engagement geometry (Adamy 20010) [45]. These types of radars are also described as "quiet" radars (Fig. D.16).

In order to hide itself from the interception of electronic surveillance (ES) systems and RWRs, the detection range of radar R_R should be longer than that of intercept receiver R_I. From Fig. D.15, a range factor α can be defined as. If $\alpha > 1$, the radar will be detected by the intercept receiver. On the contrary, if $\alpha \leq 1$ the radar can detect the platform, while the intercept receiver platform cannot detect the radar. In fact, so-called LPI performance is a probability event [46].

As part of characteristics of low probability of intercept (LPI) radar, we should state that LPI is that property of an emitter that, because of its low power, wide bandwidth, frequency variability, or other design attributes, makes it difficult to be detected or identified by means of passive intercept devices such as radar warning, electronic support, and electronic intelligence receivers. In this paper, we present the properties and generation of some the important LPI waveforms based on phase and frequency modulation. The properties such as phase/frequency variation, autocorrelation function, and ambiguity function are shown for each waveform.

These signals are then used to test with the help of signal processing technique that detects the waveform parameters of the intercepted signals. The technique is

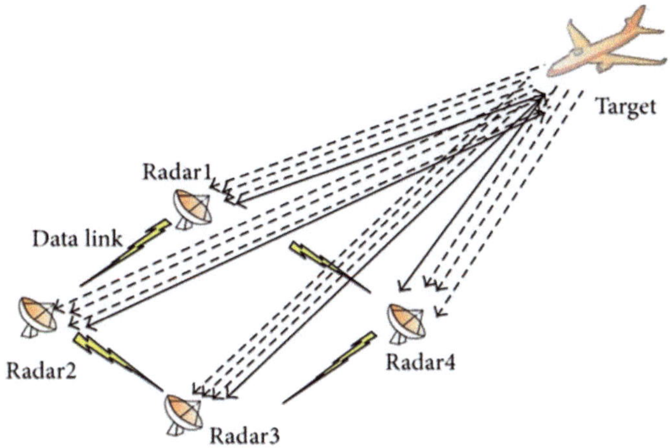

Fig. D.17 Example of an LPI radar network. (Source: www.wikipedia.com)

based on the use of parallel filter (sub-band) arrays and higher-order statistics followed by image processing methods such as image threshold binarization and mathematical morphology to recognize and extract the shape of the radar signal even when the signal-to-noise ratio (SNR) is −3 dB. Figure D.17 is a scientific diagram of example of an LPI radar network.

We can summarize the characteristics of LPI radar as follows:

- Low side-lobe antennas
- Irregular antenna scan patterns
- High-duty-cycle/wide band transmission
- Accurate power management
- Carrier frequency
- Very high sensitivity
- High processing gain
- Coherent detection
- Monostatic/bistatic configurations

Definitions of each of the above characteristics are well defined by Aytug Denk and his thesis work at the Naval Postgraduate School [47].

Appendix E: China's Stealth Fighters and Bombers Technology

Chinese industry still is struggling to manufacture arguably the most important subsystems for these new planes: their engines.

E.1 Introduction

According to a report published in *The National Interest* by David Axe, he claims that "China's Stealth Fighters and Stealth Bombers Have a Big Problem" [48].

The Chinese military is building up a meaningful force of J-20 stealth fighters, Y-20 strategic airlifters, and other high-tech military aircraft while also developing a new stealth fighter, fighter-bomber, and heavy bomber (Fig. E.1).

But for all of these advancements, Chinese industry still is struggling to manufacture arguably the most important subsystems for these new planes: their engines.

Aviation website Alert 5 spotted [49] a stock exchange filing by the Hebei subsidiary of China's Central Iron & Steel Research Institute. The filing includes production projections for military engines for the next decade and reveals some startling shortfalls. The Alert 6 claims that China still struggling to develop new military turbofan engines according to a regulator stock exchange filing by the Hebei subsidiary of China's Central Iron & Steel Research (CISRI) has disclosed the production numbers of military engines for the next decade. See Table E.1.

Production and development gaps could result in the latest Chinese warplanes flying with older engine models, including imported Russian motors that might be underpowered and unreliable. The mismatch between airframes and engines could be a drag on the overall performance of Chinese military aircraft.

Data provided by Hebei Cisri Dekai Technology Co. Ltd. shows a maximum of only five WS-15 and WS-19 engines each year from 2020 till 2026 (see Table E.2). The WS-15 will power the J-20 stealth fighter, while the WS-19 is being developed for the FC-31 fighter. See Fig. E.2.

B. Zohuri, *Radar Energy Warfare and the Challenges of Stealth Technology*,
https://doi.org/10.1007/978-3-030-40619-6

Fig. E.1 Chinese stealth fighter of fifth generation

Table E.1 Chinese Hebei Cisri Dekai Technology data

公司公告

24/234

国家**-15 型号发动机是重点中的重点，未来预计每年需要 5 台份发动机，同一类别的**-19 型号也是如此，15 型号的钛合金铸件单台份 190 万元，19 型号 120 万元，德凯均已经成为合格供应商，预计的产值如下：

时间	2020 年	2021 年	2022 年	2023 年	2024 年	2025 年	2026 年
15 产量正常预估	5	5	5	5	5	5	5
15 产量保守预估	3	3	3	3	3	3	3
19 产量正常预估	5	5	5	5	5	5	5
19 产量保守预估	3	3	3	3	3	3	3
市场份额	20%	30%	30%	50%	50%	50%	50%
德凯产值正常预估	234	351	351	585	585	585	585

**-18 型号，国家有非常迫切的需求，但是由于种种原因，一直进展不畅，目前处于半停滞状态，其铝镁钛轻质合金铸件德凯都在开展研制生产工作，每台份德凯产值 102 万元，预计情况如下：

20

Table E.2 Chinese Hebei Cisri Dekai Technology data for FC-31

**-18型号，国家有非常迫切的需求，但是由于种种原因，一直进展不畅，目前处于半停滞状态，其铝镁钛轻质合金铸件德凯都在开展研制生产工作，每台份德凯产值102万元，预计情况如下：

20

时间	2020年	2021年	2022年	2023年	2024年	2025年	2026年
产量正常预估	5	5	10	10	10	20	20
产量保守预估	3	3	6	6	6	15	15
市场份额	50%	50%	50%	50%	50%	50%	50%
德凯产值正常预估	255	255	510	510	510	1020	1020

Fig. E.2 Chinese FC-31 image

The WS-18 engine is running into trouble with development half-suspended as the company research into new materials. The WS-18 is designed for the H-6K bomber and Y-20 airlifter. See Fig. E.3 and Table E.3 for its data.

Fig. E.3 Chinese Y-20 airlifter image

Table E.3 Chinese Hebei Cisri Dekai Technology data for Y-20 airlifter

**-20 型号反推装置，该装置生产为双流水，两家单位把流水都放到德凯，内含包括上下导轨梁在内的近 50 种铸件产品，目前的价格为研制阶段价格，单台分 230 万元。预计价格会随着稳定批产价格下降 30%，按照正常情况，预计 2020 年生产 8 台份，2021 年 8 台份，2022 年 4 台份，2023 年 20 台份，2024 年 30 台份，2025 年 40 台份。该型号发动机用铝合金和钛合金铸件大约分别是 1.5 万元、75 万元每台份，镁合金铸件均在 120 厂自行生产，没有外协采购。高纳已经在积极开展双流水工作。该型号德凯产值规划如下：

17

时间	2020 年	2021 年	2022 年	2023 年	2024 年	2025 年	2026 年
生产量正常预估	8	8	6	4	20	30	40
生产量保守预估	6	6	6	4	12	18	24
反推份额	100%	80%	75%	75%	75%	70%	70%
发动机铝件份额	30%	40%	50%	60%	60%	60%	60%
发动机钛合金份额	10%	10%	20%	30%	35%	35%	35%
德凯产值预估	1904	1537	1130	784	2786	3954	5272

Perhaps the biggest shortfall is in the production of WS-15s and WS-19s, the custom motors, respectively, for J-20 stealth fighters and FC-31 export stealth fighters. "Data provided by Hebei Cisri Dekai Technology Co. Ltd. shows a maximum of only five WS-15 and WS-19 engines each year from 2020 'til 2026,' Alert 5 reported.

The first few combat-capable J-20s reportedly entered service in 2017. Flight Global's survey of all the world's military aircraft for 2020 listed 15 J-20s in frontline use. J-20s usually appear in public with Russian-made AL-31 motors, which experts consider to be inadequate for the heavy, long-range, supersonic fighter.

Even the uprated 117S version of the AL-31F "would likely not be sufficient to extract the full performance potential of this advanced airframe," wrote Carlo Kopp and Peter Goon, analysts with the Air Power Australia think tank.

A dearth of WS-15s could force J-20 regiments to continue flying with AL-31s. Meanwhile it could be difficult for Chinese industry to find buyers for the FC-31 if the plane lacks a custom engine. Prototype FC-31s fly with what appear to be Russian-made RD-93s.

Chinese industry has been trying to develop the WS-18 engine for heavy subsonic aircraft. The type could power Y-20 airlifters and H-6K bombers. But Hebei "is running into trouble with development," Alert 5 reported. Work on the new turbofan is "half-suspended as the company researches into new materials."

Another alternative engine for the Y-20, the WS-20, will also enter limited production starting from 2024. See Table E.4.

The WS-10, which powers the J-10, J-11, J-15, and J-16 fighters, is having a successful production run. The company sees gradual increase in annual production numbers starting from 320 engines in 2020 till 450 engines by 2026. See Table E.5.

"Another alternative engine for the Y-20, the WS-20, will also enter limited production starting from 2024," Alert 5 explained. But for now, the Y-20 and H-6K fly with Russian D-30 turbofans.

The D-30 however is a low-bypass model that's better suited for supersonic fighters than for an efficient, slow-flying cargo-hauler.

Troubles with the WS-15, WS-19, and WS-18 should worry Chinese military planners. Shortfalls could prevent new aircraft types from performing to their maximum potential. Beijing perhaps should worry the most about a new "sixth-generation" stealth fighter that Chinese officials want to develop in order eventually to replace the J-20.

As demanding as the J-20 is on its engines, a successor fighter likely would be even more demanding.

The PLA is expected to induct a large number of Z-20 utility helicopters into service in the next decade as demands for the WZ-10 turboshaft engine are being increased from 65 for 2020 till 205 engines each year by 2026.

Table E.4 Another engine data for alternative Y-20 engine and WS-20

公司公告

21/234

在三代机**-10 型号中，德凯独家提供内环前段等 7 种铝镁合金铸件，每台份 2.3 万元，主机厂需要开辟双流水以确保自身供应安全，保守估计德凯可以提供 70%的铝镁合金件，在钛合金领域，德凯批产提供了轴承座铸件，占据市场份额 40%，单台份 7 万元；德凯正在开展后通风器主体、中央齿轮机匣壳体、内环前段、安装座、中介机匣等铸件的研制工作，目前进展顺利，该型号单台份钛合金铸件采购金额约 53 万元。整体上德凯 2020 年占据市场份额 5%，2021 年 15%，2022 年后达到 20%。发动机生产量稳定在 300-500 台份，该型号德凯产值规划如下：

时间	2020 年	2021 年	2022 年	2023 年	2024 年	2025 年	2026 年
生产量正常估计	320.0	320.0	400.0	400.0	400.0	450.0	450.0
生产量保守预估	320.0	320.0	320.0	320.0	350.0	350.0	350.0
铝镁件份额	100%	90%	80%	70%	70%	70%	70%
钛合金份额	5%	15%	20%	30%	35%	35%	35%
德凯产值预估	1584.0	3206.4	4976.0	7004.0	8064.0	9072.0	9072.0

Table E.5 New data for engines build in 2020–2026

五代涡轴型号发动机和**-10 型号涡轴发动机，主机匣和前机匣的四流水在德凯，铸件研制难度很大，两种铸件售价 40 万元，五代涡轴未来在 2025 年为研制阶段，每年 5 台份，**-10 型号发动机需求很大，2019 年 60 台份，2020 年 100 台份，2021 年 150 台份，2022 年后每年 200 台份。上述两个型号德凯规划产值如下：

时间	2020 年	2021 年	2022 年	2023 年	2024 年	2025 年	2026 年
生产量正常预估	65	105	155	205	205	205	205
生产量保守预估	50	100	100	100	150	150	150
市场份额	10%	25%	25%	25%	25%	25%	25%
德凯产值预估	468	1890	2790	3690	3690	3690	3690

Fig. E.4 Chinese H-20 stealth bomber image

Furthermore, the H-20 stealth bomber (see Fig. E.4) that is reportedly under development likewise probably cannot adequately perform with the same hand-me-down motors that power the Chinese military's current heavy aircraft. For the next generation of warplanes, China needs custom engines [50].

But it is not impossible for Chinese industry to overcome engine-development problems or to scale up production.

References

1. S. S. Vinogradov and P. D. Smith, ' Radar Cross-Section Studies of Lens Reflectors', *Progress In Electromagnetics Research, PIER 72, 325–337, 2007*
2. Luneberg, R. K. *(1944). Mathematical Theory of Optics. Providence,* Rhode Island: Brown University. pp. 189–213.
3. Brown, J. (1953). Wireless Engineer. 30: 250.
4. Gutman, A. S. (1954). "Modified Luneberg Lens". J. Appl. Phys. 25 (7): 855. Bibcode:1954JAP....25..855G. doi: https://doi.org/10.1063/1.1721757.
5. Morgan, S. P. (1958). "General solution of the Luneberg lens problem". *J. Appl. Phys.* ***29*** *(9): 1358–1368.* Bibcode:1958JAP....29.1358M. doi: https://doi.org/10.1063/1.1723441.
6. "Solutions of problems (prob. 3, vol. VIII. p. 18*8)". The Cambridge and Dublin mathematical journal.* Macmillan. **9**: 9–11. 1854.
7. "Problems (3)". *The Cambridge and Dublin mathematical journal.* Macmillan. **8**: 188. 1853.
8. Niven, ed. (1890). *The Scientific Papers of James Clerk Maxwell.* New York: Dover Publications. p. 76.
9. Lo, Y. T.; Lee, S. W. (1993). *Antenna Handbook: Antenna theory.* Antenna Handbook. Springer. p. 40. ISBN 9780442015930.
10. http://thediplomat.com/2018/12
11. Pai, Shih, "Magnetogasdynamics and Plasma Dynamics", Springer-Verlag, Vienna, Prentice-Hall, Inc., Englewood Cliffs, NJ 1962.
12. Rosa, R. J., "Magnetohydrodynamic Energy Conservation", McGraw-Hill Book Company, NY, 1968.
13. Gurijanov, E. P., and Harsha, P. T. : AJAX: New Directions in Hypersonic Technology", AIAA-96-4609, Seventh Aerospace Planes and Hypersonic Technology Conference, Norfolk, VA, 1996.
14. Bruno, C., Czysz, P.A., and Murthy, S.N.B., "Electro-magnetic Interactions in a Hypersonic Propulsion System," AIAA 97-3389, 33rd AIAA/ASME/SAE/ASEE Joint Propulsion Conference, Seattle, WA, July 6–9, 1997.
15. Litchford, R.J., Cole, J.W., Bityurin, V.A., and Lineberry, J.T., "Thermodynamic Cycle Analysis of Magnetohydrodynamic-Bypass Hypersonic Airbreathing Engines," NASA/TP—2000-210387, July 2000.
16. Murthy, S.N.B. and Blankson, I.M., "MHD Energy Bypass for Turbojet-Based Engines," IAF-00-5-5-05, 51st International Astronautical Congress, Rio de Janeiro, Brazil, October 2-6, 2000.

B. Zohuri, *Radar Energy Warfare and the Challenges of Stealth Technology*,
https://doi.org/10.1007/978-3-030-40619-6

17. Sergey O. Macheret, Mikhail N. Schneider, and Richard B. Miles, "Magnetohydrodynamic and Electrohydrodynamic Control of Hypersonic Flows of Weakly Ionized Plasmas", AIAA JOURNAL, Vol. 42, No. 7, July 2004
18. Zohuri, B. "Scalar Wave Driven Energy Applications" Sep 4, 2018, Published by Springer Publishing Company, New York, NY.
19. Zohuri, B. "Principle of Scalar Electrodynamics Phenomena Proof and Theoretical Research", Journal of Energy and Power Engineering 12 (2018) 408-417. doi: https://doi.org/10.17265/1934-8975/2018.08.005.
20. Zohuri, B. "Directed Energy Beam Weapons, The Dawn of New Age Defenses" will be published By Springer Publishing Company, April 2019.
21. Horst Eckardt, "What are "Scalar Waves"?, www.aias.us, www.atomicprescision.com, www.upitec.org, 02 January, 2012.
22. Georgetown Journal of International Affairs, March 29, 2018.
23. https://en.wikipedia.org/wiki/DF-ZF
24. MacDonald, A. D., Microwave Breakdown in Gases, Wiley Publisher 1966, New York.
25. https://en.wikipedia.org/wiki/Paschen%27s_law
26. U.S. Patent 512,340 "Coil for Electro-Magnets", Nikola Tesla (1894).
27. H. Eckardt, D. W. Lindstrom, "Solution of the ECE vacuum equations", in "Generally Covariant Unified Field Theory", vol. 7 (Abramis, Suffolk, 2011), pp. 207-227 (see also www.aias.us, section publications).
28. Bearden, T. E. (1981), Solutions to Tesla's Secrets and the Soviet Tesla Weapons. Tesla Book Company, Millbrae, California.
29. Michael T. Borkowski, Raytheon Company, Chapter 5 of the book "RADAR HANDBOOK" edited by Merrill I. Skolnik, 2nd edition McGraw-Hill, 1990, New York, NY.
30. https://www.tek.com/blog/fundamentals-radar-measurement-and-signal-analysis-part-1
31. Dimitris V. Dranidis, Airborne Stealth In A Nutshell Part II, Countering Stealth – Technology & Tactics,http://www.google.com/url?sa=t&rct=j&q=&esrc=s&source=web&cd=1&ved=2ahUKEwj8j56ch83lAhUYrZ4KHdffCRAQFjAAegQIAxAC&url=http%3A%2F%2Fwww.harpoonhq.com%2Fwaypoint%2Farticles%2FArticle_021.pdf&usg=AOvVaw3tvhiMWdEHOaeWmA-XnI8k (Accessed November 2019), *The magazine of the computer Harpoon community - http://www.harpoonhq.com/waypoint/*
32. Rabideau, D.J. (2003). "Ubiquitous MIMO multifunction digital array radar". The Thirty-Seventh Asilomar Conference on Signals, Systems & Computers. **1**: 1057–1064. doi:https://doi.org/10.1109/ACSSC.2003.1292087. ISBN 978-0-7803-8104-9
33. Bliss, D.W.; Forsythe, K.W. (2003). "Multiple-input multiple-output (MIMO) radar and imaging: degrees of freedom and resolution". The Thirty-Seventh Asilomar Conference on Signals, Systems & Computers, 2003. Pacific Grove, CA, USA: IEEE: 54–59. doi:10.1109/ACSSC.2003.1291865. ISBN 9780780381049
34. K. W. Forsythe, D.W. Bliss, and G.S. Fawcett. Multiple-input multiple output (MIMO) radar: performance issues. Conference on Signals, Systems and Computers, 1:310–315, November 2004.
35. Gao, Xin, et al. "On the MUSIC-derived approaches of angle estimation for bistatic MIMO radar." Wireless Networks and Information Systems, 2009. WNIS'09. International Conference on. IEEE, 2009.
36. Li, Jian, and Petre Stoica. "MIMO radar with collocated antennas." IEEE Signal Processing Magazine 24.5 (2007): 106-114.
37. Sturm, Christian, et al. "Spectrally interleaved multi-carrier signals for radar network applications and multi-input multi-output radar." IET Radar, Sonar & Navigation 7.3 (2013): 261-269
38. Chen, Chun-Yang, and P. P. Vaidyanathan. "MIMO radar ambiguity properties and optimization using frequency-hopping waveforms." IEEE Transactions on Signal Processing 56.12 (2008): 5926-5936

39. Anastasios Deliggianis, Anastasia Panoui, S. Lambotharan, and Jonathon Chambers, “Game-Theoretic Power Allocation and the Nash Equilibrium Analysis for a Multistatic MIMO Radar Network”, IEEE Transactions on Signal Processing, September 2017
40. J. Li and P. Stoica, *MIMO Radar Signal Processing*. Hoboken, NJ, USA: Wiley, 2009.
41. J. Li and P. Stoica, “MIMO radar with colocated antennas,” *IEEE Signal Process. Mag.*, vol. 24, no. 5, pp. 106–114, Sep. 2007.
42. A. M. Haimovich, R. S. Blum, and L. J. Cimini, “MIMO radar with widely separated antennas,” *IEEE Signal Process. Mag.*, vol. 25, no. 1, pp. 116–129, Dec. 2007.
43. A. Deligiannis, S. Lambotharan, and J. A. Chambers, “Beamforming for fully-overlapped two-dimensional phased-MIMO radar,” in *Proc. IEEE Radar Conf.*, Arlington, VA, USA, 2015, pp. 599–604.
44. https://www.radartutorial.eu/02.basics/Frequency%20Modulated%20Continuous%20Wave%20Radar.en.html
45. Hou Jiangang, Tao Ran, Shan Tao, and Qi Lin. 2004. A novel LPI radar signal based on hyperbolic frequency hopping combined with barker phase code. 2004 7th International Conference on Signal Processing Proceedings (IEEE Cat.no.04TH8739) (2004): 2070; 2070-3 vol.3; 3o.3.
46. *GuoSui Liu, Hong Gu, WeiMin Su, and HongBo Sun. 2001. The analysis and design of modern low probability of intercept radar. 2001 CIE International Conference on Radar Proceedings (Cat no.01TH8559) (2001): 120; 120-124; 124.*
47. Aytung Denk, “DETECTION AND JAMMING LOW PROBABILITY OF INTERCEPT (LPI) RADARS” Thesis work, Naval Postgraduate School, Monterey, California, September 2006.
48. https://nationalinterest.org/blog/buzz/chinas-stealth-fighters-and-stealth-bombers-have-big-problem-108571.
49. https://alert5.com/2019/12/26/china-will-struggle-to-develop-new-military-turbofan-engines-at-least-till-2026/#more-79574.
50. https://nationalinterest.org/blog/buzz/china-building-its-very-own-stealth-bombers-meet-h-20-and-jh-xx-102507

Index

B. Zohuri, *Radar Energy Warfare and the Challenges of Stealth Technology*,
https://doi.org/10.1007/978-3-030-40619-6

MIX
Papier aus verantwortungsvollen Quellen
Paper from responsible sources
FSC® C105338
FSC
www.fsc.org

If you have any concerns about our products,
you can contact us on
ProductSafety@springernature.com

In case Publisher is established outside the EU,
the EU authorized representative is:
Springer Nature Customer Service Center GmbH
Europaplatz 3, 69115 Heidelberg, Germany

Printed by Libri Plureos GmbH
in Hamburg, Germany